DIRECT-CONTACT HEAT TRANSFER

Edited by

Frank Kreith
Solar Energy Research Institute
Golden, Colorado
and University of Colorado, Boulder

R. F. Boehm
University of Utah
Salt Lake City

⊙ **HEMISPHERE PUBLISHING CORPORATION**
A subsidiary of Harper & Row, Publishers, Inc.
Washington   New York   London

DISTRIBUTION OUTSIDE NORTH AMERICA
SPRINGER-VERLAG
Berlin   Heidelberg   New York   London   Paris   Tokyo

DIRECT–CONTACT HEAT TRANSFER

1 2 3 4 5 6 7 8 9 0   B R B R   8 9 8

Library of Congress Cataloging-in-Publication Data

Direct contact heat transfer.
    Presentations made at a workshop on direct contact heat transfer at the Solar Energy Research Institute, Golden, Colorado, in the summer of 1985.

    Bibliography: p.
    Includes indexes.
    1. Heat exchangers—Congresses. 2. Heat—Transmission—Congresses. 3. Two-phase flow—Congresses. I. Kreith, Frank. II. Boehm, R. F.
    TJ263.D55   1987   621.402′2   87-12058
    ISBN 0-89116-635-1   Hemisphere Publishing Corporation

DISTRIBUTION OUTSIDE NORTH AMERICA:
ISBN 3-540-18209-8   Springer-Verlag   Berlin

CONTENTS

CONTRIBUTORS

D. BHARATHAN
Solar Energy Research Institute
1617 Cole Boulevarde
Golden, CO 80401

R. F. BOEHM
Department of Mechanical Engineering
University of Utah
Salt Lake City, UT 84112

MARK S. BOHN
Solar Energy Research Institute
1617 Cole Boulevarde
Golden, CO 80401

M. Q. BREWSTER
Department of Mechanical Engineering
University of Illinois
 at Urbana-Champaign
1206 West Green Street
Urbana, IL 61801

JOHN C. CAMPBELL
Professional Engineer
1217 Derbyshire Drive
Ballwin, MO 63021

MICHAEL M. CHEN
Department of Mechanical
 Engineering
University of Illinois
 at Urbana-Champaign
1206 West Green Street
Urbana, IL 61801

C. T. CROWE
Engineering International
Suite 200
301-116 Avenue, S.E.
Bellevue, WA 99004; and
Mechanical Engineering
 Department
Washington State University
Pullman, WA 99164

JAMES R. FAIR
Department of Chemical Engineering
The University of Texas
Austin, TX 78712

HAROLD R. JACOBS
Mechanical Engineering Department
The Pennsylvania State University
University Park, PA 16802

JAMES G. KNUDSEN
Chemical Engineering Department
Oregon State University
Corvallis, OR 97331; and
HTRI
1499 Huntington Drive
South Pasadena, CA 91030

FRANK KREITH
Solar Energy Research Institute
1617 Cole Boulevarde
Golden, CO 80401

R. LETAN
Mechanical Engineering Department
Ben-Gurion University of the Negev
Beer-Sheva, Israel

E. MARSCHALL
Department of Mechanical
   and Environmental Engineering
University of California
Santa Barbara, CA 93106

A. F. MILLS
Department of Mechanical Engineering
University of California
Los Angeles, CA 90024

J. J. PERONA
Department of Chemical Engineering
University of Tennessee
Knoxville, TN 37996-2200

JAMES R. WELTY
Division of Engineering
   and Geosciences
Office of Basic Energy Sciences
U.S. Department of Energy
Washington, DC 20545

JOHN D. WRIGHT
Solar Energy Research Institute
1617 Cole Boulevarde
Golden, CO 80401

PREFACE

It is common engineering practice to use a closed heat exchanger to transfer heat between two streams. In a closed heat exchanger each fluid has its own flow channel and the fluids are separated by a solid wall. The variety of closed heat exchanger types marketed is increasing each year. They range from simple devices made of plastic or paper for low temperatures to the sophisticated ceramic configurations specifically for high temperatures. Since each type of exchanger has a number of different geometries available, several choices are at the designer's fingertips. Moreover, simplified design techniques are available for most closed configurations either by a Modified Log Mean Temperature Difference approach or by the use of the Effectiveness—NTU technique. Hence, the typical engineer generally uses closed devices without considering alternatives.

However, other options for heat exchange do exist: direct-contact processes. As implied by the name, this approach accomplishes heat transfer by bringing a higher temperature stream and a lower temperature stream into contact. Specific examples of direct-contact heat transfer devices are cooling towers, open feed water heaters, distillation units, and barometric condensers. These devices are often viewed as special situations, and empirical design techniques for them have been developed over time but without the underpinnings of a basic physical understanding of direct-contact phenomena.

As the costs of energy and industrial facilities increase, direct contact devices are being given new consideration because these devices offer the possibility of increased performance and decreased first cost. In addition, they may have application to processes in which closed exchangers encounter problems such as corrosion and high initial cost.

In order to explore the potential of direct heat transfer processes, and establish a pool of knowledge summarizing our current understanding, as well as to delineate what needs to be done in research and development to provide information necessary to increase the use of direct contact processes, the National Science Foundation supported a workshop on direct contact heat transfer at the Solar Energy Research Institute in the summer of 1985. We served as organizers for this workshop, which emphasized an area of thermal engineering that, in our opinion, has great promise for the future, but has not yet reached the point of wide-spread commercial application. Hence, a summary of the state of knowledge at this point is timely.

The workshop had a dual objective:

1. To summarize the current state of knowledge in such a form that industrial practitioners can make use of the available information.
2. To indicate the research and development needed to advance the state-of-the-art, indicating not only what kind of research is needed, but also the industrial potential that could be realized if the information to be obtained through the proposed research activities were available.

In order to achieve these objectives, we invited some of the leading researchers, engineers and practitioners in the field of direct-contact heat transfer for a two-day workshop to discuss the key issues in the field. The workshop consisted of 8 lectures and 4 discussion sessions. Each discussion session dealt with two related lecture topics. The lectures were tutorial in nature, thus presenting the best available correlation of data and summaries of techniques in the field and also showing how this information could be used for practical applications in industry. We have attempted to summarize all the information on research needs from this workshop as the final chapter in this book for the use of the engineering research community and funding agencies.

After completing the workshop, we decided that it would be helpful if in addition to the technical summaries, there would also be appendixes that would illustrate the manner in which direct-contact heat transfer devices can be designed and evaluated. Searching for readily available information, we found that a majority of applications of direct-contact processes have been made in areas in which energy use and efficiency of utilization is at a premium. These areas are in the evolving fields of energy conservation and renewable energy conversion, where often low grade energy sources must be utilized and high second law thermodynamic efficiencies are required in order to make these processes viable. Hence, the examples in the appendixes are chosen from energy production in a geothermal installation, energy conversion in a solar pond application, energy conversation in open cycle OTEC, and in a cooling tower. However, these illustrative examples are not intended to be exhaustive, but rather to provide a background for other applications that imaginative engineers and designers can utilize.

We hope that the presentations from eminent authorities in the field that we have collected here will serve the engineering community and that the appendix material will help in generating more applications of direct-contact heat transfer processes.

Frank Kreith
R. F. Boehm

DIRECT–CONTACT
HEAT TRANSFER

DIRECT-CONTACT HEAT TRANSFER PROCESSES

R. F. Boehm and Frank Kreith

1 INTRODUCTION

A key challenge in the design of efficient energy conversion systems is to achieve effective heat transfer at temperatures that can extract the maximum thermodynamic potential of the system's heat source. In classical heat exchangers, heat transfer takes place through a wall separating the hot and the cold fluid streams. Thus, conventional heat exchangers are limited in their ability to tap the maximum thermodynamic potential because they have built-in thermal losses associated with the separation of the fluid streams by an intervening solid wall. This type of configuration also leads to a deterioration of the heat transfer effectiveness as the heat transfer coefficients decrease with time due to fouling. In the case of high-temperature applications, thermal stress and corrosion problems are imposed on the wall materials themselves. These situations are present irrespective of whether the two fluid streams are solid, liquid, vapor, gaseous, or some mixtures of them.

Despite these potential problems the transfer of heat between two streams is achieved in most engineering applications by closed heat-exchange devices in which a solid surface separates the two streams, and the variety of closed heat-

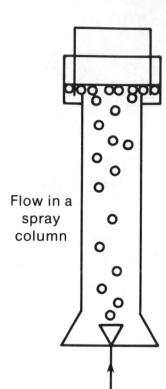

Flow in a
spray
column

Figure 1.1 Schematic diagram of spray tower.

exchanger types available is increasing each year. These exchangers range from
simple devices made of plastic or paper for low-temperature applications to the
sophisticated ceramic exchangers for high-temperature applications, such as gas
turbine regenerators. Since each type of exchanger comes in a variety of different
geometries from several manufacturers, a great deal of flexibility and choice is
available at a designer's fingertips. More importantly, however, simplified design
techniques have been developed for almost any closed heat-exchanger
configuration, either with the modified log mean temperature difference (LMTD)
method or with the effectiveness-NTU approach. Both of these methods have
been available in the form of graphs or simple computer programs for many years,
and, as a result, most engineers have usually selected closed heat exchange devices
without considering alternatives.

In addition to their thermodynamic limitations, closed-type heat exchangers
may also have problems related to first cost and operating expense. To improve
the performance of closed-type heat exchangers the usual approach is to increase
the surface area. This, of course, leads to increased initial cost. If the properties
of one or both steams produce corrosion or fouling of the surface, maintenance and
operational costs are incurred. To combat these problems, special materials may

have to be used in the construction of the exchanger, and frequent cleaning may be necessary. Both of these measures lead to increased operational cost.

In view of the limitations of closed-type heat-exchanger devices, increasing interest in another type of heat-exchange approach has developed in recent years: *direct-contact heat transfer*. The term direct contact encompasses a wide range of devices, but all of them have one thing in common: heat transfer is achieved through intimate contact of two material streams without the presence of an intervening surface. More efficient and lower-cost heat transfer equipment is the promise of these types of processes.

Direct-contact processes have been used for many years for operations in which the primary goal was mass transfer, and initial applications to heat transfer processes have emphasized situations where heat and mass transfer are intimately coupled. However, the inclusion of a chapter on heat transfer in packed and fluidized beds in the 1985 *Handbook on Heat Transfer Applications* is a recognition of the increasing importance of direct-contact heat transfer to engineering practice in applications such as energy conversion and combustion of fossil fuels.

A problem facing the engineer who wants to use direct-contact heat transfer is the limited general understanding of the processes and a lack of a reliable methodology for predicting the thermodynamic performance. Current practice usually relies on empirical correlations for a given geometry that are difficult, if not impossible, to transfer to another geometric configuration. Except for some limited use of mass transfer data to predict heat transfer characteristics by means of the analogy between heat and mass transfer, there exist no generalized approaches to predict the performance of direct-contact heat transfer devices similar to the LMTD or effectiveness-NTU method in closed heat exchangers. A few examples of direct-contact heat transfer applications will illustrate the situation.

2 DIRECT-CONTACT DEVICES AND APPLICATION

As an example of a simple countercurrent flow, direct-contact heat transfer device, consider the spray tower shown in Fig. 1.1. It consists essentially of a cylindrical vessel in which one fluid moves downward by gravity dispersed in droplets within a continuous immiscible second fluid stream flowing upward. It is known that this type of contact device has a high capacity but low efficiency. The low efficiency is largely the result of the low dispersed phase holdup (less than 40%) due to the loose packing flow.

To combine the high capacity with a good efficiency, chemical engineers in the early 1960s placed inside the cylinder a packing capable of increasing the dispersion of droplet to 90% holdup as shown in Fig. 1.2. In addition, the dense-packing flow provided a large interfacial area and a low value of axial dispersion. But since there are many different packing materials that can be used, the kind of correlation will depend on the specific geometry of the device. Figure 1.3 shows several kinds of commercial packings that are widely used in industry. Other types of internal configurations have been used to improve the contactor performance.

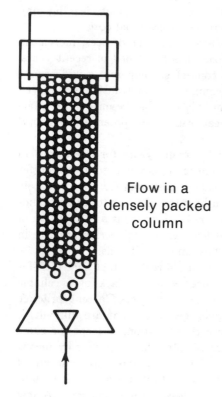

Flow in a
densely packed
column

Figure 1.2 Schematic diagram of a packed-bed device.

Most of the direct heat-exchanger applications are commercially accomplished in one of the following devices: spray columns, baffle tray columns, packed columns, crossflow tray columns, and pipeline contactors. Figure 1.4 illustrates a perforated plate contactor. Figure 1.5 illustrates the heat transfer section of a heavy hydrocarbon fractionator that uses both baffles and crossflow trays. Figure 1.6 illustrates a spray chamber in which pyrolysis gases are cooled in a device similar to that in Fig. 1.1. Figure 1.7 is a special packing condenser reflux in a vacuum steam fractionator, and Fig. 1.8 shows a barometric type of steam condenser. Figure 1.9 shows a type of agitated column. All of these devices will be discussed in more detail later; at this point they are shown mainly to illustrate the kind of applications that direct-contact heat transfer offers.

Another major application of direct-contact heat transfer is in closed cycle cooling of various types of power plants. Prior to 1970, more than 75% of the power plants used open circuit water cooling in their condensers, but in response to a shortage of water and specific environmental regulations from the clean water act, use of closed cycle cooling has increased dramatically. There are essentially three types of cooling devices in common use: cooling towers, cooling ponds, and spray systems. All three use direct-contact heat transfer.

(a) Raschig ring   (b) Lessing ring   (c) Berl saddle

(d) Intalox saddle   (e) Tellerette   (f) Pall ring

Source: Chemical Engineers Handbook

Figure 1.3 Sketches of some commercial random packings.

Cooling towers may be wet or dry. In wet cooling towers, water is pumped to the top of the cooling section and allowed to splash down through the packing region while air flows counter- or crosswise to the water. As a fraction of the water evaporates, the air is heated and humidified while the remainder of the water is cooled. Air flow may be accomplished by buoyancy or by fans. Figure 1.10 shows a crossflow and counterflow cooling tower, both with natural draft.

Dry cooling towers may be direct or indirect. In a direct system the turbine steam is ducted to the tower and condensed there. In an indirect system a secondary circulating loop is cooled by the tower while the steam is condensed in a conventional shallow tube condenser. Combinations of wet and dry systems have also been used. Figure 1.11 shows an indirect dry cooling system with a direct-contact condenser, and Fig. 1.12 depicts a combined system.

Cooling ponds are artificial impoundments into which heated cooling water is discharged and from which cooled water is withdrawn. Heat is rejected from the pond surface to the atmosphere by radiation, convection, and evaporation. Figure 1.13 shows a stratified pond with an entrance mixing zone, a surface layer in which heat transfer to the environment takes place, and a colder subsurface region that feeds the plant intake.

Spray ponds or spray channels are cooling ponds in which the transfer processes to the atmosphere are enhanced by spraying the water into the air to

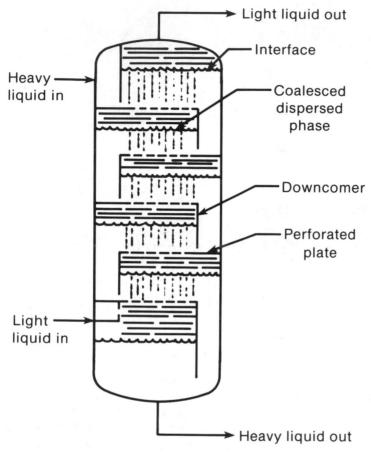

Figure 1.4 Schematic diagram of a perforated-plate column.

form a large interfacial area for heat and mass transfer. Spray channels consist of a series of spray modules distributed along a canal between the intake and discharge. Figure 1.14 is a typical power spray channel system with power spray module details. There is a similarity between spray systems and towers in that a dispersed liquid phase is created. In the former, however, the contact region is not enclosed, and no positive control of air flow is therefore possible. Methods of analysis for all of these systems are presented in Rohsenow et al. (1985). Since condensers for power plants have been widely used, a great deal of empirical data for their design is available.

High-temperature applications of direct-contact heat transfer are used in pulverized coal combustion and have been proposed for solar central receiver systems. As is shown in more detail in Chapter 9, radiation to or from small particles can be used effectively to absorb very high heat fluxes in relatively small volumes. This process is widely used in burning coal at high efficiency using small

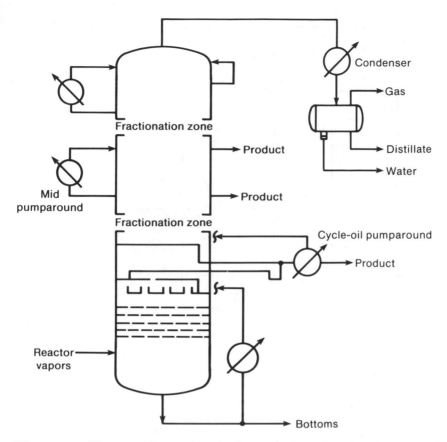

Figure 1.5 Heat transfer section of a heavy hydrocarbon fractionator with baffles and crossflow trays.

particles in devices for the pulverized combustion of coal. It has also been proposed for solar energy conversion systems.

Figure 1.15 shows a schematic diagram for a direct-contact solar flux receiver. In this scheme a thin film of nitrate salt flows down an incline, while solar radiation is reflected from a field of heliostats onto its exposed surface. This eliminates the need for an intervening surface as would be required if the salt were contained in tubes. Thus, the thermal resistance between the source (the sun) and the sink (the salt film) is minimized and the thermodynamic efficiency is maximized.

About 15 years ago, Fair (1972) provided a summary of design recommendations for direct-contact cooler/condensers. He characterized the performance of this type of equipment in terms of the flooding, entrainment, and pressure drop characteristics, in addition to the thermodynamic performance, and also noted the lack of a generalized approach in that a particular set of equations is required to describe each situation. Fair divided direct-contact gas-liquid heat transfer into four general classifications: simple gas cooling, gas cooling with vaporization of

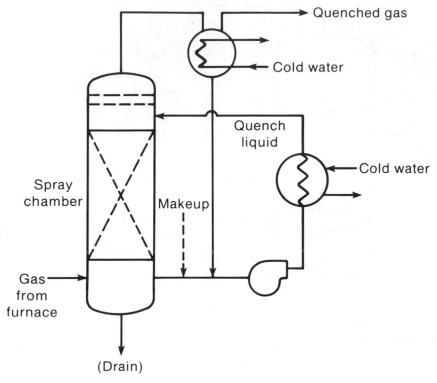

Figure 1.6 Spray quencher for pyrolysis gases.

coolant, gas cooling with partial condensation, and gas cooling with total condensation.

Although the design methodologies presented by Fair are for direct-contact cooler/condensers employing the devices shown in Figs. 1.1-1.6, the design relations are approximate. Many are derived by analogy from mass-transfer data or correlations, and none is applicable directly to tower diameters exceeding 6 ft. (usually restricted to $D \leq 2$ ft). Equations are based on data obtained with air-water, air-oil, and H_2-light hydrocarbon-oil.

To account better for all the influential variables, more comprehensive correlations are needed. Areas to be addressed should include the degree of back-mixing and of entrainment in large-diameter spray chambers and the axial drop-size distribution in pipeline contactors with turbulent mist flow. Data obtained in studies of water injection in line desuperheaters for steam and with other gas-liquid combinations indicate that the uniformity of initial phase dispersion is quite important because thermal effectiveness decreases rapidly with nonuniformity of dispersal. Although there is a large amount of absorption/desorption data for different packings, the validity of the heat transfer correlations is still subject to uncertainty and much of the design information is proprietary.

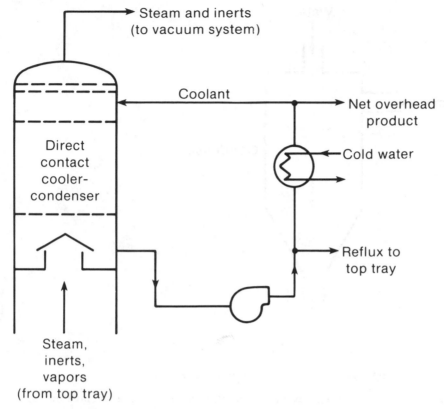

Figure 1.7 Condenser for vacuum steam fractionator.

3 CLASSIFICATIONS OF DIRECT-CONTACT HEAT TRANSFER PROCESSES

As we have seen, there exists a wide variety of processes and devices for transferring heat by direct contact. We will now attempt to classify these processes. Classifications of direct contact devices could be made according any of several criteria: the relative flow directions (cocurrent or countercurrent flow are the most common); the driving forces for the flows (gravity, pressure, and centrifugal forces are most widely used); or phase characteristics of the contacting streams (e.g., liquid-liquid, solid-gas, gas-liquid).

Classification according to flow direction is seldom meaningful because the multipass arrangements and its variations, as used in closed exchangers, are neither feasible nor are they normally needed in direct-contact devices. Generally, parallel- or counterflow schemes are most convenient, although a form of crossflow can also be used. In direct-contact devices, just as in closed heat exchangers, counterflow is thermodynamically more effective than parallel flow. But the flow direction is immaterial when one fluid in the heat transfer process undergoes an

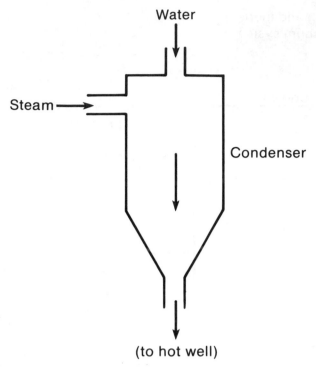

Figure 1.8 Schematic of a barometric steam condenser.

isothermal phase change. For example, the relative flow directions have no meaning in an open feedwater heater where the condensation of steam is used to warm feedwater. This could be considered as a limiting case for either cocurrent or countercurrent flow.

The general situation is that one fluid transfers heat with another, distinct from the first, and does not undergo a phase change during this process. In this situation a counterflow arrangement is often preferred because it is efficient and also facilitates the separation of the fluids after the heat transfer process has been accomplished. Examples of both cocurrent and countercurrent flow are shown in Fig. 1.16.

The most frequently used driving force for the separation of two immiscible streams relative to one another is gravity, although sometimes the separation is accomplished by other body-force means; for example, for applications in space an electrostatic field has been proposed. But if a force other than gravity is employed, the situation is considered a special case. Hence, categorization of direct-contact processes by the forces that cause the flow is not a useful way of making clear distinctions.

While in theory heat can be transferred in direct contact between virtually any combinations of substance streams (e.g., solid, liquid, or vapor), in practice only some distinct groupings have been used. Historically, perhaps the most

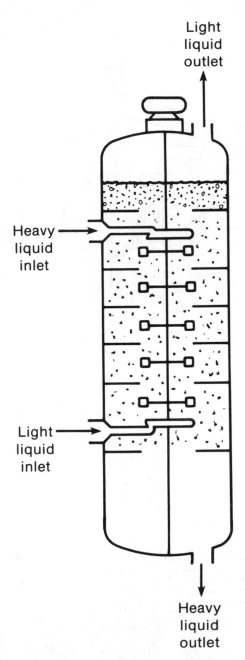

Figure 1.9 Schematic diagram of an agitated column.

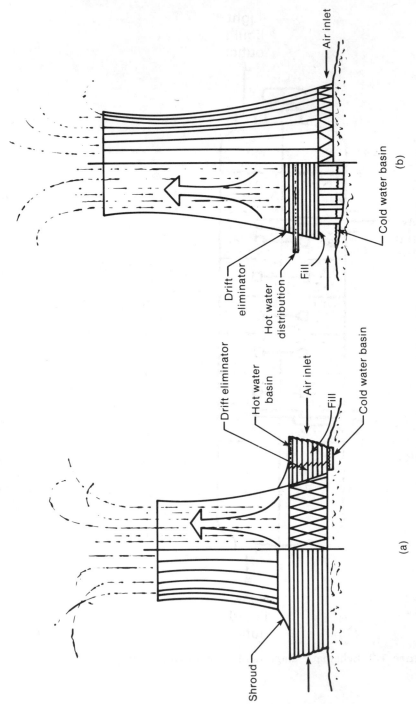

Figure 1.10 Direct cooling tower configurations with natural draft (a) crossflow, and (b) counterflow.

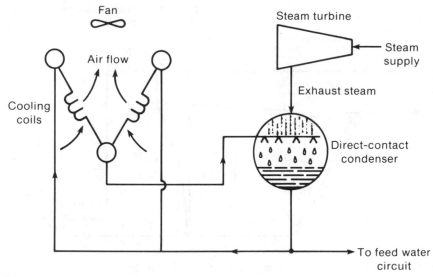

Figure 1.11 Indirect dry cooling system with direct-contact condenser.

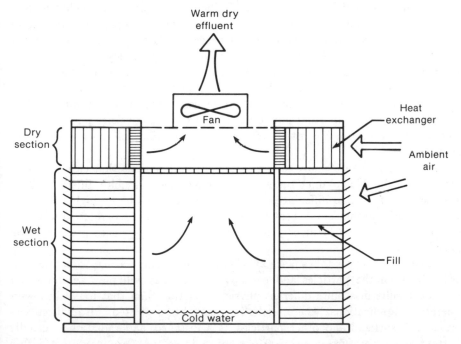

Figure 1.12 Hybrid wet-dry cooling tower.

important device has been the bubble or spray column, which was used to transfer heat between a gas and a liquid. But heat transfer in direct contact between two

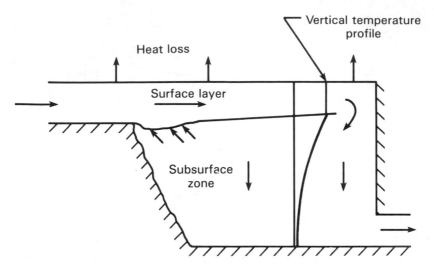

Figure 1.13 Schematic view of deep stratified cooling pond.

immiscible liquids of different densities falls in this category, and the transfer of heat between solid particles moving through a liquid or a gas can be included also.

These applications are closely related to mass transfer operations that have been used for many years in the chemical process industry. In fact, equipment similar to that used in mass transfer operations has been employed for some heat transfer processes, and design techniques have usually drawn on the analogy between heat and mass transfer. But complete analogies between mass transfer and heat transfer in bubble columns are not realistic because of a number of factors. One of the most important differences is that the heat transfer phenomena do not occur at the (usually) constant temperature conditions of the mass transfer analogs.

Although there are many solid-gas stream combinations that closely resemble the flow in bubble columns there are also cases that are quite different. Two factors account for these differences: the temperatures might be high enough to change the basic mechanism of heat transfer from convection/conduction to radiation, and there may be artificially induced heat transfer surfaces present, as in fluidized beds.

If the heat transfer takes place at relatively high temperature, radiation heat transfer between the particles and perhaps also the gas will play an important role. This results in a quite different physical situation than that found in lower-temperature applications where radiation can be neglected. Radiation can interact with surfaces and other particles at a distance, and radiation is usually important at temperatures above $800^{\circ}C$ when solid particles are in contact with a gas.

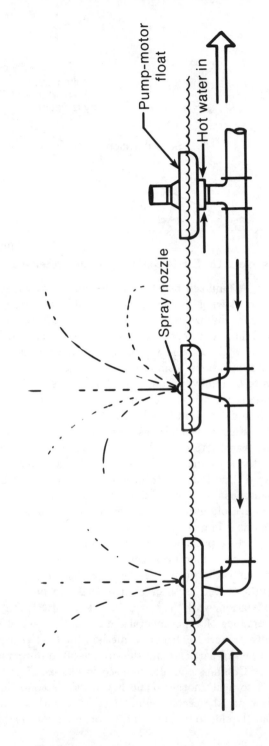

Figure 1.14 Typical power spray canal system.

Pump-motor float

Hot water in

Spray nozzle

15

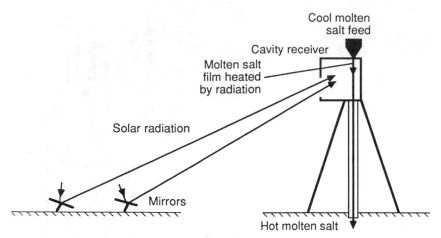

Figure 1.15 Direct-contact solar flux receiver.

A fluidized bed heat exchanger is shown schematically in Fig. 1.17. A gas stream flowing through a bed of solid particles (sand, coal, or catalyst, for example) can "fluidize" this bed within a certain range of flow rates or gas velocities. In the fluidized state, the gas-solid system is very well mixed due to the free mobility of the solid particles with a uniform temperature all over the bed. In many technical applications (combustion, coal gasification, or other chemical reactions) heat has to be removed from—or added to—a fluidized bed. This can be done by using immersed heating or cooling elements such as single tubes, tube bundles, or coils with liquid coolant (See Fig. 1.17).

Fluidized solids-gas and solids-liquid combinations can be used to enhance heat transfer to closed heat-exchanger surfaces. Although this arrangement is not, strictly speaking, a direct-contact heat transfer situation, the microscopic heat transfer mechanisms from a solids-gas or solids-liquid stream to a surface involve phenomena similar to those found in direct contact. For this reason, fluidized beds are often considered direct-contact heat transfer devices, even if one of the streams is in a closed circuit.

Heat transfer between fluidized beds and the surfaces of immersed heating or cooling elements has been a subject of research for more than 30 years. A huge amount of literature has been published in this field, and a lot of experimental results as well as theoretical models can be found in the numerous summaries of fluidization work (e.g., Saxena et al., 1981; Zabrodsky, 1966). But despite the abundance of experimental data and theoretical work, there seems to be no generally accepted calculation method for heat transfer between a gas or liquid fluidized bed of particles and the surface of an immersed solid body.

Columns can also be used in change-of-phase situations, where one liquid is used to heat another liquid beyond the second liquid's saturation temperature. In this case, the second fluid in contact with the first fluid undergoes a transition from liquid to vapor. This configuration is identified as a *three-phase heat*

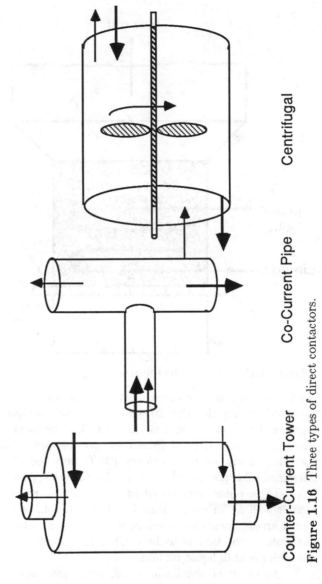

Counter-Current Tower   Co-Current Pipe   Centrifugal

Figure 1.16 Three types of direct contactors.

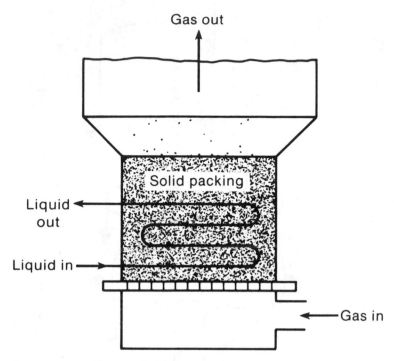

Figure 1.17 Fluidized bed heat exchanger.

exchanger in the literature. Several studies of this configuration have been reported during the last 20 years (see, for example, Sideman and Gat, 1966; Epstein, 1984; Jacobs and Boehm, 1980). This form of heat exchanger has been investigated over the last decade for possible applications to power cycles driven by temperature sources between $100^{\circ}C$ and $300^{\circ}C$ (Jacobs and Boehm, 1980). Stratified solar ponds and geothermal wells are typical applications.

Three-phase, direct-contact heat exchangers can be designed for condensation as well as for evaporation (see Fig. 1.18). In condensation a variety of operational arrangements are possible, including some where the coolant is sprayed into the bulk vapor that is to be condensed and others where the vapor is bubbled through the bulk liquid coolant.

As in all of the liquid-liquid, liquid-gas, and three-phase configurations, a density difference is generally also used to move the two fluids in opposite directions in a direct-contact condenser. If there is no subcooling or superheating of the change-of-phase fluid, then the direction of fluid motion is of no importance, but normally counterflow or crossflow is still used for ease in ultimate separation of the two streams.

Phase-change applications of direct-contact heat transfer with one of the fluids undergoing a solid-liquid transition have so far received little attention. A possible application of this situation is in phase-change storage, where a substance

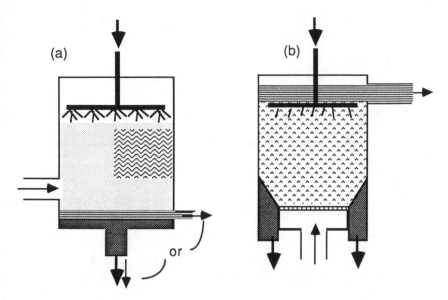

Figure 1.18 Examples of direct-contact condensers.

such as water or paraffin absorbs heat while changing phase from a solid to a liquid. Practical problems might exist in keeping the solidified substance fluidized at the flow necessary for the operation of the system.

Despite the large number of subcategories in the classification by phase, this approach seems to offer the best chance of arriving at a generalizable characterization of the direct-contact heat transfer process. Moreover, design criteria for engineers can be developed more conveniently in this categorization than in any other. Hence, we have adopted this method of classification in the organization of this book.

4 HEAT TRANSFER MECHANISMS

While direct-contact processes are most conveniently categorized according to the physical characteristics and phases of the fluids involved, it is sometimes of value to make distinctions also according to the details of the heat transfer mechanisms present. In all cases it is useful to consider the physical situation to be one of a *continuous* fluid in direct contact with a *dispersed* fluid. The dispersed fluid will be assumed to be in distinct masses surrounded by a bulk, continuous phase. These masses can approximate spherical forms under special circumstances, but normally they are of a nonspherical geometry. They will be referred to as *particles* in what follows.

It is helpful to consider the heat transfer mechanisms and in some situations also the fluid dynamics present. This approach will give an additional way of visualizing details of the direct-contact processes. To do this we consider the *internal* and the *external* heat transfer mechanisms separately. Internal and

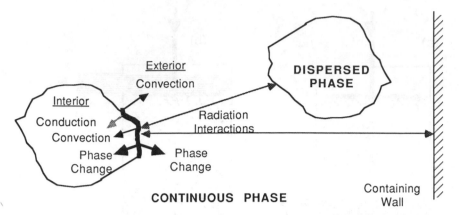

Figure 1.19 Schematic diagram for classifying direct-contact processes.

external refer here, respectively, to inside of dispersed phase particles and outside of the dispersed phase, that is, in the adjacent continuous phase. Figure 1.19 illustrates the processes schematically.

First, consider the external mechanisms. These are the processes by which the particle transmits energy to, or receives energy from, the bulk fluid or surrounding surfaces. Typically, this is a convection process, but under special circumstances it can be by conduction, phase change, or radiation. A forced flow in fluidized beds can be accompanied by intense turbulence, but in gravity-driven systems the external flow is usually laminar or only moderately turbulent. Conduction dominates when fluid in the external flow is laminar and has a large thermal conductivity.

The convective process in the continuous medium, as in bubble columns, is a conventional process that has been extensively studied. To describe this transfer mechanism and, in many cases, the overall performance, it is important to understand the convection/conduction process in flow over a single particle of arbitrary shape. Consideration of the enhancement or degradation of heat transfer to the continuous phase due to the interaction of an ensemble of dispersed particles is the second most important aspect.

As mentioned previously, radiation in the external region is a special case and is important when the particles are at a high enough temperature to radiate appreciably to the surrounding walls and to other particles. Radiation can also be important when the surrounding walls are at high enough temperature to radiate appreciably to the particles. If the continuous phase is a liquid, radiation is not important. Because of their "special case" characteristics, configurations that involve radiation as the major means of transport usually require a quite different analysis than any of the other situations.

Phase change of the continuous phase fluid due to heating or cooling by the particles occurs in a small percentage of cases. This includes situations where water (as the continuous fluid) evaporates into air bubbles (as the dispersed fluid) that are traveling through the water. An example of this is the injection of air streams into a cooling pond to enhance the removal of waste heat by both convection and phase change.

Evaporation is just one example of a whole group of important phenomena that involve heat and mass transfer simultaneously. As noted earlier, mass transfer operations have been the genesis for much of the equipment used in direct-contact heat transfer applications. In all cases of direct-contact heat transfer, the possible effects of mass transfer must be considered. Sometimes, mass transfer effects will be negligible, and sometimes they will have a profound impact on the overall process. As shown in Chapter 3, the proper handling of the mass transfer phenomena in heat transfer situations can be very critical.

Inside the particles one of three possible situations is found to take place during the heat transfer process: conduction, convection, and phase change. While more than one process can occur at a given time, the relative magnitudes of the heat transfer rates of the three mechanisms are usually so different that one dominates.

Conduction always dominates inside solid particles but can also be significant in liquid and sometimes in vapor particles when the particle is small. Convection can occur inside the particle under certain conditions and, when it does, it will be the dominant mechanism compared with conduction. For convection to occur the particle must be liquid or vapor, and there must be a driving force for the convective flow. This driving force can be viscous shear over the particle surface due to the particle flow relative to the continuous phase, or it can be a density-driven force due to gravity or centrifugal motion acting on the particle, or it can be a gradient with interfacial tension. Usually the magnitude of the convection heat transfer is smaller in the latter situation than when the convection is driven by viscous shear when fluid particles can take on irregular shapes and can oscillate, agglomerate, or break up. The convection phenomena can become very complex under these circumstances.

Finally, and often most importantly, phase change can occur inside the particle. When evaporation or condensation occurs, these modes of heat transfer result in the highest transfer coefficients. When this fact is coupled with the high surface areas implicit in direct-contact heat transfer, generally these kinds of processes are desirable for achieving high performance or low-temperature-difference heat transfer.

Special attention must be given to the possible influence of heat transfer on the fluid mechanics present (e.g., the relative velocities) and vice versa. While many of the combinations of fluids and containment geometries result in an essentially constant situation throughout, the acceleration or deceleration of particles due to changes in size or shape can have a major effect on the performance in a specific application. Size effects on performance are better understood than shape effects of the dispersed phase. Also particle breakup or agglomeration is often present, but currently these are not well understood and not exploited in design.

5 ORGANIZATION AND OVERVIEW

Direct-contact heat transfer devices offer many benefits over closed-type heat exchangers. However, so far there have only been limited examinations of the field, and all of them for specialized applications (Sideman, 1966; Jacobs and Boehm, 1980; Sideman and Moalem-Maron, 1982; Vallario and DeBellis, 1984). It is therefore timely that a careful assessment be made of the strengths and shortcomings of this field and that the type of information needed to increase commercialization of direct-contact heat transfer be spelled out.

During the summer of 1985 a workshop was held at the Solar Energy Research Institute under the sponsorship of the National Science Foundation. Attendees included many key engineers and researchers in the direct-contact heat transfer field. Eight invited papers were given on a wide breadth of topics in this area, and these are included in this book. In addition, a discussion among the attendees was held following each pair of related papers. The discussions were presided over by experts in direct-contact heat transfer work. Summaries of the discussion and the conclusions reached by the group are included in five additional chapters. Numerical examples illustrating typical applications and engineering designs are presented in the Appendix. In all cases the emphasis was placed on summarizing what we know and where the key problem areas are that impede design and industrial utilization of divert contact devices.

Chapter 2 summarizes typical industrial applications of direct-contact heat transfer, with particular emphasis on the chemical process industry. In that field there is a heavy reliance on the analogies between heat and mass transfer, and for many years engineers have been working to make the most of the analogy. Numerical modeling of direct-contact heat transfer systems has gained importance over the years. As understanding increases of the basic physical processes, this approach should yield even larger benefits. Chapter 3 gives insights to fundamental aspects of this approach as well as some examples of previous work in this area. Chapter 4 summarizes the key problems delineated by the authors of Chapters 2 and 3 as well as aspects that were brought out in the discussions.

Mass transfer effects during direct-contact heat transfer processes are the emphasis of Chapter 5. Sometimes mass transfer can be used beneficially, but sometimes a significant amount of mass transfer renders a direct-contact heat transfer process undesirable. Hence, the potential impacts of mass transfer need to be considered during the design of direct-contact heat transfer devices. Liquid-liquid or liquid-gas processes are discussed in Chapter 6. In these cases the bubble size remains essentially constant during its travel through the heat exchanger, and these configurations have been subjected to experimental measurements and theoretical modeling. Although some widely applied design approaches are also available, several aspects of the performance of these devices are clearly not yet well understood. Chapter 7 is a discussion of the information given in Chapters 5 and 6, and a summary of the key points brought out in the discussion.

The next three chapters deal with various aspects of solids-gas transfer systems. Chapter 8 examines the basic formulation and previous work on high-temperature problems where radiation heat transfer is either dominant or at least important. High-temperature applications of direct-contact heat transfer have been fewer than those at lower temperatures, but interest has been increasing for solar energy and coal combustion applications as well as for thermal storage. A reason for the lack of past applications of these devices has been inadequate understanding of the basic heat transfer mechanisms in high-temperature situations. Chapter 9 deals with important problems of gas-solids fluidization and enhancement of heat transfer at a boundary. As noted earlier, the fact that a heat transfer surface is imbedded in the solid-gas flow makes this, strictly speaking, a closed heat-exchanger process. However, the fundamentals of the gas-solids interactions, including the driving forces for the convection and radiation, are direct-contact processes. The flow in this particular kind of application is typically very turbulent, and a study of it provides better understanding of direct-contact processes generally. A summary of the discussions that took place about material presented in Chapters 8 and 9 is given in Chapter 10.

In Chapter 11 the fundamentals of evaporation are examined. This is one of the most promising applications of direct-contact heat transfer processes because excellent efficiencies can be achieved. It includes devices such as distillation columns and evaporative coolers. The understanding gained from these processes permits applications of direct-contact evaporation to conditions previously not thought possible. Condensation heat transfer is the focus of Chapter 12. An interest in devices using direct-contact condensation developed late in the last century with applications of barometric condensors. Recently, there has been an interest to work in this area for a variety of other applications, some using immiscible fluids. As with evaporation, this mode offers a very effective means of transferring heat. Chapter 13 is a summary of additional points that were made about evaporation and condensation following the presentation of the original papers. In the concluding chapter of the book, the editors summarize the research required to bring direct-contact heat transfer to its appropriate place in the overall spectrum of thermal equipment and design.

In an effort to assist the heat transfer designer, six appendices have been added to the body of the book. Each of them illustrates by numerical example worked out in detail thermal design of a direct contact heat transfer device or process. Included are a combined heat and mass transfer process, a high temperature direct contact air-molten salt heat exchanger, a preheater/boiler for a solar pond power plant, a direct contact liquid-liquid heat transfer device, a spray column for extracting heat from geothermal brines, and a water cooling tower for a power plant.

We hope that this book can serve as the starting point for the design of direct contact heat transfer devices and will also stimulate the research needed to gain the understanding necessary to develop engineering design methods for direct contact heat processes that will make it possible to build heat transfer devices that use less material and operate more effectively than current industrial equipment.

REFERENCES

Borodulya, V. A., and Kovensky, V. I., (1983), "Radiative Heat Transfer Between a Fluidized Bed and a Surface," *Int. J. Heat Mass Transfer*, Vol. 26, No. 2, pp. 277-287.

Chan, C. K., and Tien, C. L., (1974), "Radiative Transfer in Packed Spheres," *Trans. ASME Ser. C.*, Vol. 96, p. 52.

Epstein, N. (1984), "Hydrodynamics of Three-Phase Fluidization," Chapter 23 in *Handbook of Fluids in Motion*, Ann Arbor Science.

Fair, J. R. (1972), "Designing Direct-Contact Coolers/Condensers," *Chemical Engineering*, June 12, pp. 91-100.

Flamant, G. (1982), "Theoretical and Experimental Study of Radiant Heat Transfer in a Solar Fluidized-Bed Receiver," *AIChE Journal*, Vol. 28, No. 4, pp. 529-535.

Jacobs, H., and R. Boehm (1980), "Direct-Contact Binary Cycles," Section 4.2.6 in *Source book on the Production of Electricity From Geothermal Energy* (J. Kestin et al., Eds.), U.S. Department of Energy, Report DOE/RA/4051-1, pp. 413-471.

Rohsenow, W., J. Harnett, E. Ganic, eds. (1985), Chapter 10, *Handbook of Heat Transfer Applications*, McGraw-Hill, New York.

Saxena, S. C., and Gabor, J. D. (1981), "Mechanisms of Heat Transfer Between a Surface and a Gas-Fluidized Bed for Combustor Applications," *Prog. Energy Combust. Sci.*, Vol. 7,, pp. 73-102.

Saxena, S., N. Grewal, J. Gabor, S. Zabrodsky, and D. Galershtein (1981), "Heat Transfer Between a Gas-Fluidized Bed and Immersed Tubes," in *Advances in Heat Transfer*, Vol. 14.

Sideman, S. (1966), "Direct-Contact Heat Transfer Between Immiscible Liquids," in *Advances in Chemical Engineering*, Vol. 6, pp. 207-286.

Sideman, S., and Y. Gat (1966), "Direct Contact Heat Transfer with Change of Phase: Spray-Column Studies of a Three-Phase Heat Exchanger," *AIChE Journal*, March, pp. 296-303.

Sideman, S., and D. Moalem-Maron (1982), "Direct-Contact Condensation," in *Advances in Heat Transfer*, Vol. 15, pp. 227-281.

Vallario, R., and D. DeBellis (1984), "State of Technology of Direct-Contact Heat Exchanging," Pacific Northwest Laboratory, Report PNL-5009, UC-95, May.

Zabrodsky, S. (1966), *Hydrodynamics and Heat Transfer in Fluidized Beds*, MIT Press.

INDUSTRIAL PRACTICES
AND NEEDS

James R. Fair

The transfer of heat between immiscible phases occurs in many steps of the typical chemical process. Many of the occurrences result simply from the contacting of phases having different temperatures and are handled by energy balances without regard to rate of thermal equilibration. Other occurrences may be contrived; there is a particular need to cool one phase or to heat the other phase, and special efforts are required to ensure that the amount of heat to be transferred and the rate at which it transfers are taken into account in the design of a contacting device that we shall call a direct-contact heat exchanger. It is with this latter case, the contrived case, with which this chapter is concerned.

Process industries applications of direct-contact heat transfer are many and varied. Table 1, from the paper by Fair (1972a), lists typical applications. To limit the scope of the present paper and to cover the most prominent of the applications in the table, we shall further limit the scope of this chapter to *gas-liquid contacting* for direct heat transfer. In other words, we shall be concerned with the need to heat or cool a gas, or to heat or cool a liquid. Thus the problem is a common one in the field of process heat transfer *design* (what type and dimensions of device are required?) or *rating* (given the type and dimensions of the device, what temperatures of terminal streams might be expected?)

Table 1 Direct-contact process heat transfer applications.

	Examples
A. Fluid	
1. Gas-liquid	
a. Gas cooling	Hot gas quenching; steam desuperheating; condensation in barometric condensers; fractionator reflux/overhead condensing
b. Gas heating	Cooling tower operation
2. Liquid-liquid	
a. Immiscible systems	Hot oil cooling; preheating water for desalination
3. Liquid-liquid-gas	Desalinating water; quenching water-containing gas with oil
4. Gas-gas	Blending streams of unequal temperatures
B. Fluid-solid	
1. Gas-solid	
a. Solids heating/gas cooling	Calcining; adsorbent regeneration
b. Solids cooling/gas heating	Regenerative heating of gas; moving bed contacting; fluid-bed contacting
2. Liquid-solid	Metal quenching
3. Liquid-gas-solid	Prilling; spray drying; hot gas quenching by a slurry
C. Solid	Blending streams of unequal temperatures

Advantages of direct-contact heat transfer include low cost of equipment, low pressure drop for gas flow, and ability of the device to handle fluids that might foul conventional heat transfer surfaces. Disadvantages arise when the fluids are not compatible or when material transfer between the phase is not desirable. Another possible disadvantage is that design methods for determining required dimensions of direct-contact equipment are not as well established as they are for shell-and-tube equipment, thus placing a burden on the process designer that he often wishes to avoid.

It is toward this concern over the adequacy of design or rating methods for direct-contact heat transfer that the present chapter is directed. An assessment of available technology will be made, and suggestions for further research and development in the area will be offered.

1 COMPUTATIONAL APPROACH

A gas-liquid direct-contact exchanger may be represented as shown in Fig. 2.1. The countercurrent mode is typical, but cocurrent or crosscurrent modes may also be used. For a differential element of height dZ and cross section S, the total required volume may be obtained by integrating across the device (herein represented as a vertical column, with or without special contacting internals):

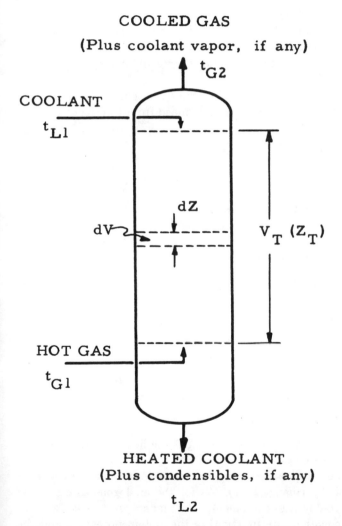

COOLED GAS

(Plus coolant vapor, if any)

t_{G2}

COOLANT

t_{L1}

dZ

dV

V_T (Z_T)

HOT GAS

t_{G1}

HEATED COOLANT
(Plus condensibles, if any)

t_{L2}

Figure 2.1 Direct-contact heat exchanger. The example system is for cooling a hot gas stream, with provisions for condensing a portion of the gas.

$$V_T = \int_0^{Z_T} S \, dZ = \int_0^{Q_T} \frac{dQ}{Ua(t_g - t_L)} \tag{1}$$

As in the analogous case of countercurrent mass transfer in applications such as gas absorption in packed columns, the height of the contacting zone is usually the objective. If a set of effective values on the right side of Equation (1) can be found, then the required height is

$$Z_T = V_T/S = \frac{Q_T}{\overline{Ua} \, \Delta t_M} \cdot \frac{1}{S} \tag{2}$$

A difficulty in the use of Equation (2) is that the volumetric heat transfer coefficient, Ua, is a function not only of individual phase coefficients but also of the degree and direction of mass transfer between the phases. Also, the mean temperature difference Δt_M is dependent on the degree of axial mixing of the flowing streams.

There are alternative approaches to the determination of heat transfer volume or height required. The latter may be determined from transfer units:

$$Z_T = N_{og,h} \cdot H_{og,h} = N_{g,h} \cdot H_{g,h} \tag{3}$$

where

$$H_{og,h} = H_{g,h} + \lambda \, H_{L,h} . \tag{4}$$

The analogy with countercurrent mass transfer operations will be obvious. The several analogs with mass transfer are shown in Table 2.

Still another approach utilizes enthalpy as a driving force. For humidification of gas, as in a cooling tower,

$$Z_T = \frac{L \, c_L}{K'a} \int \frac{dt_L}{i^* - i} \tag{5}$$

For dehumidification of gas, as in a cooler-condenser,

$$Z_T = \frac{G}{K'a} \int \frac{di}{i - i^*} \tag{6}$$

Still other approaches are available when there is departure from the counter-current contacting mode. For example, when trays are used,

$$Z_T = (N_{s,h}/E_h) \, TS \tag{7}$$

where $N_{s,h}$ is the number of heat transfer stages, E_h is the heat transfer efficiency and TS is the spacing between trays.

More details on these approaches may be found in the papers by Fair (1972a,b) and in the thesis by Huang (1982). As for the analogous case of mass transfer in equivalent contacting equipment, the mass and heat transfer coefficients in the height equations are functions of the system properties and the contacting device geometries. The geometries are usually subdivided according to type of internals in the contacting column: packing, spray, and tray.

Table 2 Heat/mass transfer analogies

$$N_{g,d} = \int \frac{dy}{y - y_i} \qquad N_{g,h} = \int \frac{dt_g}{t_g - t_i} \qquad N_g{}' = \int \frac{di}{i_i - i}$$

$$N_{og,d} = \int \frac{dy}{y - y^*} \qquad N_{og,h} = \int \frac{dt_g}{t_g - t_L} \qquad N_{og}{}' = \int \frac{di}{i^* - i}$$

$$H_{g,d} = \frac{G_m}{k_g \, a \, P} \qquad H_{g,h} = \frac{G \, c_g}{h_g a} \qquad H_g{}' = \frac{G}{k' a}$$

$$H_{L,d} = \frac{L_m}{\rho_m k_L a} \qquad H_{L,h} = \frac{L \, c_L}{h_L a} \qquad H_L{}' = \frac{L \, c_L}{Ua}$$

$$\lambda = m \frac{G_m}{L_m} \qquad \lambda = \frac{\int dt_L}{\int dt_g} \qquad \lambda = \frac{c_g G}{L \, c_L}$$

$$H_{og,d} = H_{g,d} + \lambda H_{L,d} \qquad H_{og,h} = H_{g,h} + \lambda H_{L,h}$$

$$H_{og}{}' = H_g{}' + \lambda H_L{}' \qquad Z_T = H_{og,d} \cdot N_{og,d} = H_{og} \cdot N_{og,h}$$

2   TRANSFER MODELS

2.1   The Case of Sensible Transfer Only

This is the simplest case for analysis, since no mass transfer is involved. An example is the cooling of a hot gas by a cool, nonvolatile oil. For the differential element,

$$G \ c_g \, dt = h_g(t_g - t_i) \, da$$
$$= h_L(t_i - t_L) \, da = L \ c_L \, dt \tag{8}$$

The total heat transferred Q_T is obtained by integrating the appropriate portions of Equation (8), and for assumed constant coefficients and driving forces,

$$Q_T = Ua \ V_T \ (t_g - t_L)_{\text{mean}} \tag{9}$$
$$= h_g a \ V_T \ (t_g - t_i)_{\text{mean}} \tag{10}$$
$$= h_L a \ V_T \ (t_i - t_L)_{\text{mean}} \tag{11}$$

with the overall volumetric coefficient being represented by

$$1/Ua = 1/h_g a + 1/h_L a \tag{12}$$

It then remains for the individual phase coefficients to be evaluated on the basis of geometry and contacting conditions.

2.2   The Case of Cooling with Dehumidification

This is the familiar cooler-condenser case in which a hot gas is cooled and a portion of it is condensed and removed with the liquid coolant stream. It has been shown by Fair (1961) that on an overall column basis, Equation (12) is modified to

$$\frac{1}{Ua} = \frac{1}{\alpha h_g a} + \frac{1}{h_L a} \frac{Q_g}{Q_T} \tag{13}$$

where Q_g is the duty for sensible cooling of the gas that leaves the contactor and Q_T is the total duty. Thus, $Q_T = Q_c + Q_g$, where Q_c represents the enthalpy change of condensed components from temperature of entry to the contactor to temperature of discharge with the exit liquid. The multiplier α in Equation (13) is the Ackerman factor (Sherwood et al., 1975), which accounts for the influence of mass transfer on heat transfer rate:

$$\alpha = \frac{C_o}{1 - e^{-C_o}} \tag{14}$$

where

$$C_o = \frac{N_A \, c_{g,A}}{h_g a \ V_T} \tag{15}$$

When heat and mass are transferred in the same direction, N_A is positive and $\alpha > 1$. Coefficients are determined on the basis of geometry and contacting conditions. When the condensing species is the same as the coolant species (e.g., water from a gas being condensed by a water stream), there is relatively little resistance to transfer in the liquid phase and the gas phase heat transfer tends to control. In an extreme case, very little fixed gas remains after the contacting, and the high rate of diffusion of molecules to the interface gives high effective rates of heat transfer. An example of this extreme case is the barometric condenser with few inerts present.

An alternative design method for the dehumidification case takes into account mass transfer coefficients. For a differential amount of transfer surface,

$$L \ c_L \ dt_L/da = k' \ (i' - i_i') = h_L \ (t_i - t_L) \tag{16}$$

$$= K' \ (i' - i'^*) \tag{17}$$

$$Q_T = \int L \ c_L \ dt_L = \int k' \ (i' - i_i') \ da \tag{18}$$

From an energy balance around the gas stream,

$$\int G \ di' = \int k'(i' - i_i') \ da \tag{19}$$

from which

$$Z_T = \frac{G}{k'a} \int \frac{di'}{(i' - i_i')} = \frac{G}{K'a} \int \frac{di'}{(i' - i'^*)} \tag{20}$$

In Equations (16)-(20), i' is a modified enthalpy of the gas, defined below:

$$i' = \phi \ c_g t_g + \lambda \ H \tag{21}$$

$$i_i' = \phi \ c_g t_i + \lambda \ H_i \tag{22}$$

with ϕ being the dimensionless Lewis number ($=h_g/(k' \ c_g)$). For the air-water system, $\phi \simeq 1.0$. For air-benzene, for example, $\phi \simeq 1.7$.

2.3 The Case of Gas Cooling with Humidification

This case is represented by the adiabatic humidifier in which a hot gas is cooled by the vaporization of a coolant liquid operating approximately at its adiabatic saturation temperature. On an overall column basis (Fair, 1961), total duty is

$$Q_T = Q_L + Q_v \tag{23}$$

$$= w_{L2}c_L \ (t_{L2} - t_{L1}) + N_A \ (i_{A2} - i_{A1}) \tag{24}$$

The overall coefficient is

$$\frac{1}{Ua} = \frac{1}{\alpha h_g a} + \frac{1}{h_L a} \frac{Q_L}{Q_T} \tag{25}$$

For an adiabatic humidifier, $t_{L2} = t_{L1}$ and $Q_T = Q_v$. Accordingly,

$$Ua = \alpha h_g a \ . \tag{26}$$

Since heat and mass are transferred in opposite directions, $\alpha < 1.0$. The transfer unit concept gives

$$H_{og,h} = H_{g,h} + \frac{G\ c_g}{L\ c_L} \frac{Q_L}{Q_T} H_{L,h} \tag{27}$$

For the adiabatic humidifier, $H_{og,h} = H_{g,h}$

2.4   The Case of Gas Heating with Humidification

The usual case of interest here is the cooling tower wherein a circulating water stream is cooled through the process of humidifying air. The heat balance gives

$$Q_T = Q_g + Q_v = Q_L$$

$$= w_{g1} c_g \left(t_{g2} - t_{g1}\right) + N_A(i_{A2} - i_{A1}) \tag{28}$$

$$= w_{L2}\ c_L\ (t_{L2} - t_{L1}) \tag{29}$$

For this combination of simultaneous heat and mass transfer, enthalpy and humidity driving forces are convenient to use.

For a differential amount of transfer surface,

$$L\ c_L\ dt_L/da = k_g'\ (i_i' - i')$$

$$= K_g'\ (i'* - i') = h_L(t_L - t_i) \tag{30}$$

$$L\ c_L\ dt_L = K_g'a\ (i'* - i')dZ = G\ di' \tag{31}$$

$$Z_T = \frac{L\ c_L}{K_g'a} \int \frac{dt_L}{(i'* - i)} = H'_{og,i} \cdot N'_{og,i} \tag{32}$$

$$\frac{1}{K_g'a} = \frac{1}{k'a} + \frac{i'* - i'}{t_L - t_i} \frac{1}{h_L a} \tag{33}$$

Also, from Equation (31),

$$Z_T = \frac{G}{K'_g a} \int \frac{di'}{i'* - i'} = H_{og,i} \cdot N_{og,i} \tag{34}$$

As for the case of dehumidification, i' is a modified enthalpy that accounts for non-air-water situations where the Lewis number is not unity (see Equations (21) and (22)).

The material balance relationships for the gas heating/humidification case are often handled graphically, using enthalpy-temperature plots. Equations (31) and (34) are used by manufacturers of cooling towers, with the number of transfer units from the plot and the heights of transfer units from large-scale test data. The approach follows closely that of determining required heights for absorbers based on numbers and heights of diffusional transfer units.

3 PERFORMANCE CHARACTERISTICS OF EQUIPMENT

3.1 Packed Columns

These columns, conventionally used for mass-transfer-controlled operations, can also be used for direct-contact heat exchangers. The literature and practice dealing with packed columns is vast, and it is beyond the scope of the present paper to describe the many different packing materials and their performance characteristics. The heat transfer/mass transfer analogy is useful for dealing with direct-contact heat exchangers; the analogy of Chilton and Colburn (1934) was found by Sherwood et al. (1975) to represent well experimental data for flow in pipes. Huang (1982) found it to work well also for direct-contact heat transfer measurements in a column packed at different times with Raschig rings, ceramic Intalox saddles, metal Pall rings, and metal Hypac rings. A useful form of the analogy is

$$h_g a = \frac{c_g\, G}{H_{g,d}} \left[\frac{Sc_g}{Pr_g}\right]^{2/3} \tag{35}$$

$$h_L a = \frac{c_L L}{H_{L,d}} \left[\frac{Sc_L}{Pr_L}\right]^{2/3} \tag{36}$$

Thus if values of $H_{g,d}$ and $H_{L,d}$ are available, values of the volumetric heat transfer coefficients can be estimated. In turn, these can be converted to overall coefficients and the required height obtained from the appropriate relationships based on the mode of heat transfer (cooling with dehumidification, sensible cooling only, etc.).

Various generalized methods are available for estimating $H_{g,d}$ and $H_{L,d}$. These include the correlation of Bolles and Fair (1979, 1982) for random packings (rings, saddles) and the correlation of Bravo et al. (1985) for structured packings such as Koch-Sulzer BX and Glitsch Gempak 4BG. Reviews of experimental heat transfer data for columns filled with conventional random packings have been provided by Fair (1972a) and Huang (1982).

As indicated earlier, the special area of cooling tower analysis and design has been handled by the use of enthalpy driving forces, taking advantage of the well-known properties of the air-water system. Cooling towers represent a form of packed column, and many studies have been made of the performance of packings (mostly of a structured type, such as wooden slats arranged carefully) for this service. Representative studies have been by Kelly and Swenson (1956) and Lichtenstein (1943). More recently, Barile et al. (1974) have reported on the performance of a "turbulent bed cooling tower" employing a fluidized bed of hollow spheres.

3.2 Spray Columns

Spray chambers are convenient for direct-contact heat transfer operations and are simple in design, containing as internals only one or more banks of spray nozzles. Pressure drop for flow of gas is minimal. Heat-transfer efficiency suffers, however, from deterioration of spray pattern as distance from the nozzles increases and the large degree of backmixing of the gas caused by the impact of the spray droplets.

The earlier work of Pigford and Pyle (1951) has been useful for the development of general approaches to estimating volumetric heat transfer coefficients. A typical presentation of data from the Pigford and Pyle work is given in Fig. 2.2. On the basis of distance from spray nozzles to gas inlet, and numbers of transfer units indicated, heights of transfer units can be determined. These, in turn, can be used with the analogy [Equations (35) and (36)] to obtain values of the heat transfer coefficients. Figure 2.2 shows ammonia/air/water mass transfer data, equivalent to gas-phase coefficients. The authors also studied the oxygen/air/water system, obtaining data for the liquid-phase rate of transfer.

On the basis of the Pigford and Pyle work plus spray column heat transfer data by others, Fair (1972b) presented the following design equation:

$$h_g a = 867 \ G^{0.82} \ L^{0.47}/Z_T^{0.38} \ , \tag{37}$$

where Z_T is the height of a single spray contact zone. While this relationship applies only to the gas phase, it is often assumed that because of circulation patterns inside the droplets there is relatively little resistance to heat transfer in the liquid phase. Reasonable confirmation of the equation was obtained experimentally by Wang (1984).

3.3 Baffled Columns

These columns contain simple segmental or disc/donut-type baffles and are often used when the liquid contains solids or materials that can deposit on surfaces and cause undesirable fouling. The scrubbing action of the cascading liquid prevents excessive buildup. While such columns have had fairly widespread use for mass transfer service, no definitive studies of their efficiency in such service have been reported. Thus, the heat transfer/mass transfer analogy is not particularly useful for determining heat transfer coefficients for baffled columns.

A number of commercial and pilot plant tests of heat transfer in baffled columns have been reported, however, and were summarized by Fair (1972a,b). He recommended the following approximate design equation:

$$Ua = 585 \ G^{0.7}L^{0.4} \tag{38}$$

for baffles spaced at about 0.6 meters, and window areas of 40% to 50% of the column cross-sectional area. For more background the papers of Fair (1961, 1972) should be consulted.

3.4 Crossflow Tray Columns

These columns are not ordinarily used for services that involve primarily heat transfer. If they should be used, the analogy is useful because of the wealth of mass transfer data available for them. While most of the tray efficiencies have been determined for distillation conditions (equimolar counterflow between

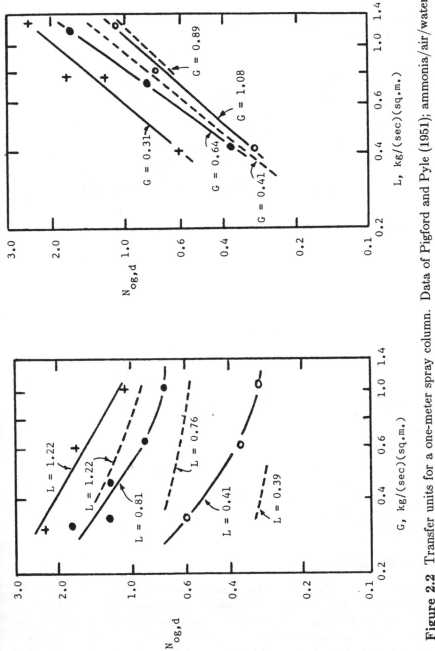

Figure 2.2 Transfer units for a one-meter spray column. Data of Pigford and Pyle (1951); ammonia/air/water; 6 Sprayco nozzles.

    ————   1.71 m high column
    - - - - -   0.85 m high column

35

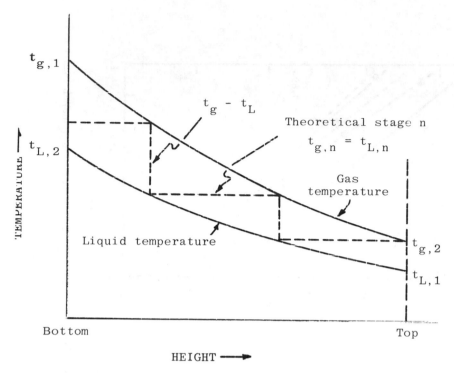

Figure 2.3 Determination of heat transfer stages in a counterflow direct-contact heat exchanger.

phases), those determined for absorption or stripping are probably more appropriate for use with the analogy.

It is possible to use a heat transfer efficiency for a crossflow tray:

$$E_{h,n} = \frac{t_{g,n} - t_{g,n-1}}{t_{L,n} - t_{g,n-1}} \tag{39}$$

Thus, when the temperature of the exit gas equals the temperature of the exit liquid, the efficiency is 100%. Many years ago, Carey (1934) proposed that numerically, $E_{mv} \simeq E_h$, where E_{mv} is the Murphree tray efficiency for mass transfer. A typical plot of temperature profiles in a countercurrent column is shown in Fig. 2.3.

4 RESEARCH NEEDS

This necessarily cursory review of direct-contact process heat transfer principles and practice has been intended to underscore several aspects of the technology. First, the idea of transferring heat directly between phases is not only workable but also economically attractive in many cases. Second, the principles of application are reasonably well worked out and at least for gas-liquid systems cover all of the cases likely to be encountered. Third, the development of methods of predicting the performance characteristics of the several equipment types involved has not progressed nearly to the point where highly reliable designs can be implemented. It is in this last area that much more research is needed.

Specific research needs include the following:

1. Better characterization of the new "high efficiency packings," such as the structured packings of the gauze and sheet-metal type, with respect to heat and mass transfer capabilities. Some of this work is in progress, but is restricted to mass transfer.

2. Further work on the applicability of the heat transfer/mass transfer analogy to gas-liquid contacting that includes significant amounts of heat transfer, taking into account varying system properties as well as device geometries.

3. Measurement of temperature and concentration profiles in the various devices used for direct-contact heat transfer, in order to evaluate proper driving forces for use in heat and mass flux equations.

4. Study of process systems to determine points of application of direct-contact heat transfer as an economically attractive replacement for conventional heat-exchange equipment.

NOMENCLATURE

a	interfacial area, m^2/m^3
c	heat capacity, $J/kg{\cdot}K$
C_o	coefficient [Eq. (15)]
D	molecular diffusion coefficient, m^2/s
e	base for Napierian logarithms
E_h	heat transfer efficiency, fractional
E_{mv}	Murphree tray efficiency, fractional
G	gas mass velocity, $kg/m^2{\cdot}s$
G_m	molar mass velocity of gas, kg moles/$m^2{\cdot}s$
h	heat transfer coefficient, $J/(m^2{\cdot}s{\cdot}K)$
$h_g a$	volumetric heat transfer coefficient for gas phase, $J/(m^3{\cdot}s{\cdot}K)$
$h_L a$	volumetric heat transfer coefficient for liquid phase, $J/(m^2{\cdot}s{\cdot}K)$
$H_{g,d}$	height of a gas-phase diffusional-transfer unit, m
$H_{g,h}$	height of a gas-phase heat transfer unit, m
$H_{og,d}$	height of an overall gas-phase diffusional transfer unit, m
$H_{og,h}$	height of an overall gas-phase heat transfer unit, m
$H_{L,d}$	height of a liquid-phase diffusional transfer unit, m
$H_{L,h}$	height of a liquid-phase heat transfer unit, m
i	enthalpy, J/kg
i^*	enthalpy of gas phase based on bulk liquid condition, J/kg
i'	modified enthalpy, J/kg [Eq. (21)]
k'	mass transfer coefficient for an individual phase, based on humidity driving force, $kg/m^2{\cdot}s$

k_t thermal conductivity, J/m·s·K

K' overall mass transfer coefficient, based on humidity driving force, kg/m²·s

$k'a$ volumetric mass transfer coefficient for an individual phase, based on humidity driving force, kg/m²·s

$K'a$ overall volumetric mass transfer coefficient, based on humidity driving force, kg/m³·s

$k_g a$ mass transfer coefficient for the gas phase, based on partial pressure driving force, kg moles/m²·s·atm

$k_L a$ mass transfer coefficient for the liquid phase, based on concentration driving force, 1/s

L liquid mass velocity, kg/m²·s

L_m liquid molar mass velocity, kg moles/m²·s

N_A mass flow of species A, kg/s

$N_{g,d}$ number of gas-phase diffusional transfer units

$N_{g,h}$ number of gas-phase heat transfer units

$N_{og,d}$ number of overall gas-phase diffusional transfer units

$N_{og,h}$ number of overall gas-phase heat transfer units

$N_{s,h}$ number of equilibrium heat transfer stages

P pressure, atm

Pr dimensionless Prandtl number $(=c\mu/k_t)$

Q rate of heat transferred, J/s

Q_c rate of heat transferred in condensation, J/s

Q_g rate of heat transfer in sensible cooling or heating of gas, J/s

Q_v rate of heat transferred in vaporization, J/s

Q_T total rate of heat transferred, J/s

S cross-sectional area, m²

Sc dimensionless Schmidt number $(=\mu/\rho D)$

t temperature, C or K

t_g bulk gas temperature, °C or K

t_i interface temperature, °C or K

t_L bulk liquid temperature, °C or K

TS tray spacing, m

U overall heat transfer coefficient, J/m²·s·K

Ua volumetric overall heat transfer coefficient, J/m³·s·K ($\overline{Ua}$ = mean value)

V volume, m³

V_T total volume for heat transfer, m³

w mass flow rate, kg/s

Z height, m (Z_T = total height)

Subscripts

A species A

g gas phase

i interface

L	liquid phase
m	molar
M	mean value
$n,n\text{-}1$	stages n and $n\text{-}1$ ($n\text{-}1$ below n)
T	total

Greek letters

α	Ackerman factor [Eq. (13)]
Δt_M	mean temperature difference (taken as logarithmic mean)
λ	stripping factor (Table 2)
μ	viscosity, $kg/m\ s$
ρ	density, kg/m^3
ϕ	dimensionless Lewis number ($= h/k'c$)

REFERENCES

Barile, R. G., Dengler, J. L., Hertwig, T. A., *AIChE Symp. Ser. No. 128, 70*, 154 (1974).

Bolles, W. L., Fair, J. R., *I. Chem. E. Symp. Ser. No. 56*, 3.3/35 (1979).

Bolles, W. L., Fair, J. R., *Chem. Eng. 89* (14) 109 (July 12, 1982).

Bravo, J. L., Rocha, J. A., Fair, J. R., *Hydrocarbon Proc. 64* (1) 91 (1985).

Carey, J. S., in *Chemical Engineers' Handbook*, first edition, J. H. Perry, ed., pp. 1191-2, McGraw-Hill, New York, 1934.

Chilton, T. H., Colburn, A. P., *Ind. Eng. Chem. 26*, 1183 (1934).

Fair, J. R., *Petro/Chem Engineer 33* (3) 39 (1961).

Fair, J. R., *Chem. Eng. Progr. Symp. Ser. No. 118, 68*, 1 (1972a).

Fair, J. R., *Chem. Eng. 79* (13) 91 (1972b).

Huang, C.-C., *"Heat Transfer by Direct Gas-Liquid Contacting,"* MS thesis, University of Texas at Austin, 1982.

Kelly, N. W., Swenson, L. K., *Chem. Eng. Progr. 52*, 263 (1956).

Lichtenstein, J., *Trans. ASME 65*, 779 (1943).

Pigford, R. L., Pyle, C., *Ind. Eng. Chem. 43*, 1649 (1951).

Sherwood, T. K., Pigford, R. L., Wilke, C. R., *Mass Transfer*, McGraw-Hill, New York, 1975.

Wang, T., *"Mass Transfer Characterization of a Countercurrent Spray Column"*, MS Thesis, University of Texas at Austin, 1984.

COMPUTATIONAL TECHNIQUES FOR TWO-PHASE FLOW AND HEAT TRANSFER

C. T. Crowe

ABSTRACT

Computational models for two-phase flow and heat transfer are reviewed. The computational models are categorized according to the inertial and thermal states of the system. The computational techniques applicable to each system are outlined and discussed. Directions for further model development are suggested.

1 INTRODUCTION

Many direct-contact heat-exchange processes involve heat exchange in two-phase flows. The purpose of this paper is to review numerical models that have been developed for heat exchange in multiphase, multicomponent systems, and suggest future directions for model development.

Direct-contact heat-exchange systems can be classified according to multicomponent or multiphase systems. Multicomponent systems relate to the transfer of heat between two substances in the same phase. The classic example is the transfer of heat between two solid materials for which the well-known models for conductive heat transfer are applicable. The direct-contact heat-exchange between two gases is not of interest here because the "direct contact" results in mixing and the loss of identity of each component. The multicomponent direct-contact heat exchange of primary interest is the liquid-liquid systems in which droplets of one liquid pass through another immiscible liquid and heat transfer takes place between the liquids. Another configuration would be the heat transfer between two streams of liquids. Liquid-liquid heat transfer systems are found in sea water desalination systems and units for extracting heat from geothermal brines. A discussion of liquid droplets immersed in another liquid is provided by Letan in Chapter 6.

Multiphase direct-contact heat-exchange systems are numerous. These systems fall under the general categories of gas-liquid, gas-solid, and liquid-solid systems as shown in Table 3-1. Gas-liquid systems include vapor bubbles in a liquid as well as droplets in a gas stream. There is a bulk of literature on the flow and heat transfer in steam-water systems (Hetsroni, 1982). These flows are identified by flow regimes such as annular-mist flows, stratified flows, slug flows, and so on. The appropriate flow regime has to be identified by reference to a flow map before the appropriate heat transfer correlation can be selected. Steam-water heat transfer is important in the nuclear power industry.

Gas-solid systems include solid particles in a gas stream as well as gas flowing through a porous medium. Fluidized beds fit into the general category of gas-solids flows. Liquid-solid systems include the flow of liquids through a porous solid or packed bed. The heat transfer from solid particles suspended in a flowing liquid also constitutes a liquid-solid system.

Thus there is a wide variety of fundamental direct-contact heat-exchange systems. The controlling heat transfer mechanism is not the same for all systems, which precludes the development of a general numerical model. The purpose of this paper is to review the numerical models available for these systems and discuss the needs for future model development.

Table 3.1. Multicomponent and multiphase heat transfer systems.

Multicomponent		Multiphase	
Liquid-Liquid	Gas-Solid	Gas-Liquid	Liquid-Solid
Drops in a Liquid	Gas in Porous Media	Bubble Flow	Liquid in Porous Media
	Packed Beds	Slug Flow	Liquid in Packed Beds
	Fluidized Beds	Stratified Flow	Fluidized Beds
	Particles in a Gas	Annular-Mist Flows	Particles in a Liquid
		Sprays	

2 NUMERICAL MODEL CLASSIFICATION

The term "computational (or numerical) model" is used to describe numerical models with varying degrees of sophistication. It is helpful for discussion to subdivide the models into more descriptive categories.

2.1 Global Models

A global model is one in which the whole process is modeled by a single equation. Generally global models are based on the parametric groupings that result from dimensional analysis. A global model may also be used to describe a physical model in which the system is divided into computational elements. An example of such a model would be the liquid-vapor system shown in Fig. 3.1. Bubbles of hot vapor rise in the central core region and transfer heat to the fluid, which moves downward in the outer annulus. The system could be divided into the core region and the annular region. Models for the heat transfer in each region would be developed. Expressions for the heat transfer between the two regions would be used to complete the model.

Liquid in

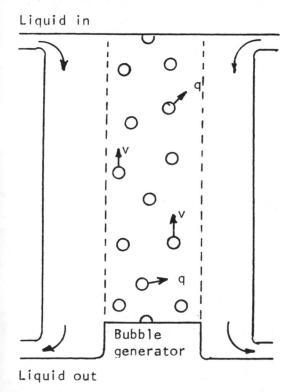

Liquid out

Figure 3.1 A bubble-liquid heat transfer system.

2.2 Local Models

Local models are based on processes that occur locally in the flow. These models may use differential equations for the fluid mechanics and heat transfer of each phase and empirical correlations for the transfer between phases. For the example of the liquid-vapor system shown in Fig. 3.1, differential equations would be written for the fluid mechanics and heat transfer in the liquid phase. The vapor bubbles would be treated as vapor bubbles would be treated as vapor cavities with constant temperature and uniform velocity. Empirical expressions would be used for the heat transfer between phases.

The resulting equations would be written in finite difference form or finite element form with the appropriate boundary conditions. The solutions to these equations provide the velocity and temperature fields and the heat exchange between phases.

2.3 Point Models

A point model is one in which the governing differential equations are applied at every point in the flow field. In the example used here, the differential equations for the heat transfer in the liquid as well as the differential equations for the heat transfer within the vapor of the bubble would be utilized. This model would require fewer assumptions and empiricism than the local model. However, the grid dimensions would have to be small enough to represent the detail of the system. Thus the local grid system would have to be smaller than the bubbles. If the flow is turbulent, the grid would have to be smaller than the smallest turbulence scale in the flow. This grid size and the ensuing number of nodal points may be impractical.

In general the local models are the most practical for two-phase systems. They give sufficient detail of the velocity and temperature field and utilize a minimum of empiricism.

3 GROUPING ACCORDING TO PHYSICAL MODELS

The development of a useful numerical model is a two-step process. First, one must identify the dominant transfer mechanisms and establish the differential or empirical equations that describe these mechanisms. Second, one must select the most appropriate computational scheme to solve the equations. In this section a generalization of models according to dominant transfer mechanisms is discussed. The significance of dilute and dense systems will be outlined and the assumption of one way and two-way coupling will be explained. Both momentum and energy transfer mechanisms are of interest here.

3.1 Inertially Dilute vs. Dense

Consider a cloud of particles being conveyed by a gas stream as shown in Fig. 3.2. The particles are at a higher temperature than the gas so heat is transferred from

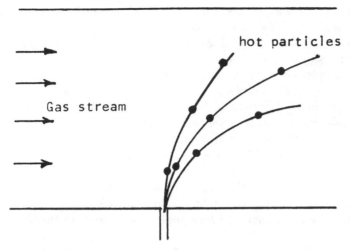

Figure 3.2 Particles injected into a gas-stream.

the particles to the gas. In an "inertially dilute" flow, the motion of the particles is completely controlled by the aerodynamic and body forces acting on the particles (Crowe, 1982). That is, the particle motion is that of an isolated particle with modified transfer properties to reflect the effect of neighboring particles. On the other hand, if the particle-particle collisions are important in controlling the particle motion, the flow is "inertially dense."

The criterion for establishing whether a multiphase flow is inertially dilute or dense is the ratio of aerodynamic response of the particle to the time between particle-particle collisions. The aerodynamic response time of a particle in Stokes flow is

$$\tau_A = \rho_d d^2 / 18 \, \mu \,, \tag{1}$$

where ρ_d is the particle's material density, d is the particle diameter, and μ is the dynamic viscosity of the gas. This is the time it takes a particle released from rest in a uniform flow to achieve 63% of the free stream velocity, assuming Stokes drag coefficient is valid.

The time between particle-particle collisions is τ_c and depends on the relative velocity between particles, the particle diameter, and the particle number density in the field.

If the ratio of particle response time to the time between collisions is much less than unity, the flow is inertially dilute.

$$\tau_A/\tau_c \ll 1 \quad \text{inertially dilute.}$$

However, if this ratio is much larger than unity, the flow is inertially dense, and the motion of the particle cloud is controlled by particle-particle collision.

$$\tau_A/\tau_c \gg 1 \quad \text{inertially dense.}$$

The particle motion in a packed bed is obviously the limiting case of a dense flow. The fluidized bed is also an inertially dense flow. The identification of inertially dilute or dense is significant in the choice of the computational model.

A complication with the definition of dilute and dense can arise for droplets which coalesce during transit through the system. A time interval is required for the formation of a new droplet after the initial contact. If this time is short compared with the time between droplet-droplet contact, then the formation time can be neglected and the conventional definition of dilute and dense applies. If the formation time is comparable to the time between contact, then the system motion is controlled by the droplet contact and formation process and the system is dense. In this situation only point models would be adequate to analyze the flow.

In the case of bubbles the fluid mechanic response time is very small because the vapor density is much smaller than the fluid density. Thus the flow bubbles tends to be dilute. Once again, if the time for bubble formation is comparable to the time between bubble-bubble contact, the formation process controls the bubble motion and the flow is dense.

3.2 Thermally Dilute or Dense

A thermally dilute or dense flow depends on the mechanism that controls particle temperature. The particle temperature can depend on the convective (or radiative) heat transfer from the particle to the surrounding liquid or gas. Or, the particle temperature can depend on the energy transferred from particle to particle by direct exchange through physical contact or radiative transfer.

The rate of convective heat transfer from the particle to the surrounding liquid or gas is given by

$$\dot{q}_T = Nu \; \pi k \; d \; \Delta T \tag{2}$$

where k is the thermal conductivity, Nu is the Nusselt number, d is the particle diameter, and ΔT is the particle-fluid temperature difference. If radiation is the controlling heat transfer mechanism between the particle and the fluid then the heat transfer rate during particle transit through the system will be a function of the local particle and fluid temperatures, particle number density as well as the particle's size, and thermal properties.

Particle-particle heat exchange can occur by physical contact or by radiative exchange between particles. Define q_c as the heat transferred by a physical contact. The rate of heat exchange by radiation between particles, $\dot{q}_R$, is a function of the particles' physical and thermal properties as well as the particle temperature and number density. Calculation of $\dot{q}_R$ could be a formidable task.

A system is thermally dense if the ratio of heat transfer rate by particle-particle radiation to the heat transfer rate to the surrounding fluid is much greater than unity.

$$\dot{q}_R / \dot{q}_T \gg 1 \qquad \text{thermally dense.}$$

Also if the ratio of the heat transferred by physical contact to that transferred to the surrounding fluid is larger than unity, the system is thermally dense.

$$q_c/\dot{q}_T \tau_c \gg 1 \qquad \text{thermally dense}$$

A porous medium is an example of a thermally dense medium. A packed bed may also be thermally dense. The two conditions necessary for a thermally dilute system are

$$q_c/\dot{q}_T \ \tau_c \ll 1 \ \text{and}$$

$$\dot{q}_R/\dot{q}_T \ll 1 \qquad \text{thermally dilute}$$

In the case of droplets, one must consider the formation time for new droplets after collision and coalescence. If the formation time is small compared with the time between collisions, then the droplet system can be considered thermally dilute. Typically, droplet temperatures are not so high that droplet-droplet radiation need be considered. Droplet-droplet radiation would only be important in low-density environments such as outer space.

The same observation applied for bubbles in a liquid. If the formation time is small compared with the time between bubble-bubble contact then the bubble system can be considered thermally dilute.

3.3 One-Way and Two-Way Coupling

Another descriptor used to describe the physical models for two-phase property transfer is one-way and two-way coupling (Crowe, 1982). Consider the gas-particle flow system shown in Fig. 3.2. The hot particles are cooled as heat is transferred by convection to the gas. A model that accounts for the energy transferred from the particle but neglects the energy added to the gas is a model with one-way thermal coupling. A model that accounts for the energy both removed from the particles and added locally to the gas is a model with two-way thermal coupling. The same definitions apply to momentum and mass transfer. A model that accounts for two-way coupling for mass, momentum, and energy is "fully" coupled.

The use of fully coupled models may not always be necessary. Obviously mass coupling is not needed for nonreacting particles. Also the particle mass concentration may be so small that the effect of the particles on the gas velocity field is negligible. In this situation only thermal one-way coupling need be included.

4 BASIC EQUATIONS FOR MULTIPHASE MOMENTUM AND ENERGY TRANSPORT

There have been several approaches used to develop the fundamental equations for mass, momentum, and energy transport in multiphase and multicomponent systems. The earlier approaches were mostly heuristic and led to many discrepancies in the final form of the equations. The first attempts to introduce mathematical rigor to the equation development were those of Drew (1971) and Ishii (1975). These approaches are based on volume and temporal averaging. Some lingering doubts still persist over the correct form of the equations.

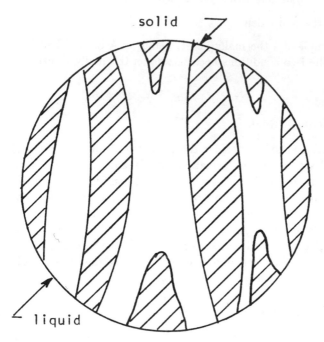

Figure 3.3 Section of porous material.

Consider the section of porous material shown in Fig. 3.3. Liquid or vapor pass through the pores of the solid material. This system is inertially and thermally dense because the solid and fluid are contiguous. One begins with the differential (point) equations for mass, momentum, and energy conservation; namely,

$$\partial \rho/\partial t + \partial(\rho\ V_i)/\partial x_i = 0 \tag{3}$$

$$\partial(\rho V_i)/\partial t + \partial(\rho V_i V_j)/\partial x_j$$
$$= -\ \partial p/\partial x_i + \mu \partial^2 V_i/\partial x_j \partial x_j + \rho g_i \tag{4}$$

$$\partial(\rho C_p T)/\partial t + \partial(\rho V_i C_p T)/\partial x_i$$
$$= K \partial^2 T/\partial x_i \partial x_i \tag{5}$$

These equations represent a system with no chemical reaction between phases. For the application to porous media, the acceleration terms in the momentum equation, Equation (4), are neglected because the Reynolds number based on pore size is small. The equations can be further specialized to reflect the properties of each phase. For example, the velocity of the solid phase is zero and the density is constant. This leaves only the energy equation for the solid phase.

The objective of the volume averaging method is to average the point equations over a volume and obtain a set of differential equations for the averaged properties. The spatial average of a property ϕ is defined as

$$<\phi> = (1/V)\int_V \phi dV \tag{6}$$

where V is the averaging volume. The value of $<\phi>$ is associated with the point at the center of the averaging volume. The phase average is defined as

$$<\phi_k> = (1/V) \int_V \phi_k \, dV \tag{7}$$

where k designates the liquid or solid phase, and the integration is performed over the entire averaging volume. The integrand is zero at points where the phase is nonexistent.

The primary tool in volume averaging is the spatial averaging theorem (Slattery, 1972), which relates the average values of the gradient $<\nabla\phi>$ to the gradient of the average value $\nabla<\phi>$ and integrals of the property ϕ over the area

$$\nabla<\phi_k> = <\nabla\phi_k> + \frac{1}{V} \int_A \mathbf{n} \; \phi_k \; dA \tag{8}$$

where A is the area of the liquid-solid interfaces within the averaging volume and $\mathbf{n}$ is the unit outward normal vector from the k-phase.

The energy equation that results from volume averaging is quite complicated and assumptions must be made to arrive at an equation that is amenable to a computational solution. The assumption of local thermal equilibrium is commonly used to simplify the equations (Whitaker, 1980). Additional deletion of terms based on order of magnitude analyses yields an energy equation with an effective thermal conductivity.

$$<\rho> C_p \partial<T>/\partial t + \rho_\ell C_{p_\ell} <V_\ell>\nabla<T>$$

$$= \nabla \cdot (K_{e\!f\!f}^T \cdot \nabla<T>) \tag{9}$$

where $<V_\ell>$ is the average velocity of the liquid phase. In general the thermal conductivity is a tensor with nine components (Whitaker, 1977). Note that the equation is similar to the point equations but with averaged properties. Averaging the point equations yields a local model because parameter values at a point are no longer obtainable.

One might expect that the equations for heat, mass, and momentum transfer in a porous media will differ from those for a system in which a discontinuous (dispersed) phase is carried or suspended by a fluid as shown in Fig. 3.4. Bubbles in a liquid, particles or droplets in a gas, or droplets in an immiscible fluid represent such systems. The most common approach is to treat the dispersed and continuous phases as two interpenetrating fluids; that is, two fluids that coexist at the same physical point. The resulting equations can be obtained by application of averaging techniques with approximations used in the evaluation of the integrals in the spatial averaging theorem.

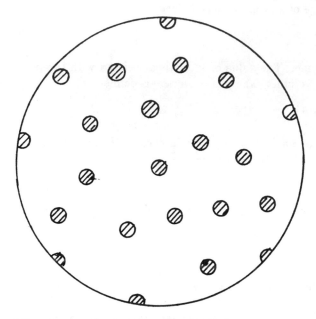

Figure 3.4 Section of a dispersed phase system.

The continuity equation for the fluid (continuous) phase is

$$\partial(\alpha_f\rho_f)/\partial t + \partial(\alpha_f\rho_f U_i)/\partial x_i = 0 \tag{10}$$

where ρ_f is the material density of the fluid, α_f is the volume fraction, and U_i is the velocity of the continuous phase. The corresponding equation for the dispersed phase is

$$\partial(\alpha_d\rho_d)/\partial t + \partial(\alpha_d\rho_d V_i)/\partial x_i = 0 \tag{11}$$

where ρ_d is the material density, α_d is the volume fraction, and V_i is the velocity of the dispersed phase. The continuity equations shown here do not include chemical reaction or mass exchange between phases.

The momentum equation for the continuous phase is written as

$$\partial(\alpha_f\rho_f U_i)/\partial t + \partial(\alpha_f\rho_f U_i U_j)/\partial x_j = -\alpha_f\partial P/\partial x_i$$
$$+ \lambda(V_i - U_i) + \partial(T'_{i,j})/\partial x_j + \alpha_f\rho_f g_i \tag{12}$$

where the term $\lambda(V_i - U_i)$ represents the drag force of the dispersed phase on the continuous phase and g_i is the body force due to gravity. The term $T'_{i,j}$ is the viscous shear tensor for the continuous phase. This should be a function of the volume fraction and strain rate of the fluid. The correct formulation for this tensor has not been established.

The momentum equation for the discontinuous phase is

$$\partial(\alpha_d\rho_d V_i)/\partial t + \partial(\alpha_d\rho_d V_i V_j)/\partial x_j = -\alpha_d\partial P/\partial x_i$$

$$+ \lambda(U_i - V_i) + \partial(T^d_{i,j})/\partial x_j + \alpha_d \rho_d g_i \tag{13}$$

where $T^d_{i,j}$ is the shear stress tensor for the discontinuous phase. This tensor represents the momentum transfer due to interactions of the dispersed phase. For example, it would represent the momentum transfer due to the particle-particle collisions. The form of this tensor is not known. In an inertially dilute flow this tensor is set equal to zero.

The energy equation for the continuous phase is given by

$$\partial(\alpha_f \rho_f e_f)/\partial t + \partial(\alpha_f \rho_f U_i e_f)/\partial x_i$$
$$= h(T_d - T_f) + K^f \partial^2 T_f/\partial x_i \partial x_i$$
$$- P[\partial \alpha_f/\partial t + \partial(\alpha_f U_i)/\partial x_i] \tag{14}$$

where e_f is the internal energy of the continuous phase and $h(T_d - T_f)$ represents the energy transferred by convection from the dispersed phase to the fluid (continuous) phase. It is assumed that the conduction of energy in the fluid can be expressed with Fourier's law. The appropriate thermal conductivity must depend on the volume fraction and particle properties. This equation also neglects the heat generation by dissipation and the energy transferred to the fluid by radiation.

The corresponding energy equation for the dispersed phase is

$$\partial(\alpha_d \rho_d e_d)/\partial t + \partial(\alpha_d \rho_d V_i e_d)/\partial x_i = h(T_f - T_d)$$
$$+ \partial Q^d_j/\partial x_j - P[\partial \alpha_d/\partial t + \partial(\alpha_d V_i)/\partial x] \tag{15}$$

where the tensor Q^d_j represents the energy transferred by particle-particle contact or interparticle radiative transfer. Once again the exact form for this tensor is unknown. For a thermally dilute system the value of this tensor is zero.

5 COMPUTATIONAL MODELS

The computational models will be discussed according to the dilute or dense nature of the systems. The models for the various flow regimes in gas-liquid flows will not be included. The reader is referred to Duckler (1978) and Michiyashi (1978) for discussion of heat transfer models in these regimes.

5.1 Inertially and Thermally Dense Systems

The flow of mass and heat transfer in a porous material (or packed bed) is an example of a inertially and thermally dense system. The earlier models for heat and mass transfer in porous media were global models based on Darcy's and Fourier's laws (Eckert and Pfender, 1978). For example, the porous media would be treated as a material with many holes that extend through the material (Schhuster, 1974). Darcy's equation, or modifications thereof, would be used to establish the flow rate. The temperature distribution and heat transfer would be modeled as a one-dimensional system for heat transfer from the pores to the fluid.

The more recent models have been based on the equations resulting from volume averaging. Certain simplifications have to be made to handle the thermal conductivity tensor such as assuming isotropy in certain planes. The equations that result can be solved by application of standard methods for numerical solutions of partial differential equations. Whitaker (1984) has obtained solutions for the drying of a wet sand bed using the equations resulting from volume averaging.

The current difficulties in modeling heat transfer in porous media center on the formulation of the equations and physical models. More effort is needed to establish methods to evaluate the thermal conductivity from physical properties and experimental measurements. Still, the use of the averaged equations appears to be the most fruitful direction for continued model development.

5.2 Inertially Dilute, Thermally Dense

An example of an inertially dilute, thermally dense system would be particles or droplets in a rarefied atmosphere where the particles move without collisions and heat is transferred from particle to particle by radiation. Computational solutions to this problem are not readily available. If the dominant heat transfer mechanism is radiation to the surroundings, the system would be thermally dilute and readily amenable to solution.

5.3 Inertially Dense, Thermally Dilute

An example of an inertially dense, thermally dilute system is a fluidized bed. In this case the particle motion is controlled by particle-particle collision and the particle-particle heat transfer is negligible (Wen and Chang, 1967).

A comprehensive review of fluidized beds is provided by Chen in Chapter 8 of this volume.

The early models were global models. Perhaps the best-known model for temperature distribution in the bed is that of Baeyens and Goosens (1973), in which the heat transferred from the particles to the gas was related to the bubble temperature. The other global model for heat transfer from the bed to the wall by Mickley and Fairbanks (1935) has been used for many years and has been the forerunner to several other global models.

Recently Gidaspow and associates (1985) have used the two-fluid equations to model the transient bubble motion in a fluidized bed. They neglect the viscous stress tensor for the fluid phase, $T'_{i,j}$, and assume that particle-particle heat transfer can be represented by Fourier's law,

$$Q_j^d = K_d \partial T_d / \partial x_j \qquad (16)$$

where K_d is the effective thermal conductivity through the dispersed phase. Inclusion of this term makes the system thermally dense, which is in contrast to Wen and Chang's findings. The particle shear stress tensor is expressed empirically as a function of the fluid volume fraction; namely,

$$\partial (T_{i,j}^d) / \partial x_j = G(\alpha_f) \, \partial \alpha_f / \partial x_i \qquad (17)$$

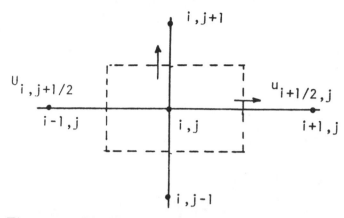

Figure 3.5 Displaced grid system used KFIX numerical code.

where

$$G(\alpha_f) = 10^{(-8.76\,\alpha_f + 5.43)} N/m^2$$

The gas-particle drag force is modeled using Ergun's equation up to fluid volume fractions of .8 and using an expression that approaches the isolated particle drag for higher fluid volume fractions.

The equations are integrated in time using the KFIX numerical code developed at Los Alamos (Rivard and Torrey, 1977). The code uses a staggered grid system, shown in Fig. 3.5, where the velocities are located on the edges of the computational cells and all the other variables are evaluated at the cell center. The grid arrangement is convenient for setting up the continuity equation and locating velocities between pressure nodes so that a pressure gradient is associated directly with a velocity.

The equations are integrated explicitly in time. The momentum equations are integrated in two parts; the first part includes the pressure gradient and the momentum coupling term and the second part includes the convection terms, gravity and viscous forces. The pressure field is adjusted implicitly each step to provide a velocity field that satisfies continuity. This procedure parallels the early work in code development done at Los Alamos (Harlow and Amsden, 1975).

Gidaspow et al. (1985) report solutions for bubble motion in a fluidized bed that qualitatively agree with experimental observations. Syamlal and Gidaspow (1985) report predictions for the average heat transfer coefficient in a fluidized bed, shown in Fig. 3.6, which agrees reasonably well with experimental data (Ozkaynak and Chen, 1980). The two numerical predictions correspond to setting the thermal conductivity for particle-particle heat transfer equal to the thermal conductivity of the material of the particle phase and equal to the value given by the Zehner and Schlunder model (Bauer and Schlunder, 1978). No information is given on the initial conditions used for the model.

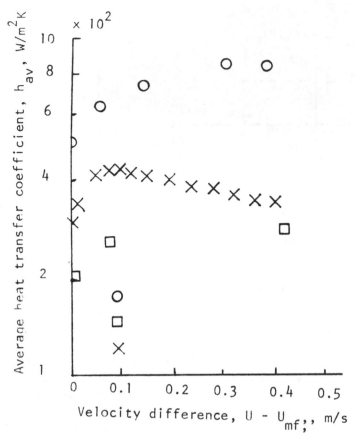

Figure 3.6 Comparison of predicted and measured heat transfer coefficients for a fluidized bed. (Ref: Syamlal and Gidaspow, 1985).

Based on the work of Gidaspow et al. it appears that the two-fluid model is promising for model development in dense phase systems. Of course there is considerably more work that has to be done to better quantify the particle-particle shear stress and heat transfer tensor for such flows. This is imperative before the models can be used reliably to predict scale up effects and changes in operational performance.

5.4 Inertially and Thermally Dilute Systems

An example of an inertially and thermally dilute system is particles (or droplets) in a spray. Other examples include bubbles in a liquid or droplets immersed in another immiscible fluid.

The early computational models for dilute gas-particle flows were based on one-way coupling (Crowe, 1982). Typically the velocity and thermal histories were calculated along trajectories without regard for the effect of the particles on the velocity and temperature field of the gas.

For inertially and thermally dilute systems, the momentum and energy equations for the dispersed phase reduce to those for individual particles (or droplets or bubbles) passing through the fluid. The dispersed phase momentum equation reduces to

$$mdV_i/dt = \beta(U_i - V_i) + mg_i - V_p \, \partial P/\partial x_i \tag{18}$$

where m is the particle mass and V_p is the particle volume. The energy equation becomes

$$mc_p dT_d/dt = \gamma \left(T_f - T_d \right) \tag{19}$$

where c_p is the specific heat of the particle material. Of course, the values used for β and γ should account for the influence of neighboring particles if the particle-particle spacing is such that this correction is necessary. This simplified form of the equation permits one to use the "trajectory" approach.

In the trajectory approach the flow field is divided into a series of computational cells as shown in Fig. 3.7. These computational cells are the nodal points for the finite difference formulation of the fluid-phase equations. The computation is begun by first solving for the velocity and temperature of the gas flow field with no particles. Particle trajectories are then calculated starting from the point where the particles are introduced into the field and continued until the particles exit from the computational field. The change in mass, momentum, and energy as the particles cross cell boundaries is stored as a source of mass, momentum, and energy for the gas phase. The gas flow field is recalculated using the source terms accumulated from the trajectory calculations. New trajectories are calculated, new source terms are re-evaluated, and the cycle is continued until convergence is achieved.

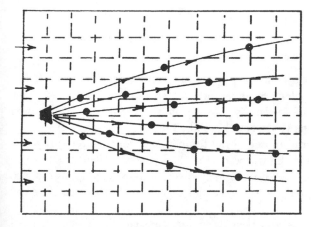

Figure 3.7 Computational field for trajectory approach.

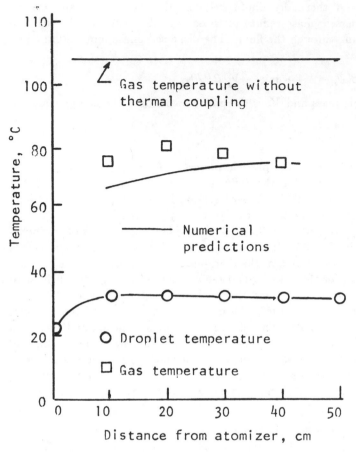

Figure 3.8 Predicted air and droplet temperatures for a water spray injected into a hot gas. (Ref: Crowe, 1980).

The trajectory approach is conceptually simple and straightforward. It has been used in a variety of problems such as droplet evaporation in gas streams (Crowe, 1980), spray drying of foodstuffs (Crowe et al., 1984), fire suppression (Alpert, 1980), and spray cooling (Palaszewski et al., 1981).

The predicted air temperature and spray temperature for a water spray in a hot air stream as obtained using the trajectory method (Crowe, 1980) is shown in Fig. 3.8. The air temperature with no coupling effects (one-way coupling) is also shown on the figure. The predictions agree quite well with the experimental results.

The trajectory method has also been used for transient liquid jets sprayed into a gas (Dukowicz, 1981). Forms of this code have been used in atomization models.

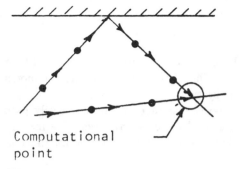

Figure 3.9 Mechanism for nonunique particle velocity and temperature at a computational point.

The two-fluid model has also been used for inertially and thermally dilute flows. The computational method for steady flows has first introduced by Spalding (1977) with the IPSA code. With this method the two fluid equations for the fluid and dispersed phases are solved in the same way using finite difference equations. The pressure field is corrected by adjusting the velocities to have the sum of the volume fractions of each phase equal unity. The application of this method to heat transfer problems has not appeared frequently in the open literature.

Problems of inertially and thermally dilute systems can also be handled by the KFIX code. This code was originally developed for steam bubbles in water.

The two-fluid model is based on the assumption that the velocity and temperature of the dispersed are unique at each point (or in each computational volume). However, there are physical situations for which this is not valid. Consider a particle impinging and rebounding from the wall as shown in Fig. 3.9. Another particle moves past the wall without collision. The temperatures and velocities of these two particles will be different momentum and thermal histories. Yet, they pass through the same point so the velocity and temperature are not unique at this point. The trajectory approach has no difficulty in treating this problem.

Although there has been considerable effort reported in the literature to develop local models for gas-particle and gas-droplet systems, there appear to be no papers for local models for heat transfer in bubble-liquid and liquid-liquid systems. It seems feasible that the modeling ideas generated for gas-particle and gas-droplet flows may apply equally well to bubble-liquid and liquid-liquid systems.

The majority of the codes that have been developed for multiphase flow and heat transfer are based on finite difference methods. There have been few codes based on finite element methods. Of course, there is no reason that the equations resulting from volume could not be solved using finite element methods. Finite element methods for convection-dominated single phase flows are now beginning to appear more frequently in the literature. In the future it is likely that finite element methods will be used to solve the two-fluid equations. Finite element methods could also be used to solve the fluid phase equations in connection with the trajectory approach.

6 CONCLUSIONS

Several computational models are available for modeling multiphase flow and heat transfer in direct-contact heat exchangers. The choice of the numerical model depends on the dense or dilute nature of the system.

The models currently available are poorly documented and not easily used by others practicing in the field. There is a need to review the available models, establish their range of applicability, and develop good documentation for the general user.

The primary need, however, is to have improved physical models for the numerical codes. These physical models include such items as the heat transfer between solids in a fluidized bed, the effective stress associated with solids flow in a dense system, and the quantitative role of turbulence in both dilute and dense systems. The local models for dilute systems require the least empiricism but data are needed for the drag on and heat transfer to a particle surrounded by other particles. The ultimate utility and reliability of computational models depend uniquely on the validity of the physical models used in the codes.

REFERENCES

Alpert, R. L., "*Calculated interaction of sprays with large-scale, buoyant flows,*" ASME Paper No. 82-WA/HT-16, 1982, Winter Annual Meeting.

Baeyens, J. and Goosens, W. R., "Some aspects of heat transfer between a vertical wall and a gas fluidized bed," *Powder Technology*, 81, 1975, pp. 91-96.

Bauer, R. and Schlunder, E. U., "Effective radial thermal conductivity of packings in gas flow. Part II: Thermal conductivity of the packing fraction without gas flow," *Intl. Chem. Eng.*, 18, 1978, pp. 189-204.

Crowe, C. T., "Modeling spray-air contact in spray-drying systems," *Advances in Drying*, ed. A. Mujumdar, Hemisphere, New York, 1980, pp. 63-99.

Crowe, C. T., "Review: Numerical models for dilute gas-particle flows," *Journal of Fluids Engineering*, 104, No. 3, 1982, pp. 297-303.

Crowe, C. T., Chow, L. C. and Chung, J. N., "*An assessment of steam operated spray dryers,*" Proc. Fourth International Drying Symposium, Kyoto, 1984, pp. 369-377.

Drew, D. A., "Averaged field equations for two-phase media," *Studies in Applied Mathematics*, MIT, Vol. L, 1971, pp. 133-166.

Duckler, A. E., "*Modeling two-phase flow and heat transfer,*" Proc. Sixth International Heat Transfer Conference, Toronto, 6, 1978, pp. 541-557.

Dukowicz, J. K., "A particle-fluid numerical model for liquid sprays," *Journal of Computational Physics*, 35, 1980, pp. 229-253.

Eckert, E. R. G. and Pfender, E., "*Heat and mass transfer in porous media with phase change,* Proc. Sixth International Heat Transfer Conference, 6, 1978, pp. 1-12.

Gidaspow, D., Syamlal, M. and Seo, Y. C., "Hydrodynamics of fluidization: supercomputer generated vs. experimental bubbles," *Proceedings of Powder and Bulk Handling and Processing*, Rosement, 1985, pp. 111-117.

Harlow, F. H. and Amsden, A. A., "Flow of interpenetrating phases," *Journal of Computational Physics*, 18, 1975, pp. 440-464.

Hetsroni, G. (ed.), *Handbook of Multiphase Systems*, Hemisphere, New York, 1982.

Ishii, M., *Thermo-Fluid Dynamic Theory of Two Phase Flow*, Eyrolles, France, 1975.

Michiyashi, I., *"Two-phase, two-component heat transfer,"* Proc. Sixth International Heat Transfer Conference, 6, 1978, pp. 219-233.

Mickley, H. S. and Fairbanks, D. F., "Mechanics of heat transfer to fluidized beds," *AIChE Journal*, Vol. 1, 1935, pp. 374-384.

Ozkaynak, T. and Chen, J. C., "Emulsion phase residence time and its use in heat transfer models in fluidized beds," *AIChE Journal*, 26, 1980, pp. 544-550.

Palaszewski, S. J., Jiji, L. M. and Weinbaum, S., "A three-dimensional air-vapor-droplet local interaction model for spray units," *Journal of Heat Transfer*, 103, 1981, pp. 514-521.

Rivard, W. C. and Torrey, M. D., *"K-FIX: A computer program for transient, two-dimensional, two-fluid flows,"* LA-NUREG-6623, Los Alamos, 1977.

Schuster, J. R., *"Nonisothermal porous flow in transpiration cooled nose tips,"* Proc. Fifth International Heat Transfer Conference, 5, Tokyo, 1974, pp. 93-97.

Slattery, J. C., *Momentum, Energy and Mass Transfer in Continua*, McGraw-Hill, New York, 1972.

Spalding, D. B., *"The calculation of free convection phenomena in gas-liquid mixtures,"* Report No. HTS/76/11, Imperial College, London, 1976.

Syamlal, M. and Gidaspow, D., "Hydrodynamics of fluidization: prediction of wall to bed heat transfer coefficients," *AIChE Journal*, 31, 1985, pp. 127-135.

Wen, C. Y. and Chang, T. M., *"Particle-particle heat transfer in air-fluidizing beds,"* Proc. of Intl. Symp. on Fluidization, ed. A.A.H. Drinkenburg, Netherlands University Press, Amsterdam, 1967, pp. 491-506.

Whitaker, S., "Simultaneous heat, mass and momentum transfer in porous media: a theory of drying," *Advances in Heat Transfer*, 13, Academic, New York, 1977, pp. 119-203.

Whitaker, S., "Heat and mass transfer in granular porous media," *Advances in Drying*, ed. A. Mujumdar, Hemisphere Pub., 1980, pp. 23-61.

Whitaker, S., *"Moisture transport mechanisms during the drying of granular materials,"* Proc. Fourth International Drying Conference, Kyoto, 1984, pp. 31-42.

4

INDUSTRIAL PRACTICES AND TWO-PHASE TRANSPORT

James G. Knudsen

1 SESSION 1. INDUSTRIAL PRACTICES AND NEEDS

1.1 Summary of Session

Various types of industrial direct-contact heat transfer equipment were described. Process engineers hesitate to use direct contact heat transfer in process design because they have little knowledge of it. Emphasis was on gas-liquid contactors; design equations with and without heat transfer were presented. Major problems associated with using direct-contact heat transfer in the process industries are

No reliable design methods
Need to account for mass transfer
Fogging and entrainment
Determination of correct temperature driving force

1.2 Discussion of Session

It was brought out in the discussion that more work needs to be done on complicated packings. There is also a limitation on the use of the heat transfer-mass transfer analogy for metal packings because of their high thermal conductivity.

The new metal structured packings have a high surface area and low pressure drop. There is no good method available for predicting heat and mass transfer on these.

There was a question relative to prediction of pressure drop across packings, and this is not accurately predictable for the new high-porosity structured packings.

There is a need to obtain fundamental data on the processes occurring in packings. Much work has been done by EPRI and the nuclear engineering community that should be applicable to direct-contact heat transfer in the process industries. There is a need to identify publications where useful information (both basic and applied) is published.

It was pointed out that process designers seem to favor packed columns as direct-contact devices perhaps because more is known about packings (particularly the traditional ones). Sprays should be used in systems where low pressure drop is required and also in systems containing fouling or dirty liquids. New packings are developed by vendors. They test them or have them tested on a proprietary basis but the results are not published.

It was indicated that the existing design equations are of the traditional chemical engineering form, which has been used for a long time. There is a need to move forward and solve the problem on the basis of first principles.

1.3 Research Needs

Payoff
  Inexpensive way to transfer heat (low pressure drop)
  Good way to transfer heat to or from dirty and fouling fluids
Results
  Save energy
  Save materials
  Accomplish processes not possible with conventional heat exchangers (for example: open cycle OTEC)
Major Need
  Improved design methods for direct contact heat transfer devices.
Specific Needs
  Characterize the new low pressure drop packings with respect to heat, mass, and momentum transfer.
  Investigate application and limitation of mass transfer-heat transfer analogy for packings.
  Determine temperature and concentration profiles.
  Data on direct-contact heat transfer between gases and slurries.
  A knowledge of the controlling mechanisms.
  Effect of entrance conditions; for example, liquid droplets entering a flowing hot gas.
  Provide a means by which information from diverse fields (chemical, civil, mechanical, etc.) can be consolidated.

Determine which mechanisms are important and which can be neglected.
A good design equation will include the significant mechanisms and neglect the
insignificant mechanisms.

2 SESSION 2. COMPUTATIONAL TECHNIQUES FOR TWO-PHASE FLOW AND HEAT TRANSFER

2.1 Summary of Session

Various types of numerical models are defined including global, local, and point
models as well as physical models of inertially dense and inertially dilute, thermally dense and thermally dilute systems. The basic equations for multiphase
momentum and energy transport are presented and computational models
applied to various physical systems. Two major types of models have been
applied to two-phase systems: the two-fluid model and the trajectory model.
Most models have been based on finite difference methods and few are based on
finite element methods. There is a need to document numerical codes better.

2.2 Discussion of Session

It was pointed out that many numerical codes are proprietary and there is a need
to get more information into the public domain about numerical codes.

An edited, abridged comment on A. T. Wassel (Science Applications International, Inc.) follows:

> Some of the problems of these codes are the poor technical documentation. It
> seems that 1) the communication skills among engineers are not good, and 2)
> funding agencies do not put emphasis on proper documentation, so we spend
> a lot of time developing codes and in the last week or two we try to docu-
> ment them. Sometimes also we try to generalize the applicability of codes. I
> do not think that we will be able to have one code that will predict all types
> of flow regimes and ultimately some fine tuning has to be done.

An edited, abridged comment of A. F. Mills (UCLA) follows:

> I think the objective of writing the codes we are talking about is to develop
> the design tool and at any given instant in time, to do the best we can to
> design the required equipment because we have a company wishing to design
> a piece of equipment today, or a government agency wanting a design today.
> So, we we put together the code, we know that each piece of numerical input
> has uncertainties associated with it. We then take bench scale data of what-
> ever we have and see how well the code does. If we can vary our numerical
> input within a reasonable range so as to get a good match with the experi-
> mental data available that is the best we can do.

An edited, abridged comment of John Chen (Lehigh University) follows:

> The point I was trying to make was to illustrate Dr. Crowe's number one
> need. My concern is that at this current stage of science in multiphase

design, we certainly do the best we can with what we know and numerical techniques are another tool that is available. My own conclusion is that the uncertainty in the constitutive relationships required to obtain closure of the specified problem is so large compared to the other unknowns, that it is dominant. Going back to the illustration that is indicated, i.e., to predict the heat transfer in dispersed flow, the two-fluid model is a six conservation equation model. The constitutive input relationships fall into two categories, the hydraulic and the thermal. The hydraulic information basically is the interfacial drag functions on the droplets that needs some numerical model for it. The thermal process is basically the boiling curve. When there is a disagreement, changing either one can make the results fit the data. But what becomes worse is that we now find that you don't have a magic droplet size; what you have, of course, is a spectrum of droplet sizes and they all behave differently. In fact, the small drops may be carried along, but the big drops fall down. A very different behavior. The uncertainty in the physics, in many cases, is overwhelming. I think that should be where the dominant effort needs to be—the mechanistic physics.

An edited, abridged comment by Mike Chen (University of Illinois) follows:

In a sense there are two approaches, one is to be fairly specific; the result of that kind of model can be extremely useful, it can be immediately used by someone. Another approach is to make the model so general that it can fit anything. The modeler can say my model is always correct; it's your problem to fit it to whatever you need. The fact is, I would say that kind of model probably is not going to be very useful. I think we are running into this kind of problem; in a sense you can defend your model on the basis of its versatility but if the model is that versatile it probably isn't a model at all. Therefore, I am coming back to the point that has been made. At this point probably we need input to make the model more specific so we have some form of a constitutive equation that would be at least useful over a wide range of conditions. It's perhaps not really fair to say we can go on modeling because even if our results aren't right we can always change it and make it fit. Most of us who are good engineers can fit something with almost any kind of equation given enough parameters to fit the data.

It was commented that if the numerical model can tell what's important and what's not important, that's an important contribution.

2.3 Research Needs

Payoff—good computational models would provide a good means to design and scale up direct-contact devices.

Result—make reliable design and means of design optimization.

Major need—good documentation of computer models and identification of model applicability.

Specific needs—good knowledge of physics of the process (to provide constitutive equations for model). This includes (a) heat, momentum, and mass transfer during particle-particle contact, (b) heat, mass, and momentum transfer between particle and fluid as affected by the presence and proximity of other particles.

5

MASS TRANSFER EFFECTS IN HEAT TRANSFER PROCESSES

J. J. Perona

1 INTRODUCTION

Mass transfer processes are commonly direct-contact processes. Such chemical process operations as gas absorption, solvent extraction, distillation, and gas adsorption are examples. Gas-liquid, liquid-liquid, or fluid-solid systems are passed through devices such as packed towers, plate towers, fixed or fluidized beds, spray, and bubble towers designed to promote physical contact of the phases.

Direct-contact heat transfer is less common. The inevitable, accompanying mass transfer may or may not be significant or desirable. In any case its extent must be estimated in the evaluation of a proposed direct-contact heat transfer operation.

2 EXAMPLE 1 - GEOTHERMAL POWER CYCLE

A typical geothermal fluid at the well-head consists of a hot brine liquid phase, steam, and noncondensable gases such as carbon dioxide and hydrogen sulfide. If this stream is mixed with a working fluid such as isobutane in a direct-contact heat exchanger, the prediction of the compositions of the effluent streams from the heat exchanger is a challenging problem in nonisothermal mass transfer and gas-liquid reactions. Furthermore, water and noncondensable species will be carried with the working fluid through all other parts of the power cycle.

A representative power cycle is illustrated in Fig. 5.1. The following mass-transfer processes affect significantly the design and economics of such a plant:

1. The geothermal fluid entering the direct-contact heat exchanger is in contact with boiling working fluid. The working fluid vapor will strip some of the dissolved noncondensable gases from the geothermal brine. The equilibrium solubilities of the noncondensable gases in the brine may depend on the extent of several chemical reactions involving carbonate, bicarbonate, sulfide, and hydrogen ion concentrations, as well as the boiler pressure and brine temperature distribution. The boiler contains two liquid phases (the aqueous phase and the evaporating organic phase) and a vapor phase. The vessel is probably well mixed due to the vigorous generation of a large volume of vapor, but the geometrical characterization of such systems as a basis for transport analyses is not well developed. The volume requirements for a direct-contact boiler are relatively small, as is its impact on the economics of such a plant.

2. The gases passing through the turbine into the condenser will be a mixture of working fluid vapor, steam, and noncondensable gases. As condensation begins, two liquid phases will be produced: an aqueous liquid and an organic liquid. Noncondensable gases will dissolve in the liquids to an extent depending on condenser temperatures, pressure, and mass-transfer characteristics.

If a surface condenser is used, the literature contains evidence that a continuous organic condensate film is formed, with water droplets as a discontinuous phase (Bernhardt et al., 1971). An average film coefficient based on volume fractions of the two phases in the condensate works fairly well. For the typical geothermal system, the fraction of water may be negligibly small.

3. A gas-liquid separator follows the condenser so that noncondensables may be vented from the system. Some working fluid vapor will go with the gases, and some gases will remain dissolved in the two liquid phases.

4. A liquid-liquid separator will recover the working fluid for recycle to the preheater and boiler. Some dissolved noncondensables will remain in the working fluid.

5. In the preheater the working fluid will be contacted with brine. Some working fluid will dissolve in the spent brine, and dissolved noncondensables may be transferred between the liquids.

Contacting in the preheater takes place between droplets of the organic phase dispersed in the hot aqueous brine. This part of the system is probably the best understood in terms of mass-transfer analysis at the present time.

It is desirable that the noncondensables remain with the brine phase so that the amount going to the condenser is small. The loss of working fluid with the spent brine is a major consideration in the economics of such facilities.

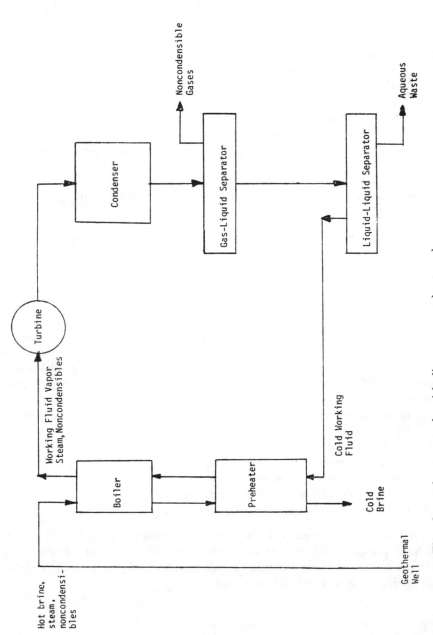

Figure 5.1 Flowsheet of power cycle with direct-contact heat exchange.

This is an example of a case where even a small amount of mass transfer can have significant effects on the economic feasibility of a process. Under typical process conditions, the equilibrium solubility of isobutane in the brine phase is only approximately 200 ppm; however, if saturation is approached in the spent brine flow, the isobutane make-up cost is severe. For example, a 1000 kw plant would require typically a brine flow of 200,000 lb/hour. If the spent brine carried an isobutane concentration of 200 ppm, the loss would be about 8 gal/hour. The value of the isobutane loss would represent a significant fraction (of the order of 20% to 30%) of the gross earnings of the plant.

A mass-transfer model of a direct-contact heat exchanger for brine-isobutane was developed by Knight and Perona (1981). Their study showed that effluent brine concentrations near saturation are obtained with one mm drops (because of high surface area) and with 5 mm drops (because of internal drop circulation and oscillation). Much lower brine concentrations can be achieved with drop sizes around 3.5 to 4 mm (Fig. 5.2). Field tests at East Mesa agreed with the latter result.

3 EXAMPLE 2 - FISCHER-TROPSCH SYNTHESIS IN A FLUIDIZED BED

The first example dealt with a system in which mass-transfer effects were incidental to the heat transfer process and had undesirable consequences. In the second example, this is not the case. Hydrocarbons may be produced by the reaction of carbon monoxide and hydrogen in the presence of an iron catalyst:

$$CO + 2H_2 \rightarrow (-CH_2-) + H_2O$$

The reaction is highly exothermic. Much of the aviation gasoline used by Germany during World War II was produced by this reaction called Fischer-Tropsch.

The heat of reaction must be removed rapidly to effect good temperature control, and fluidized beds have been investigated for that purpose. The steps involved in the reaction are

1.   mass transfer of carbon monoxide and hydrogen to the catalyst particle surface
2.   sorption of carbon monoxide onto the catalyst
3.   surface reactions to form products and liberate heat
4.   desorption of products from the surface
5.   mass transfer of produces to the bulk gas
6.   transfer of heat from catalyst particle to the gas.

In the Fischer-Tropsch example, heat and mass transfer occur between a gas and fluidized particles. These rate processes generally depend on relative velocities between gas and particle and thus require detailed knowledge of the complex flow and mixing patterns within fluidized beds.

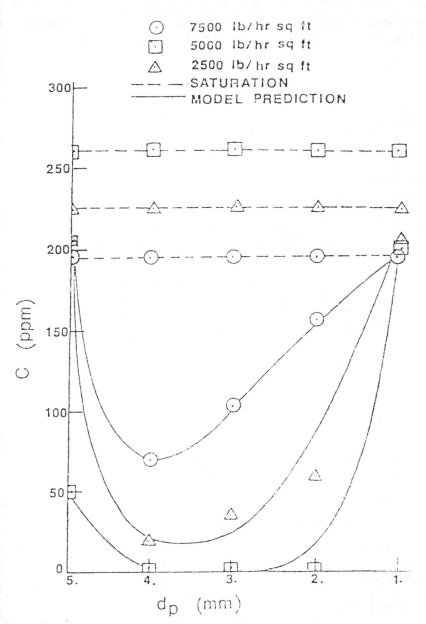

Figure 5.2 Effects of drop size and isobutane flow rate on isobutane exit concentration.

4 FORMULATION OF RATE EQUATIONS

A brief discussion of heat and mass-transfer rate expressions is set down with emphasis on the similarities and dissimilarities in the transport processes.

Diffusive mass transfer is analogous to conductive heat transfer. The rate equation derives from random molecular motion in the presence of a concentration gradient:

$$N_A = -DA\ \partial C_A / \partial Z \tag{1}$$

The molecular diffusivity D is analogous to thermal diffusivity.

Film theory is commonly applied in convective mass-transfer systems, just as it is in convective heat transfer. For diffusion through a stagnant film of thickness δ:

$$N_A = (D/\delta)A\ \Delta C = kA\ \Delta C \tag{2}$$

Convective mass transfer has been represented by theoretical models other than film theory (e.g., penetration theory, surface renewal models), giving rise to alternative interpretations of the mass-transfer coefficient k. Nevertheless, the similarities in the rate equations for diffusive and convective heat and mass transport are clear.

Dissimilarities arise for applications to interphase transport primarily because of two considerations: (a) concentrations in two separate phases cannot be added or subtracted, as can temperatures, and (b) saturation concentrations must be dealt with. As an illustration, consider convective transport between a liquid drop and a surrounding continuous fluid phase. If both phases are taken to have individual film coefficients h_d and k_d for the drop, and h_c and k_c for the surrounding continuous phase, the heat transfer rate can be expressed as

$$q = h_c A\ (T_c - T_i) = h_d A\ (T_i - T_d) \tag{3}$$

$$= UA\ (T_c - T_d) \tag{4}$$

For the mass-transfer case, the concentrations in the two phases must be related through an equilibrium expression. At the interface the two fluids are in immediate contact and are generally taken to be in equilibrium. Rate expressions may be written as

$$N_A = k_c A\ (C_A - C_{A_{eq}})_c = k_d A\ (C_{A\ eq} - C_A)_d \tag{5}$$

Equilibrium concentrations are commonly related through phase equilibrium constants. Historically, these are often designated by special names and symbols for different pairs of phases (e.g., Henry's law constant for gas-liquid systems).

$$\left[C_{A_c} / C_{A_d} \right]_{eq} = H \tag{6}$$

An overall mass-transfer coefficient can be derived as follows:

$$\frac{N_A}{A} \left(\frac{1}{k_c} \right) = C_{A_c} - C_{A_{eq}\ c} \tag{7}$$

$$\frac{N_A}{A} \left(\frac{H}{k_d} \right) = HC_{A\ eq_d} - HC_{Ad} = C_{A\ eq_c} - C^*_{A_c} \tag{8}$$

where $C_{A_c}^*$ is the calculated (fictitious) concentration of component A in the continuous phase in equilibrium with the bulk concentration of A in the drop. Adding the two equations, (7) and (8), yields the overall coefficient K_o:

$$\frac{N_A}{A} \left(\frac{1}{k_c} + \frac{H}{k_d} \right) = \frac{N_A}{A} \cdot \frac{1}{K_o} = C_{A_c} - C_{A_c}^* \tag{9}$$

In contrast, the overall heat transfer coefficient U is simpler because the nature of temperature is not different for different phases. Clearly, the mathematical analogies between heat and mass transfer cannot be applied to overall coefficients.

Accurate values of the transport area are required for both heat and mass-transfer systems. These are often difficult to estimate in direct-contact operations. Drops and bubbles may vary from near-spherical to ellipsoidal to quite irregular shapes, and they may change with time and position depending on hydrodynamic conditions. The flow of two phase fluids through packed beds is even more difficult to characterize geometrically.

The area parameter A is often expressed in units of area per unit volume of contacting vessel. In older literature, the product of the mass-transfer coefficient and the area are generally treated as a single quantity, referred to as a "volumetric coefficient." Since the mass-transfer coefficient depends heavily on molecular diffusion properties, and the interfacial area depends on hydrodynamic and surface properties, the use of volumetric coefficients has not been very successful.

In mass-transfer operations the diffusing component may react chemically as it passes from one phase to another. If the reaction is relatively fast, it may enhance the rate of mass transfer by steepening the concentration gradient of the diffusing species. A good analysis of this effect is presented by Danckwerts (1970).

5 ANALOGIES

Similarities in transport mechanisms have led to the use of mathematical analogies between heat and mass (and momentum) transfer. For similar hydrodynamic situations, heat transfer coefficients can be estimated from mass-transfer coefficients, and vice versa.

Perhaps the most common and useful are the j-factor analogies. Dimensional analysis and empirical evidence have led to the formulation

$$j_h = j_m = f\left(N_{\text{Re}}\right) \tag{10}$$

where

$$j_h = \frac{h}{C_p G} N_{Pr}^{2/3} \left(\frac{\mu_w}{\mu} \right)^{0.14} \tag{11}$$

$$j_m = \frac{k}{u} N_{Sc}^{2/3} \tag{12}$$

As an example, for flow inside a pipe

$$j_h = j_m = 0.023 \ N_{Re}^{-0.2} \qquad (13)$$

An another example, for single-phase fluid flow through beds of spheres

$$j_h = 1.85 \ N_{Re}^{-0.51}, \quad N_{Re} < 50 \qquad (14)$$

$$j_h = 1.08 \ N_{Re}^{-0.41}, \quad N_{Re} > 50 \qquad (15)$$

Correlations for j_h, as provided by Eckert (1956), are quite close but do not agree exactly with a correlation for j_m provided by Sherwood et al. (1975):

$$j_m = 1.17 \ N_{Re}^{-0.415} \qquad (16)$$

The analogies become inaccurate in cases where phenomena occurring in one mode of transport cannot be mirrored in the other. Some examples are (a) a high rate of mass transfer that causes a significant diffusion-driven bulk flow, (b) significant viscous dissipation, (c) chemical reaction, and (d) variable physical properties.

The analogies can be applied to individual convection coefficients only. Overall mass-transfer coefficients contain an equilibrium parameter, as shown in Equation (9).

6 INTERFACIAL AREA

For surface heat exchangers, the exchanger area is the most accurately defined parameter in the design equation. The area is often difficult to estimate accurately in the case of direct contacting. Drops and bubbles can assume an infinite variety of nonspherical forms, and solid particles may be be quite irregular as well as being porous to make internal area accessible.

Data on the shapes and velocities of single drops and bubbles moving through liquids have been correlated by Grace et al. (1976). Three dimensionless groups are required in the graphical correlation:

$$N_{Re} = \rho_c \, d_e \ u / \mu_c \quad \text{(Reynolds)}$$

$$N_{Eo} = g \ d_e^2 \ \Delta\rho / \sigma \quad \text{(Eötvös)}$$

$$N_M \ = g \ \mu_c^4 \ \Delta\rho / \rho_c^2 \ \sigma^3 \quad \text{(Morton)}$$

The map is divided into three large regimes: spherical, ellipsoidal, and spherical cap. The ellipsoidal is subdivided into "wobbling" at low Morton numbers and "oblate" for high Morton numbers. The spherical cap regime contains subareas for "open turbulent wake," "ellipsoidal cap skirted," and "ellipsoidal cap dimpled." Grace (1982) provides correlations for velocities in the various regimes.

Interfacial areas are not well in hand for the nonspherical regimes. Aspect ratios have been correlated by Grace et al. (1976). The aspect ratio is defined as

$$E = \text{vertical dimension/horizontal dimension}$$

These may be used in conjunction with equations for the volumes of revolution of ellipsoids; however, it is common for shape oscillations to occur, so that area changes with time. Also, the drop or bubble may not be symmetrical. For spherical caps, area estimates are even more difficult to make. Finally, there are indications that the behavior of highly-purified systems is different from those where surface-active contaminants are ordinarily present.

Bubble and drop swarms are more complicated because of interactions such as coalescence. Interfacial areas in contactors such as packed towers defy geometrical description. A variety of physical and chemical methods for estimating areas have been developed. Physical methods generally involve light transmission and reflection techniques.

Attempts have been made to measure interfacial areas indirectly. The basic idea is to carry out an area-dependent rate process in the device of interest, and to obtain the area from the measured rate. An early noteworthy effort was that of Shulman et al. (1955), who made column packings in the shapes of Raschig rings and Berl saddles out of naphtalene, and measured rates of evaporation. A more widely used method employs a chemical reaction, and was developed largely by Danckwerts (1970).

The Danckwerts model of absorption of a gas with simultaneous reaction in the liquid phase has been used widely to determine effective interfacial areas in gas-liquid systems such as packed columns, plate columns, and stirred tanks. This same general methodology has been applied on occasion to liquid-liquid systems. The technique requires a reaction of known kinetics in the contacting device so that measurements of the extent of chemical reaction can be used to calculate the effective interfacial area.

Consider a chemical reaction between dissolved gas molecules A and a liquid phase B: $A + B \rightarrow product$. As the transferring component A diffuses into the phase where it reacts with component B, the reaction causes the concentration gradient of A near the interface to be larger than would be the case if no reaction occurred. Thus the rate of transfer is enhanced. According to the Danckwerts model, the average rate of transfer is given by

$$r_A = A_i C_A^* \left[D_A k_2 C_B + k_L^2 \right]^{1/2} \qquad (17)$$

This model requires that the reaction be pseudo-first order, i.e., that the concentration of B is not significantly depleted throughout the reaction zone. This condition is met provided the following inequality is valid:

$$\left[D_A k_2 C_B / k_L^2 \right]^{1/2} \ll 1 + C_B / C_A^* \qquad (18)$$

To estimate interfacial areas of dispersed fluids in a contactor, a series of experiments is performed in which all variables are held constant except for C_B, which is varied over a wide range subject to the constraints of the inequality. From the measured rates of mass transfer, the model equation can be used to find a best estimate of the interfacial area A_i.

A useful system for liquid-liquid heat transfer studies is the hydrolysis of methyl salicylate developed by Bruce and Perona (1985). Methyl salicylate is more dense than water, with a specific gravity of 1.179 at 25°C, and has a normal boiling point of 223°C. It is fairly nontoxic and exhibits good phase separation from water.

7 GAS-LIQUID SYSTEMS

A useful reference for gas-liquid systems is a review paper by Charpentier (1981). Correlations for interfacial area, and mass-transfer coefficients for gas and liquid phases, are presented for the following devices:

packed columns in countercurrent flow (includes 22 different sizes, shapes, and materials of packing)
packed bubble columns
packed columns in cocurrent flow
plate columns (includes bubble cap and sieve trays)
bubble columns
spray towers
stirred tanks

For packed columns with countercurrent flow perhaps the most comprehensive correlation for area and mass-transfer coefficients are those of Onda et al. (1968):

$$\frac{a}{a_t} = 1 - \exp\left\{-1.45\left(\frac{\sigma_c}{\sigma}\right)^{0.75}\left(\frac{L}{a_t\mu_L}\right)^{0.1}\left(\frac{L^2 a_t}{\rho_L^2 g}\right)^{-0.05}\left(\frac{L^2}{\rho_L \sigma_{a_t}}\right)^{0.2}\right\} \quad (19)$$

where a_t is the total dry area of the packing. Equation (19) shows that the surface tension of the liquid σ and the critical surface tension for the packing material σ_c are important parameters.

$$k_L\left(\frac{\rho}{\mu g}\right)^{1/3} = 0.0051\left(\frac{L}{a\mu}\right)^{2/3}\left(\frac{\mu}{\rho D}\right)^{-1/2}\left(a_t d\right)^{0.4} \quad (20)$$

Values of area range up to about 4 cm^2/cm^3 of bed volume, and values of k_L lie between 4×10^{-3} and 2×10^{-2} cm/sec.

Charpentier (1981) concludes that available mass-transfer and area correlations for packed, spray, and plate columns can be used with a fair degree of confidence to design columns up to 2 to 3 m in diameter. For other contactors, pilot scale experiments are recommended.

Equations (19) and (20) illustrate the importance of adopting the use of area-based transport coefficients. Gas-liquid interfacial area depends most strongly on the dry area of the packing a_t and the surface tension parameters. The mass-transfer coefficient depends strongly on Reynolds number and Schmidt number. Combining these into a volumetric coefficient $k_L a$ or the equivalent volumetric heat transfer coefficient would not enhance our understanding or our ability to correlate these variables.

8 LIQUID-LIQUID SYSTEMS

A detailed discussion of transport in liquid-liquid systems is presented in another paper in this symposium. Since that paper is oriented primarily toward heat transfer, a few correlations for mass transfer are noted here.

Calderbank and Korchinski (1956) studied heat and mass transfer in a mercury-aqueous glycerol solution and noted three different kinds of drop behavior: rigid drops, circulating drops, and oscillating drops. Their experimental and theoretical work indicated that internal drop circulation increased the apparent thermal diffusivity by a factor of 2.25 times the rigid drop value.

Von Berg (1971) surveyed the literature and recommended the following internal mass transfer correlations:

for rigid drops (Vermuelen, 1964):

$$k_d = \frac{-r_d}{3t} \ln \left[1 - \left(1 - \exp \left(-\pi^2 Dt/r_d^2 \right) \right)^{1/2} \right] \tag{21}$$

for circulating drops (Calderbank and Korchinski, 1956):

$$k_d = \frac{-r_d}{3t} \ln \left[1 - \left(1 - \exp \left(-2.25\pi^2 Dt/r_d^2 \right) \right)^{1/2} \right] \tag{22}$$

for oscillating drops (Handlos and Baron, 1957):

$$k_d = 0.00375 \, u \left[\left(\mu_c/\mu_c + \mu_d \right) \right] \tag{23}$$

Garner and Tayeban (1960) studied mass transfer from drops to the continuous phase for rigid, circulating, and oscillating drops. They were the first to note that drop wakes act as reservoirs for diffusing solute. For drops greater than 5 mm in diameter, oscillation and shedding of wakes was observed.

Heertjes and deNie (1971) reviewed various studies concerning mass transfer to and from drops, and cited the following correlations:

for rigid drops (Rowe et al., 1965):

$$N_{Sh_c} = 2 + 0.76 \, N_{Re}^{1/2} \, N_{Sc}^{1/3} \tag{24}$$

for circulating drops (Garner and Tayeban, 1969):

$$N_{Sh_c} = 0.6 \, N_{Re}^{1/2} \, N_{Sc}^{1/2} \tag{25}$$

for oscillating drops (Garner and Tayeban, 1960):

$$N_{Sh_c} = 50 + 0.0085 \, N_{Re} N_{Sc}^{0.7} \tag{26}$$

Treybal (1963) suggested a correlation by Ruby and Elgin (1955) for drop swarms. This relation is a modified form of Higbie's equation:

$$N_{Sh_c} = 0.725 \left[N_{Pe}\right]^{0.57} \left[N_{Sc}\right]^{-0.15} (1 - \phi) \tag{27}$$

For liquid-liquid dispersions, a review paper by Tavlarides and Stamatoudis (1981) surveys work on drop breakage and coalescence. They report 12 different correlations for Sauter mean diameter in stirred tanks.

Corresponding heat transfer coefficients can be obtained from correlations such as these through the analogies.

9 FLUID-PARTICLE SYSTEMS

Heat- and mass-transfer operations and chemical reactions involving fluid-particle contacting may be effected in fixed or fluidized beds. Heat transfer across packed beds tends to be slow unless large temperature gradients are imposed, so that fluidized beds are advantageous if a large rate of heat transfer or if close temperature control is required. A j-factor correlation for mass transfer in fixed beds was presented as Equation (16).

Liquid fluidized beds typically exhibit smooth or particulate fluidization. Mass- (and heat) transfer correlations similar to those for flow past single suspended particles apply, with an adjustment based on bed void fraction (Dwivedi and Upadhyay, 1977):

$$j_m = 1.1068 \; \epsilon^{-1} \; N_{Re}^{-0.72} \tag{28}$$

A gas fluidized bed of fine particles is usually visualized as comprising a bubble phase and an emulsion phase. The emulsion phase has properties somewhat similar to a liquid due to the interstitial gas flow. As gas bubbles move upward through the emulsion phase, a cloud of recirculating gas may accompany the bubble, depending on the bubble size and velocity. The emulsion phase is carried upward in the wakes of the rising bubbles and flows downward between bubbles. For large fast bubbles, the interstitial emulsion gas may flow downward.

In the bubbling bed model of Kunii and Levenspiel (1968), the bed is divided into three regions: bubble, cloud, and emulsion. The overall mass-transfer coefficient for the bed is expressed in terms of the volume of solids in each region (per volume of bubble), λ_b, λ_c, and λ_e, and interchange coefficients between the three regions, K_{bc} and K_{ce}. For an overall mass-transfer coefficient K_m defined by

$$-\frac{1}{V_b} \frac{dN_{Ab}}{dt} = K_m \left(C_{Ab} - C_{AS}\right) \tag{29}$$

the expression for the mass-transfer coefficient is

$$K_m = \lambda_b B_d \frac{Sh_t}{Sh_{mf}} + \cfrac{1}{\cfrac{1}{K_{bc}} + \cfrac{1}{\gamma_c B_d + \cfrac{1}{\cfrac{1}{K_{ce}} + \cfrac{1}{\gamma_e B_d}}}} \tag{30}$$

where

$$B_d = \frac{6D}{y\phi_s d_p^2} \, Sh_{mf} \tag{31}$$

and the subscripts on the Sherwood number refer to the terminal settling velocity and the minimum fluidizing velocity. Some of the difficulties in the application of the bubbling bed model are estimating the average bubble size and the volume parameter γ_b. A number of other models have been proposed based on the bubble phase-emulsion phase concept. Primary differences are whether some fraction of the solid phase is in the bubble gas, and to what extent axial mixing of each phase takes place. A recent review paper by Miyauchi et al. (1981) indicates that a successive contact model best represents catalytic reaction data.

Analogous models may be used for global heat transfer between particles and gas.

NOMENCLATURE

A area
a interfacial area/vessel volume
a_t dry packing area/vessel volume
B_d defined by Equation (31)
C_A concentration of component A
C_B concentration of component B
C_p heat capacity
ΔC concentration difference
D molecular diffusivity
d nominal packing diameter
d_e volume-equivalent drop diameter
d_p particle diameter
G mass velocity per channel cross-sectional area
g gravitational constant
h heat transfer coefficient
H phase equilibrium constant
j j-factor defined by Equations (11) and (12)
k mass-transfer coefficient
k_2 second-order reaction velocity constant
K_o overall mass-transfer coefficient
K_{bc} interchange coefficient between bubble and cloud
K_{ce} interchange coefficient between cloud and emulsion
K_m overall mass-transfer coefficient in bubbling bed
L liquid mass velocity per channel cross-sectional area
N_A rate of diffusion of component A
N_{Ab} moles of A in bubble phase
N_{Eo} Eötvös number
N_M Morton number

N_{Pe} Peclet number
N_{Pr} Prandtl number
N_{Re} Reynolds number
N_{Sc} Schmidt number
N_{Sh} Sherwood number
q rate of heat transfer
r_A rate of reaction of component A
r_d drop radius
t time
T temperature
u velocity
U overall heat transfer coefficient
V_b volume of bubble phase
y mole fraction of inert gas
Z length dimension in direction of transfer

Greek

γ_b volume of solid in bubble phase/volume bubble phase
γ_c volume of solid in cloud and wake/volume bubble phase
γ_e volume of solid in emulsion phase/volume bubble phase
σ film thickness
ϵ void fraction
μ viscosity
ρ density
σ surface tension
σ_c critical surface tension (maximum that wets packing)
ϕ volume fraction of dispersed phase
ϕ_s sphericity

Subscripts

c continuous phase
d dispersed phase
h heat transfer
i interface
L liquid phase
m mass transfer
mf minimum fluidizing condition
t terminal velocity condition
w wall temperature

REFERENCES

Bernhardt, S. H., Sheridan, J. J. and Westwater, J. W., *AIChE Symposium Series*, No. 118, Vol. 68 (1971).
Bruce, W. D. and J. J. Perona, IEC Process Design and Dev. 24:62 (1985).
Calderbank, P. H. and Korchinski, I. J. O. *Chem. Eng. Sci.*, 65 (1956).
Charpentier, J. C., *Advances in Chemical Engineering*, 11, Academic Press, New York (1981).
Danckwerts, P. V., *Gas-Liquid Reactions*, McGraw-Hill, New York, (1970).
Dwivedi, P. N. and S. N. Upadhyay, *Ind. Eng. Chem. Proc. Des. Dev.*, 16, 157 (1977).
Eckert, E. R. G., *Trans ASME*, 56, 1273 (1956).
Garner, F. H. and Tayeban, M., *Anales de Fisica Y Quimica* (Madrid), LIV-B, 479 (1960).
Grace, J. R., T. Wairegi and T. H. Nguyen, *Trans. Instn. Chem. Engrs.*, 54, 167 (1976).
Grace, J. R., Ch. 38 of *Handbook of Fluids in Motion*, edited by N. P. Cheremisinoff and R. Gupta, Ann Arbor Science (1982).
Handlos, A. E. and Baron, T., *AIChE Journal*, 3, (1957).
Heertjes, P. M. and de Nie, L. H., *Recent Advances in Liquid-Liquid Extraction*, Ed. C. Hanson, Pergamon Press, New York (1971), pp. 367-406.
Knight, J. F. and J. J. Perona, AIChE Symposium Series 208, Vol. 77, 1981, presented at the National Heat Transfer Conference, Milwaukee (1981).
Kunii, D. and O. Levenspiel, *Fluidization Engineering*, p. 201, Wiley, New York (1968).
Miyauchi, T., S. Forusaki, S. Morooka, and Y. Ikeda, *Advances in Chemical Engineering*, 11, Academic Press, New York (1981).
Onda, K., H. Takeuati and Y. Okumoto, *J. Chem. Eng. Japan*, 1, 56 (1968).
Rowe, P. N., Claxton, K. T. and Lewis J. B., *Trans. Instn. Chem. Engrs.*, 43, 14 (1965).
Ruby, C. L. and Elgin, J. C., *Chem. Engng. Prog. Symp. Ser.*, 51, (16), 17 (1955).
Sherwood, T. K., R. L. Pigford and C. R. Wilke, *Mass Transfer*, p. 242, McGraw-Hill, New York (1975).
Shulman, H. L., C. F. Ullrich, A. F. Proulx and J. O. Zimmerman, *AIChE Journal*, 1, 253 (1955).
Tavlarides, L. L. and M. Stamatoudis, *Advances in Chemical Engineering*, 11, Academic Press, New York (1981).
Treybal, R. E., *Liquid Extraction*, 2nd ed., McGraw-Hill, New York (1963).
Vermeulen, T., *Ind. Engng. Chem.*, 45, 1964 (1953).
Von Berg, R., *Recent Advances in Liquid-Liquid Extraction*, edited by C. Hanson, Pergamon Press, New York (1971), pp. 427.

6

LIQUID-LIQUID PROCESSES

R. Letan

1 CONTACTORS IN LIQUID-LIQUID PROCESSES

In the design of an industrial liquid-liquid process the decisions concern selection of the contacting phase, a solvent or a working fluid, and the type of contactor, its size and range of operating conditions.

In thermal processes where the liquid-liquid contactor is used for heating or cooling, as in a desalination process or a power plant, the final product is cheap. In such processes the equipment and the running expenses have to be low. That relates to maintenance of the equipment without clogging and deposition of solids on heat transfer surfaces. However, in such processes the main concern is directed to the working fluid losses caused by dissolution and entrainment. Insoluble organic liquids are selected to minimize dissolution, and the equipment is designed to eliminate entrainment.

In a contactor where a large distribution of droplets is produced, including micron sizes which stay in an emulsion, the losses of working fluid even in very small amounts may become critical to the process. That may be beyond the economics of electricity production in a power plant or may make the desalted water obsolete. Thus in such processes the dispersion of the liquid has to be carefully controlled to produce uniformly sized droplets sufficiently large for displacement and separation.

In other thermal processes where one of the liquids is used for heating or cooling of the other liquid to affect precipitation, dissolution, or reaction, the main concern may be directed to the properties of the product rather than to the cheapest contacting liquids or simplest equipment.

Similar considerations are applied to extraction processes. The qualities of a more expensive product may dictate the selection of a more soluble contacting liquid or a more compact contactor. In such processes the solvent losses and the cost of its recovery as well as the maintenance of the equipment may be tolerable. On the other hand the availability of a cheap insoluble solvent and simple equipment may bring upon the implementation of extraction processes that otherwise would not have been considered.

The various types of liquid-liquid contactors may be classified into two main categories: stagewise and differential equipment. The staged contactors provide discrete stages where the liquids are mixed, settled, and separately removed. Stages are usually joined in cascades. This type of equipment is generally a multiple unit of mixer-settlers. The differential equipment is usually a column with or without internal devices. Columns provide a continuous contact between the liquids. A single column is sized to perform the task of as many stages as required.

In mixer-settlers the degree of dispersion is determined by the intensity of mechanical agitation. A wide distribution of droplet sizes usually prevails and affects the size of the settler. Excessively fine dispersions produce stable emulsions that have to be separately treated for disengagement. The number of stages in a unit may be easily varied to adjust for the desired temperature or concentration change.

Columns may be operated cocurrently or countercurrently. A cocurrent operation yields at best the results of a single ideal stage. Such an operation may be conducted either vertically or horizontally. Countercurrent flow is obtained in vertical columns.

The spray column and wetted-wall column are operated without internal devices along the column proper. Internal devices improve the performance of the contactors, such as the baffle column, packed column, perforated plate column, or a combination of internal devices. Some of these devices promote transfer rates, in particular where reformation of droplets takes place. Other devices preserve the gradients along the column by restraining the turbulent vortices but reducing the flow rates in some cases.

Spray columns are the simplest contactors in which one of the liquids is dispersed. The absence of internal devices makes spray columns adaptable to processes in which solids may be deposited. However, the absence of internal devices allows the continuous liquid, with the droplets dispersed in it, to backmix freely. The study of spray columns performance illustrates, however, some basic features that characterize contactors in general.

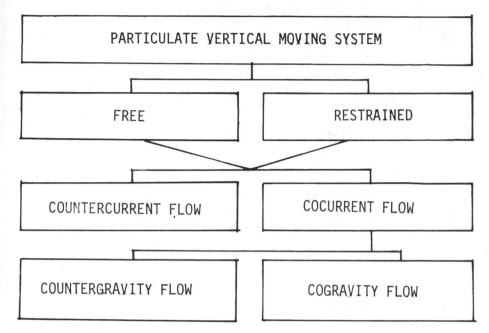

Figure 6.1 Particulate vertical moving systems.

2 MECHANICS OF VERTICAL MOVING SYSTEMS

Dispersions of solid particles, liquid drops, and gas bubbles in fluids can be classified and analyzed according to the mechanics of their systems. The object of this work is to analyze systems of liquid drops dispersed in an immiscible liquid. Drops are deformable but within some flow regions the drops behave like solid particles. Therefore, the mechanics of both rigid and deformable particles have to be considered. In the descriptions to follow the size range of particles extends between about $(10 \ \mu m) < d < (10 \ mm)$.

2.1 Operational Mechanics

Dispersions in general can be classified as heterogeneous or aggregative, and homogeneous or particulate. This part of the work relates to the particulate systems in the restricted sense of this definition, i.e., the quiescent, vertical moving, uniformly distributed systems are considered (Fig. 6.1).

This class of systems is uniquely characterized by velocity-holdup relations irrespective of the direction of flow. Thus in a system of this kind the two limiting situations correspond to a single particle in a fluid and to a packed bed. Flow characteristics of particulate systems between these two limits involve a relationship of velocities of the phases to holdup, and the velocity of a single representative particle. Studies of this relationship have followed two methods of approach: extension of the dynamics of a single particle to a multiparticle system; and

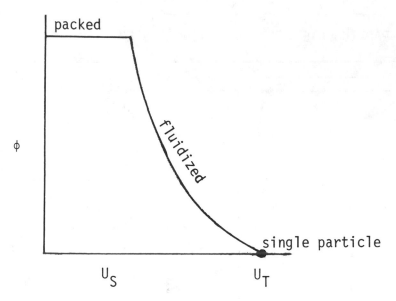

Figure 6.2 A general relationship of slip velocity-holdup.

modification of continuum mechanics of single phase fluids, accounting for the presence of particles.

Slip velocity-holdup relations. Extensive experimental and theoretical studies were carried out to establish a quantitative relationship between holdup and velocity in vertical moving particular systems [1–4]. The experimental data usually referred to fluidized and sedimenting systems of solid particles. More limited was the scope of data reported for fluid particles, such as liquid drops [5–7].

The vertical-holdup relations were formulated for superficial velocity of the fluidizing fluid [2], linear velocity [4], or slip velocity of the system [1]. Lapidus and Elgin [1] postulated that all particulate vertical moving systems are controlled by the same fundamental forces provided a relative motion, or slip velocity, existed between particles and fluid regardless of whether the particles are solid, liquid, or gaseous, and irrespective of the relative direction of motion. That postulation led to a generalized formula of slip velocity-holdup:

$$U_S = f(\phi) \tag{1}$$

It was verified experimentally [8,9] in various types of operations yielding the same functionality illustrated in Fig. 6.2.

Based on the fact that slip velocity of a single particle in a fluid is its terminal velocity;

$$U_S/U_T = f(\phi), \tag{2}$$

where $U_S = U_T$ for $\phi = 0$. The function $f(\phi)$ usually takes an exponential [6] or a power form [2,10]:

$$U_S/U_T = (1 - \phi)^n \, . \tag{3}$$

The above expression in a slightly different form was semiempirically obtained by Richardson and Zaki [2]. They considered the drag imposed on a constituent particle in a suspension and dimensionally analyzed the various groups involved. The relations were presented as

$$V_c = U_T(1 - \phi)^m \, . \tag{4}$$

The exponent, m, was experimentally obtained [2] for four regions of flow:

$$Re_o \leq 0.2, \quad m = 4.65;$$

$$0.2 \leq Re_o < 1, \, m = 4.35 \, Re_o^{-0.03};$$

$$1 \leq Re_o \leq 500, \, m = 4.45 \, Re_o^{-0.01};$$

$$500 < Re_o, \, m = 2.4$$

Applying the definition of slip velocity in batch fluidized systems [1]

$$U_S = V_c/(1 - \phi) \tag{5}$$

to Equation (4) yields $n = m - 1$.

Zenz's [3] graphical correlation is in principle based on the same dimensional groups. It presents curves of $(Re/C_D)^{1/3}$ versus $(Re^2 C_D)^{1/3}$ with voidage, $(1 - \phi)$, as a parameter. The Re–C_D groups are based on superficial velocity of the fluid, particle diameter, and the physical properties of both phases. The correlation is based on extensive data extracted from experimental works [11–13]. The other empirical or semiempirical correlations available in the literature are of a more restricted character.

The slip velocity-holdup relation was treated theoretically too. The theoretical relations and the coefficients involved in them varied with the formulation of the governing equations as well as with the particular forms of the subsidiary equations utilized in the solution.

Zuber [4] formulated the velocity-holdup relation of a particulate system as a problem of a suspension of particles of an apparent viscosity determined by the presence and motion of particles in it. He utilized in his derivations the Brinkman [14] and Roscoe [15] equation

$$\frac{\mu}{\mu_c} = \frac{1}{(1 - \phi)^{2.5}} \tag{6}$$

and obtained a relation of the form of Equation (3) with $n = 3.5$ for the Stokes region, and $n = 1.1$ for $Re = 500$. Letan [16] utilized the same procedure as Zuber to flow regions up to $Re_o \leq 2000$ and obtained:

$$\frac{U_S}{U_T} = \frac{(1 + 0.15 \, Re_o^{0.687})(1 - \phi)^{3.5}}{1 + 0.15 \left(Re_o \cdot \dfrac{U_S}{U_T} \right)^{0.687} (1 - \phi)^{1.72}} \tag{7}$$

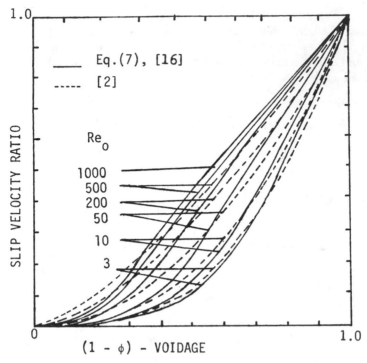

Figure 6.3 Slip velocity-voidage relationship [16].

The above relation with Re_o as a parameter is compared with Richardson and Zaki's [2] respective relation in Fig. 6.3. The applicability of the derived equation to liquid-liquid systems was also confirmed by comparison with experimental data [16] in the flow range of $100 < Re_o < 1300$.

Modes of operation: free and restrained. Lapidus and Elgin [1] postulated two basic categories of vertical moving particulate systems, free and mechanically restrained. Within each category several types of systems can be distinguished. As illustrated in Fig. 6.1 in both the free and the restrained systems the flow may be conducted countercurrently or cocurrently. In countercurrent flows the dispersed and continuous phases move in their respective directions by virtue of the difference in their densities. In cocurrent flows the particles may be moved in both counter- and cogravity directions. This classification refers to all kinds of particulate systems irrespective of the phase character.

The distinction between free and restrained systems corresponds to the controls or constraints imposed on the exits of the phases concerned. In a free system the flow rate of the particulate phase is externally controlled at the inlet, where the outlet has no constraints. The character of the flow path in the column is determined by the inlet rates and properties of the two phases. On the other hand, in a mechanically restrained system, the flow rate of the particulate phase is controlled at the outlet. In this case the flow path in the column is controlled by the exit constraints.

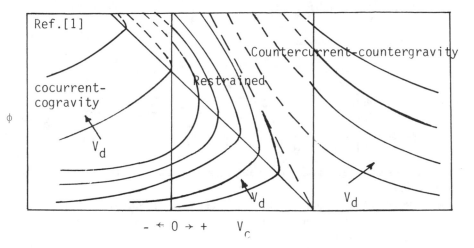

Figure 6.4 Operational diagram of a particulate system [1].

In solid-fluid systems the restrained operation is achieved by a control valve on the exit of the solids from the container. This way the solids may build up inside the column to a packed bed, and then the exit rate is again increased up to the feed rate.

The restrained operation can be achieved in fluid-fluid systems too, as for example in a liquid-liquid spray column operated with dense packing. In this case the exit constraint is provided by restricting the surface available for coalescence of the drops. As the coalescence rate is reduced the fluid particles "queue" at the interface forming a dense packing from the top of the column down. After a dense packing has been established through the column, it is possible to adjust the inlet flow rate to the rate of coalescence, i.e., to the exit rate of flow. If the rate of coalescence at the interface is increased by mechanical or chemical promoters, the dense packing breaks up and a disperse or free packing is again established. Operation of densely packed spray columns was mostly reported in thermal processes [17–19].

The operational features of free and restrained systems are schematically illustrated in Fig. 6.4, which is a generalized operational diagram for vertical moving fluidized systems [1].

Operating flow conditions. The relations between flow rates of the liquids and holdup of the dispersed phase in the column are determined by the properties of the phases, drop sizes, and mode of operation.

The slip velocity in a particulate system is defined by the difference in linear velocities of the phases;

$$U_S = U_c - U_d \tag{8}$$

while the linear velocities are related to the respective superficial velocities and holdup of particles, or drops:

$$U_d = \frac{V_d}{\phi}, \quad \text{and} \quad U_c = \frac{V_c}{1 - \phi} \tag{9}$$

Therefore, in countercurrent flow

$$U_S = \frac{V_c}{1 - \phi} + \frac{V_d}{\phi} \tag{10}$$

and in cocurrent flow

$$U_S = \pm \left[\frac{V_c}{1 - \phi} - \frac{V_d}{\phi} \right] \tag{11}$$

where the positive sign applies to countergravity flow, and the negative sign to cogravity flow.

These operational equations are combined with the slip velocity-holdup relation and used for construction of operational diagrams (Fig. 6.4) with zones of specified types of flow [1].

For a countercurrent flow Equation (1) may be combined with Equation (3) or with any other slip velocity holdup relation to yield

$$U_T (1 - \phi)^n = \frac{V_d}{\phi} + \frac{V_c}{(1 - \phi)} \tag{12}$$

which is the operational relation of the system. The operational diagram is constructed with holdup ϕ plotted against superficial velocity of the continuous phase, V_c, and the superficial velocity of the dispersed phase V_d as a parameter. The regions of the disperse and restrained packing are illustrated in Fig. 6.4. The two regions are separated by a flooding line that represents the disruption of stable flow conditions.

At flooding [1]

$$\left. \frac{\partial V_c}{\partial \phi} \right|_{V_d} = 0 \tag{13}$$

Differentiation of Eq. (12) with respect to ϕ, yields at flooding:

$$\phi_f = \frac{\left[(n + 2)^2 + 4(n + 1)(1/R - 1) \right]^{1/2} - (n + 2)}{2(n + 1)(1/R - 1)} \tag{14}$$

where $R = V_d/V_c$. The limiting case corresponds to $V_c = 0$, e.g., $R = \infty$ where the maximum holdup is obtained:

$$\left. \phi_f \right|_{max} = 1/(n+1) \tag{15}$$

The operational holdup ϕ in a system is usually determined at about 0.8 ϕ_f. Within a vertical column operated at a holdup ϕ the pressure drop in the continuous phase in all types of flow [1] is obtained as

$$\Delta P/\Delta z = (\rho_c - \rho_d)g \cdot \phi \tag{16}$$

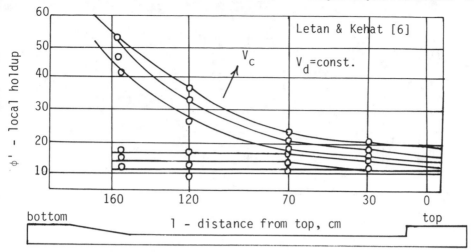

Figure 6.5 Local holdup in a liquid-liquid system [6].

The analysis of operational mechanics summarized above is the basis for design of contacting devices operated with particulate systems.

2.2 Nonuniformities

The data found in the literature on holdup and drop sizes in liquid-liquid spray columns usually refer to average quantities. Few works were concerned with the longitudinal distributions of holdup or sizes.

Holdup. Holdup was longitudinally measured by Letan and Kehat [6] in a spray column of kerosene drops in water. Typical distribution curves for one kerosene superficial velocity and several water superficial velocities in disperse packing are shown in Fig. 6.5. At low flow rates of water the holdup was constant along the column. As the flow rate of water increased, the holdup increased from the top to the bottom of the column. As the water flow rate increased further, the variation of holdup between top and bottom was considerable. At higher kerosene flow rates, the flow rate of water above which no constant region of holdup existed was lower.

In the regions of constant holdup the two phases flow at constant velocities. In the regions of changing holdup the phases accelerate in their respective directions of flow.

An equilibrium slip velocity, U_S^*, defined in the constant holdup region of the column was calculated for the countercurrent flow:

$$U_S^* = V_d/\phi' + V_c(1 - \phi') \qquad (17)$$

where ϕ' is the local holdup. The experimental data were replotted and yielded an empirical exponential function of the equilibrium slip velocity against the corresponding holdup illustrated in Fig. 6.6:

$$U_S^* = U_{S0}^* \exp(-a\phi') \qquad (18)$$

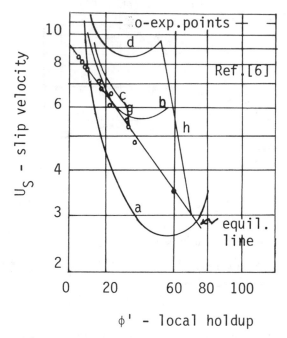

Figure 6.6 Equilibrium and operating curves [6].

where $U_{S0}^{\bullet}$ is obtained by extrapolating to $\phi' = 0$, and a is an empirical constant. Both $U_{S0}^{\bullet}$ and a, depend on the physical properties of the liquids and on the drop size.

Of particular interest in Fig. 6.6 are the operating curves, which locally obey the slip velocity-holdup relation of

$$U_S = V_d/\phi' + V_c/(1 - \phi') \tag{19}$$

The curve is designated as an operating curve. It represents all possible relations of slip velocity and holdup for one pair of flow rates of the two liquids. Four typical operating curves for the same kerosene flow rate and an increasing water flow rate from a to d are drawn in Fig. 6.6 together with the experimental equilibrium line.

Curve a (Fig. 6.6) intersects the equilibrium line at two points, i.e., at two holdups that correspond to the two modes of packing: disperse and dense. For the sets of flow rates where the holdup varied in the upper part of the column for disperse or restrained packing, the holdup distribution is represented by the upper branches of curve a.

Curve b represents an operating curve to which the equilibrium line is tangential; i.e., only one point of equilibrium is obtained. That situation corresponds to a single mode of packing obtainable in the column. The curve represents, for a given kerosene flow rate, the maximum water flow rate for which

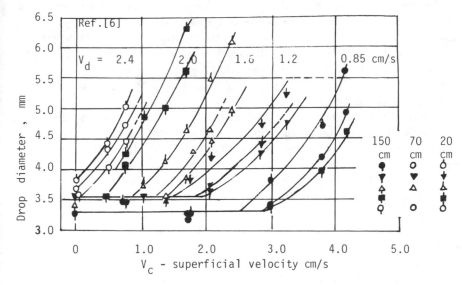

Figure 6.7 Local drop size in a spray column [6].

a stable region can be obtained in the column proper. An increase of water flow rate beyond this value will cause rejection of kerosene drops from the bottom of the column proper; i.e., flooding will commence.

Curves c and d relate to flow rates that are beyond the range of stable holdup in the column proper. However, the conical entry section of the column extends the range of flow rates of the two phases, before rejection of the drops takes place at the water outlet. The superficial velocities of the two phases are reduced in the conical section, and there an equilibrium condition is reached as shown by the right-hand branches g, h of curves c and d in Fig. 6.6.

The actual situation in the column is more complex than described above. Coalescence starts at flow rates below rejection, and the average drop size increases. That of course affects the equilibrium line which depends on the drop size.

Drop size. The drops in the column do not vary in size along the column at normal operating conditions. The size produced at the nozzles remains unchanged as long as the operating curves correspond to the type a or b shown in Fig. 6.6. The onset of coalescence of drops in the column proper coincides with a distribution of holdup that does not reach an equilibrium state in the column proper, as represented by curves c or d.

The slip velocity in the column proper for the situation represented by curve d decreases from the left with increased holdup down the column to the level of the curve minimum. Below this level the slip velocity, and hence the drag forces, increases down the column. The combination of higher holdup and higher drag creates favorable conditions for coalescence in the lower part of the column, and this starts the coalescence process throughout the whole column [6].

The drop sizes were measured at three locations in the column proper as shown in Fig. 6.7. The average drop size in disperse packing at low flow rates of the two phases was independent of the flow rate of either phase or the position in the column. With the flow rates the drop size increased considerably. The larger drops appeared earlier at the top and last at the bottom. These phenomena took place when the operating curves were of the type d, with the minimum slip velocity within the column proper.

Actually, as phenomenologically predicted, coalescence started at flow rates for which the point of minimum velocity was allocated at the bottom of the column proper. The initial average size of 3.3–3.5 mm of droplets was increased at the top of the column up to 6.5 mm at higher flow rates.

The results obtained show that the visually observed disturbances in a spray column can be associated with the mechanism of operation and used for definition of flooding: (1) the point at which coalescence commences may be identified at the flow rates of the two phases for which their operating curve has its minimum at the bottom of the column proper, or (2) the flow rates at which rejection from the column proper takes place. This point can be obtained from the locus of the tangential points of the operating curves with the equilibrium line. This definition predicts flooding at lower flow rates.

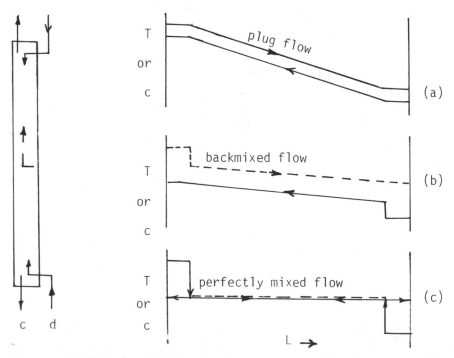

Figure 6.8 Mixing modes in a two-phase system.

2.3 Mixing Effects

The intensity of mixing of any phase in a contactor is usually very well demonstrated by the phase temperature or concentration profiles. Figure 6.8 illustrates possible profiles of the two phases along a contactor. Case (a) presents a gradual change along the contactor from inlet to outlet of each phase. Such an ideal situation is a "plug flow." The other limiting case is represented by the perfectly mixed flow (c), where a step change takes place at the inlet. The temperature or concentration is uniform along the contractor. Between the two extreme cases is found the back-mixed contactor, where the temperature or concentration changes steeply at the inlet; however, some changes also take place along the contactor. The degree of mixing in a contactor is usually studied by injection of a tracer.

Particulate countercurrent systems are characterized by a steep change, almost a step change, in the driving force at the inlet to the system. That has been demonstrated in packed beds [20], spray columns [21–24], liquid fluidized beds [25,26], and other particulate contacting devices. These phenomena were attributed to longitudinal dispersion or recirculation of the continuous phase.

The longitudinal dispersion in a system is usually represented by a longitudinal diffusivity coefficient, D_L [27],

$$D_L \frac{\partial^2 c}{\partial z^2} - U \frac{\partial c}{\partial z} = \frac{\partial c}{\partial t} \tag{20}$$

where U is an interstitial velocity. The coefficient, D_L, is experimentally determined from tracer concentration profiles. The so calculated coefficient is then presented in the form of an appropriate Peclet number, $Pe = Ud/D_L$. In some cases the length dimension used is the particle diameter, the column diameter, or the column length.

The longitudinal dispersion equation was postulated [27] for purely random processes. Hiby [20] has photographically illustrated that the process of longitudinal dispersion in the fixed particulate systems was of defined orientation. He showed that streamlines of an injected dye were well traced. Backmixing was observed in turbulent flows only. In other particulate systems the mechanism of dispersion or mixing depends on the flow regime of the two phases.

In a quiescent regime the longitudinal dispersion is primarily due to the fluid retention in the wakes of the particles. The transverse dispersion, or rather the residence time distribution of particles and their wakes, is caused by the radial velocity profile of the dispersed phase. The phenomenon of upstream backmixing is negligible or not encountered.

In turbulent systems the upstream backmixing prevails. The mixing of flows in such cases is carried along by the large-scale vortices. The radial velocity distribution, as well as the retention of fluid in wakes, has a negligible effect on the degree of mixing in such cases. The scale and intensity of the turbulent vortices increase with the diameter of the column. These phenomena can be experimentally investigated in large vessels. Theoretical studies of apparent viscosity of

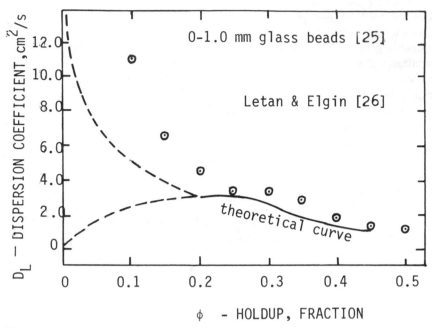

Figure 6.9 Longitudinal dispersion in a liquid fluidized bed [26].

dispersions and the turbulence in them may be very useful. In small laboratories' columns the contribution of turbulent vortices to "mixing" is relatively small. Large baffled systems perform similarly.

The various mixing mechanisms attributed to the fluids in particulate systems were sometimes studied in an integrative way, yielding a single dispersion coefficient, D_L, experimentally determined. However, by the size of the column engaged, the relative contributions may be estimated. In large vessels the turbulent eddies control. In small columns the wake and radial velocity profiles are relevant. At intermediate sizes of columns all the effects are comparable.

Letan and Elgin [26] adopted that approach in a study of the contribution of wakes to longitudinal dispersion of the continuous phase in quiescent systems. The derived dispersion coefficient related to the wake parameters and the operating flow conditions of the system. The theoretical values of the dispersion coefficient D_L and the respective Peclet numbers were compared with experimental works [25,28]. The comparison with batch expanded glass beads in water [25] is illustrated in Fig. 6.9. The predicted curves are in favorable agreement with the measured points. Thus in a quiescent particulate system the longitudinal dispersion is caused by the translation of wakes.

The longitudinal dispersion of the dispersed phase in a quiescent system is primarily due to a nonuniform velocity of droplets [29,30]. Letan and Kehat [29] measured the residence time distribution of the dispersed phase in a spray column

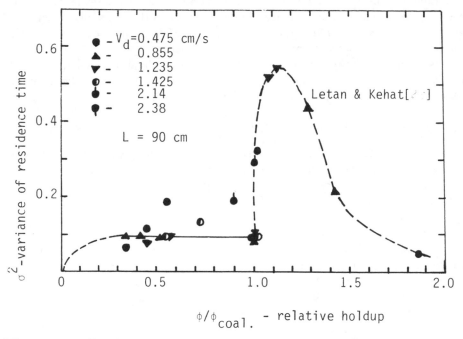

Figure 6.10 Residence time distribution in a spray column [29].

for a wide range of flow rates both in disperse and in dense packings of drops. The results have shown (Fig. 6.10) that in a disperse packing below the onset of coalescence the variance of residence time distribution was 0.1. That indicated an almost plug flow of the drops. With the onset of coalescence a variance of 0.55 was measured. It decreased at higher flow rates, as channels of kerosene were formed in the column.

In dense packings of drops the distribution of drop velocities is much wider. The variance measured there reached 0.5 at the higher holdups [29]. It decreased with holdup to 0.3 in the lower range. This distribution was attributed to wall effects and consequently the formation of a radial velocity profile in the core of the column. The respective Peclet numbers calculated from the variance were 4 to 8.5.

If indeed the wall affected the dense packing, then in larger columns the distribution will be smaller. Greskovich [31] obtained in a densely packed 3 m long spray column a 40% reduction of HTU values as the column diameter increased from 0.10 to 0.15 m. On the other hand, Mixon et al. [30] obtained in a densely packed, 1 m long spray column, Peclet numbers that increased from 0.13 to 2.1 as the diameter of the column decreased from 0.15 to 0.075 m. These results indicate a larger scatter than observed by Letan and Kehat [29]. However, as the flow rates, holdup, and drop sizes were not specified, the results of the two works cannot be compared.

Finally it may be concluded that in disperse packings the drops move uniformly up to the onset of coalescence. In dense packings the distribution of velocities is due to a radial profile in the column. These effects may change with the column size in a way that has not yet been established.

In general it may be summarized that longitudinal dispersion in small columns can be predicted and controlled. In large columns the dispersion is mostly due to bulk turbulence, which requires further study.

3 HEAT AND MASS TRANSFER

Study of the mechanism of heat and mass transfer in complex particulate systems is sometimes hampered by inadequate information and understanding of the fluid mechanics of these systems. In such systems global or integral methods are usually applied to heat- and mass-transfer processes, and the results are lumped as overall coefficients.

By measurement of local concentrations or temperatures in a particulate system a better understanding of the mechanism is gained, and theoretical or phenomenological models can be devised.

One of the most widely applied models has been the model of longitudinal dispersion. It has accounted for the more significant phenomena revealed in local measurements of temperatures and concentrations. The model has involved only one arbitrary empirical coefficient for each phase in any process examined.

Some of the particulate systems studies have been prompted in the direction of flow pattern investigation, and physical models have been sought for. One of those models tailored for quiescent particulate systems is the wake model. The model applies to particles beyond the Stokes regime. It relates the fluid circulation patterns in these systems to retention in wakes and the wake shedding behind the moving particles. The mathematical model does not incorporate arbitrary coefficients and is supposed to have all its parameters hydrodynamically determined.

3.1 Methods and Models

Integral methods are of practical interest, being applicable to engineering design in a straightforward way. These methods have been applied to particulate systems in general and to spray columns in particular.

The integral methods usually employ the logarithmic mean temperature difference (LMTD) as the driving force in the system:

$$\Delta T_m = \text{LMTD} = \frac{\Delta T_1 - \Delta T_2}{\ln(\Delta T_1/\Delta T_2)} \tag{21}$$

The methods differ in the heat transfer coefficients utilized in the calculations.

One of the approaches relates to the internal and external thermal resistance of the droplets and their surface for heat transfer. The internal resistance of the droplet is evaluated for a conducting circulating or mixed fluid inside. The external resistance or conductance is obtainable from common correlations of

$$Nu = C_1 + C_2 Re^m Pr^n \tag{22}$$

for a specified flow regime of the droplet. These two compose the heat transfer coefficient U. The specific surface area between the phases is defined for all the droplets in the column per unit volume of the column. Thus

$$a = 6\phi/d \tag{23}$$

That approach should have led to the column length, as

$$L = \frac{(V \cdot \rho \cdot c \cdot \Delta T)_c}{(6\phi/d) \cdot U \cdot \Delta T_m} \tag{24}$$

where the two phases in a plug flow along the column. That, however, does not describe correctly the phenomena governing the spray column or similar particulate systems. Therefore, an empirical correction factor is required for Eq. (24).

A more commonly practiced method employs an experimental volumetric heat transfer coefficient U_V:

$$U_V = \frac{(V \cdot \rho \cdot c \cdot \Delta T)_c}{L \cdot \Delta T_m} \tag{25}$$

In the literature the volumetric heat transfer coefficients were correlated against the operating variables of the experimental system. Woodward [18] presented his results against holdup of droplets in the liquid-liquid spray column. Plass, Jacobs, and Boehm [32] obtained a correlation with holdup and flow rate ratio,

$$\phi > 0.05, \quad U_V = 4.5 \times 10^4 (\phi - 0.05) e^{-0.75R} + 600 (\text{Btu/hr ft}^3 \, {}^\circ\text{F}) \tag{26}$$

A method similar to the volumetric heat transfer coefficient is the method of calculating the number of transfer units (NTU) and the height of a transfer unit (HTU). Here again the results are empirically correlated.

Both the volumetric heat transfer coefficients and the HTU correlations are applicable to systems of the same geometry and operating conditions as in the experiments used for the correlations. Variation of size or operation may yield considerably different results. The intensity of longitudinal mixing due to vortices, wakes, or radial distribution makes the system perform in a nonsimilar way.

Local temperature measurements in spray columns have indicated a sharp drop of temperature at the inlet of the continuous phase. A concentration discontinuity was noted in studies of extraction spray columns [21–23]. As discussed earlier in the section Mixing Effects, the model of longitudinal dispersion was developed and adopted to particulate systems including the liquid-liquid spray column. The longitudinal dispersion coefficient D_L was supposed to account for all the mixing effects.

The differential energy balance on a column of a two-phase system yielded

$$\left[\alpha_e \frac{d^2 T}{dz^2} - U \frac{dT}{dz} \right]_{d,c} = \frac{h \cdot a}{(\rho c)_{d,c}} (T_d - T_c) \tag{27}$$

Here again the thermal eddy diffusivity α_e, or the longitudinal dispersion coefficient D_L in tracer concentrations studies, had to be determined experimentally. The experimentally obtained coefficients varied considerably and illustrated the effects of the geometries and operating conditions. In fact, the considerably large discrepancies manifested the inadequacy of the randomness expressed in the eddy diffusivity model.

In the "physical" models that followed, attempts were made to analyze the fluid dynamics of the system investigated. The distinction between the "mixing" contributions in the system led to the study of the physical phenomena taking place. Among those studies was also the "wake model" developed by Letan and Kehat [33].

3.2 The Wake Model

The wake model was developed for a liquid-liquid spray column in which the bulk flow was quiescent. Thus the model applied in its original form to systems in which turbulent vortices have not been developed. The assumption in the model was that longitudinal dispersion in such systems is affected by the translation and shedding of the wakes formed behind the droplets. The model may be used in other particulate systems, and with modifications applied to turbulent flows too. Description of the physical model and the basic mathematical derivations are for convenience presented for drops rising in the column and cooling down.

The physical model describes the phenomena that govern the performance of the particulate system along the path of the two phases: The drop formed at the nozzle rises up. After a short distance the boundary layer on the droplet separates at the rear of the droplet. A toroidal vortex is formed at the stagnation point. It grows, and the separation ring moves forward. The drop starts oscillating with a highly intense turbulent mixing taking place inside the drop.

Elements of the separated boundary layer reach the temperature of the drop surface. These elements are entrained into the mixed vortices of the wake and stay there till the wake attains its final size. Within the time interval of its formation the wake accumulates all the heat lost by the drop. Within this zone of wake formation the continuous phase does not change in temperature. As the wake reaches its maximum size, elements of its substance are shed at the mixed wake temperature. The separated boundary layer is entrained into the wake to substitute for the shed elements. The shed elements mix with the continuous phase surrounding the droplet. The temperature of the continuous phase increases down the column. The temperatures of the drops and wakes decrease up the column.

At the top of the column the drops coalesce with the liquid above the interface. The wakes, detached from the drops, mix into the incoming stream of the continuous phase and flow back down. The zones described above are illustrated in Fig. 6.11.

In the formulation of the mathematical model the following assumptions are made: steady state, no heat losses from the column, constant average physical properties of the liquids, uniform holdup along the column, uniformly sized drops,

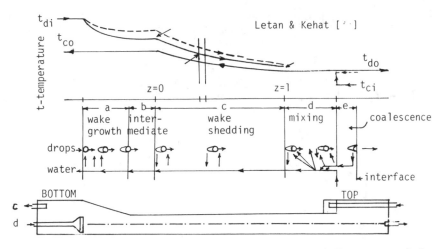

Figure 6.11 Physical model of heat transfer in a particulate system [33].

the flow rate of fluid into the wake equals the rate of shedding from the wake in the wake shedding zone.

The heat balance equations on the drop, wake, and continuous phase in the zone of shedding are, respectively, the following:

$$\frac{dT_d}{dz} + \frac{m}{r}\,(T_d - T_c) = 0 \tag{28}$$

$$\frac{dT_w}{dz} + \frac{m}{M}\,(T_w - T_d) = 0 \tag{29}$$

$$\frac{dT_c}{dz} + \frac{m}{P}\,(T_w - T_c) = 0 \tag{30}$$

where

$$r = \frac{(\rho c_p)_d}{(\rho c_p)_c} \tag{31}$$

$$R = V_d/V_c \tag{32}$$

$$P = \frac{1}{R} + M \tag{33}$$

The dimensionless volume of shed vortices per unit length of column is m. The dimensionless volume of wake is M. Eqs. (28)–(30), combined with the heat balance equations of the other zones, yield the temperature profiles of the two phases:

$$\frac{T_d - T_{co}}{T_{di} - T_{co}} = \left\{ \frac{m}{r} \left[\frac{1+S}{\alpha_1} \left[1 - \exp(\alpha_1 z) \right] \right. \right.$$

$$-\frac{S}{\alpha_2}\left[1 - \exp\left(\alpha_2 z\right)\right] + 1\right\}\exp\left(-\frac{M}{r}\right) \qquad (34)$$

$$\frac{T_c - T_{co}}{T_{di} - T_{co}} = \left\{\frac{m}{r}\left[\frac{1+S}{\alpha_1}\left[1 - \left(\frac{r}{m}\alpha_1 + 1\right)\exp(\alpha_1 z)\right]\right.\right.$$

$$\left.\left. - \frac{S}{\alpha_2}\left[1 - \left(\frac{r}{m}\alpha_2 + 1\right)\exp\left(\alpha_2 z\right)\right]\right] + 1\right\}\exp\left(-\frac{M}{r}\right) \qquad (35)$$

where

$$\alpha_{1,2} = -\frac{m}{2}\left[\left(\frac{1}{M} + \frac{1}{r} - \frac{1}{p}\right) \pm \sqrt{\left(\frac{1}{M} + \frac{1}{r} + \frac{1}{P}\right)^2 - \frac{4}{Mr}}\right] \qquad (36)$$

and

$$S = \frac{\alpha_1 + \frac{m}{r} - \frac{rm}{PM}\left[\exp(\frac{M}{r}) - 1\right]}{\alpha_2 - \alpha_1} \qquad (37)$$

In the mixing zone at the top

$$T_{do} = T_{cl} \qquad (38)$$

The overall heat balance yields

$$T_{co} - T_{ci} = Rr(T_{di} - T_{do}) \qquad (39)$$

Experimental temperature profiles [33], as well as measurements of inlet-outlet temperatures [32] in spray columns, have confirmed the validity of the model.

The same physical model and mathematical derivations were presented for transfer of solute between the phases in a spray column [36]. Previous works published in the literature were used for comparison with the developed model.

Rising drops, disperse packing, and solute transfer from drops to the continuous phase are considered for convenience in the described physical model. The basic mass-transfer equations are presented below.

At equilibrium the solute concentration in the two phases is

$$c_c = c_d/K \qquad (40)$$

where K is the distribution coefficient. It is equivalent to r, in a thermal system. In the wake shedding zone the mass on the drop, wake, and continuous phase are conducted in a way similar to the energy balance:

$$\frac{dc_d}{dz} + m\left(\frac{c_d}{K} - c_c\right) = 0 \tag{41}$$

$$\frac{dc_w}{dz} + \frac{m}{M}\left(c_w - \frac{c_d}{K}\right) = 0 \tag{42}$$

$$\frac{dc_c}{dz} + \frac{m}{P}\left(c_w - c_c\right) = 0 \tag{43}$$

where P is expressed by Eq. (33).

The concentration profiles of the two phases along the column are then expressed in a way similar to Eqs. (34)–(37). Concentration of solute in the dispersed phase:

$$\frac{c_d/K - c_{co}}{c_{di}/K - c_{co}} = \left\{ \left[\left[\frac{1+S}{\alpha_1}\left[1 - \exp\left(\alpha_1 z\right)\right] \right.\right.\right.$$

$$\left.\left.\left. - \frac{S}{\alpha_2}\left[1 - \exp\left(\alpha_2 z\right)\right] \right] + 1\right\}\exp\left(-\frac{M}{K}\right) \tag{44}$$

Concentration of solute in the continuous phase:

$$\frac{c_c - c_{co}}{c_{di}/K - c_{co}} = \left\{ \frac{m}{K}\left\{ \frac{1+S}{\alpha_1}\left[1 - \left(\frac{K}{m}\alpha_1 + 1\right)\exp\left(\alpha_1 z\right)\right] \right.\right.$$

$$\left.\left. - \frac{S}{\alpha_2}\left[1 - \left(\frac{K}{m}\alpha_2 + 1\right)\exp\left(\alpha_2 z\right)\right]\right\} + 1\right\}\exp\left(-\frac{M}{K}\right) \tag{45}$$

where

$$\alpha_{1,2} = -\frac{m}{2}\left[\left(\frac{1}{M} + \frac{1}{K} - \frac{1}{P}\right) \pm \sqrt{\left(\frac{1}{M} + \frac{1}{K} + \frac{1}{P}\right)^2 - \frac{4}{MK}}\right] \tag{46}$$

and

$$S = \frac{\alpha_1 + \frac{m}{K} - \frac{Km}{PM}\left[\exp\left(\frac{M}{K}\right) - 1\right]}{\alpha_2 - \alpha_1} \tag{47}$$

In the mixing zone at the top,

$$c_{do} = Kc_{cl} \tag{48}$$

Mass balance around the whole column yields

$$c_{co} - c_{c1} = R\left(c_{di} - c_{do}\right) \tag{49}$$

The mass-transfer model was tested against nine experimental extraction works in spray columns [34]. However, the comparison was conducted with sometimes scarce data on the operating conditions and distribution coefficients. Therefore, the wake parameters were rather arbitrarily chosen, and the deviations between the predictions and experimental results were usually large.

In long columns the formation and mixing zones are negligible compared with the wake shedding zone. The other parameters required in the wake model are the wake parameters: the wake size and the rate of shedding. Both vary with the flow regime around the droplet, i.e., slip velocity and holdup. Extensive studies are available on wake size, and therefore the order of magnitude, M, may be estimated. The rate of shedding, m, may be assessed at low holdups in disperse systems using available correlations of dimensionless groups [26]. However, study of the hydrodynamics of wakes will improve the estimates of that parameter too.

The wake model was developed for a quiescent system. It may, however, be modified for a turbulent bulk in which eddies move along the column.

4 DESIGN RELATIONS AND APPLICATIONS

Direct-contact heat exchangers have attracted attention mainly because of their applicability to fouling and corrosive fluids, clogging suspensions, and solid media of granular character. The utilization of low-grade heat sources for power production has recently renewed the interest in the direct-contact devices.

Operationally, a direct-contact heat exchanger is a container of any shape or configuration that provides contact between the process and working fluids without interfering surfaces between them. The conventional shape, configuration, and operation relate to a tubular, vertical column, in which the fluids move countercurrently, with one being particulately dispersed. The present study is concerned with liquid droplets dispersed in another immiscible liquid.

The main contact volume of the two liquids is the column proper. Therefore, the elemental geometry variables are the diameter of the column proper, and the length of the column proper. The inlets and outlets and the conical extensions, as well as the settlers at the top and bottom of the column, usually follow the well-established features of the Elgin tower [37].

4.1 Diameter and Length of the Column Proper

Particulate systems are contacted co-currently or countercurrently. In cocurrent flows the system may approach at the utmost the efficiency of a single stage or a perfect mixer; i.e. the two phases may reach equilibrium at their outlets. For such contact a container is more suited than a column. A column is usually employed for a close approach of inlet-outlet temperatures or concentrations of the phases. To achieve that goal the system has to be operated countercurrently in a quiescent state. The design relations of diameter and length for such systems can be formulated. The formulations to follow are presented, for the sake of simplicity, for uni-sized droplets, uniform holdup, and constant properties.

Diameter. The diameter of the column proper is calculated for a specified flow rate of the process liquid. For convenience the continuous phase is employed as the process liquid:

$$D = 2 \left[\frac{Q_c}{V_c} \right]^{1/2} \tag{50}$$

where Q_c is the specified volumetric flow rate. The superficial velocity V_c is to be calculated through the following relations: The superficial velocity is of the two phases are related in a counterflow by Eq. (10), which can also be expressed as

$$U_S = V_c \left[\frac{R}{\phi} - \frac{1}{1 - \phi} \right] \tag{51}$$

The slip velocity, U_S, is related to holdup by Eq. (3). Combination of these yields

$$V_c = \frac{U_T(1 - \phi)^n}{R/\phi - 1/(1 - \phi)} \tag{52}$$

with $n(Re_o)$ presented earlier [2]. The terminal velocity is obtained from the drag and gravity forces on the droplet:

$$C_D U_T^2 = \frac{4}{3} d \frac{(\rho_d - \rho_c)g}{\rho_c} \tag{53}$$

The droplet Reynolds number Re_o, defined as

$$Re_o = \frac{U_T d \rho_c}{\mu_c} \tag{54}$$

may be substituted for U_T in Eq. (53):

$$C_D Re_o^2 = \frac{4}{3} d^3 \frac{\rho_c(\rho_d - \rho_c)g}{\mu_c^2} \tag{55}$$

The **drag** coefficient C_D is a function of Re_o. That function is used to yield Re_o^2 *vs.* Re_o. Then the value of $C_D Re_o^2$ is obtained from Eq. (55), to provide the appropriate Re_o, from which U_T is calculated as

$$U_T = \frac{Re_o}{d} \frac{\mu_c}{\rho_c} \tag{56}$$

The other variable of Eq. (52) is the operational holdup. Usually,

$$\phi/\phi_f = 0.7 - 0.9 \tag{57}$$

The other variable of Eq. (52) is the operational holdup. Usually,

$$\phi/\phi_f = 0.7 - 0.9 \tag{57}$$

where ϕ_f is the holdup at flooding as expressed in Eq. (14).

The diameter of the column proper is obtained from Eqs. (50) and (52),

$$D = 2\left[\frac{Q_c(R(1-\phi)+\phi)}{\pi U_T \phi(1-\phi)^{(n+1)}}\right]^{1/2} \tag{58}$$

In the Elgin tower [37] the column proper is extended into a flared section with an angle of about $16°$ from the vertical. It gradually reduces the velocity of the continuous phase to about 0.2 of its value in the column proper. The distributor of droplets is located at the bottom of the conical section.

Length. The integral and global methods of HTU, and U_v, coefficients, as well as the dispersion and physical models were devised for the length estimates. The integral coefficients are adequate for the design of a heat exchanger of the same size and operational range as used in the experiments that yielded the correlation. Variation of size and operating conditions necessitates comprehension of the transfer mechanism between the phases. The wake model presented above yielded adequate estimates in quiescent systems. Length of the heat exchanger required to affect a temperature change ΔT_c, or ΔT_d in a quiescent column, can be calculated by Eqs. (34) and (35). The length, L, appears in the exponential form in the "wake" equations. The length is linear with the temperature change only for the specific case of $Rr = 1$

$$L = \frac{\left[\dfrac{T_{do} - T_{co}}{T_{di} - T_{co}} \cdot \exp(MR) - 1\right]}{\dfrac{r}{mS} - \dfrac{1}{\alpha_1}\left(\dfrac{1}{S} + 1\right)} \tag{59}$$

where S and α_1 are expressed by Eqs. (37) and (36), respectively.

This is the only operating condition for which the HTU, or U_v, values can be used for linear scale-up of length. For any other operating conditions, e.g., $Rr \neq 1$, we obtain within the calculated U_v, an exponential function of length. That showed that the calculated overall coefficients cannot be correlated with the operating variables only. They depend on length too.

In long columns the temperatures approach is very close. Shortening the column affects the approach very little, but changes considerably the volume of the heat exchanger, yielding much higher volumetric heat-transfer coefficients U_v [38]. Thus for a specified temperature change to be affected in the heat exchanger the length of a quiescent column has to be calculated by Eqs. (31)–(39).

The wake parameters M and m have to be established hydrodynamically. In the meantime, methods of evaluation of these parameters are adopted [39]. The relative wake volume, M is calculated in the range of $20 < Re_o \leq 500$ as follows:

for $20 < Re_s \leq 150$

$$M = 0.25(1 - \cos \alpha_s)(\cos \alpha_s - \cos 2\alpha_s) + (1.05 \, Re_s - 1.45)\sin^2 \alpha_s \quad (60)$$

for $150 < Re_s \leq 500$

$$M = 0.25(1 - \cos \alpha_s)(\cos \alpha_s - \cos 2\alpha_s) + 0.835 \sin^2 \alpha_s \quad (61)$$

where

$$Re_s = dU_s\rho_c/\mu_c = Re_o(1 - \phi)^n \quad (62)$$

The angle of boundary layer separation on a sphere α_s, was measured and graphically presented [40] as a unique function of Re_s, namely, $\alpha_s(Re_s)$. The above relations hold for

$$M < (1 - \phi)/\phi . \quad (63)$$

The other wake parameter, the relative wake shedding m, is evaluated [39] as

$$m = \frac{1}{d} \cdot \frac{F}{Re_o^{\frac{1}{2}}} \cdot \frac{1 + \phi/R(1 - \phi)}{(1 - \phi)^{(n/2)}} \cdot \frac{\sin^2 \alpha_s}{(\cos \alpha_s)^{\frac{1}{2}}} \quad (64)$$

F depends on the properties of the system. It may be obtained by a single experiment in a system in which all the other variables are well established.

Example. Application of the design relations to a preheater is illustrated. A preheater is designed for a solar pond power plant operated with a direct-contact boiler. The working fluid selected is pentane. The heating fluid is the concentrated brine from the pond introduced into the boiler at $85^{\circ}C$. Condenser is operated at $30^{\circ}C$. The turbine and pump efficiencies are 0.75 and 0.70, respectively [38]. The experimental data indicated that an approach temperature of $1^{\circ}C$ is the lowest achievable in a boiler [38].

Performance of the whole system is determined by specifying two parameters. The parameters selected for the example are the temperature difference and the approach temperature. The numerical values assigned to these parameters are $\Delta T = 9 \,^{\circ}C$, and $\Delta T_{app} = 1.5 \,^{\circ}C$. The example is illustrated in Fig. 6.12. Thus the two selected parameters uniquely define the system efficiency (6.9%) and the mass flow rate ratio, $\dot{m}_B/\dot{m}_R \cong 19$.

With the mass flow rate ratio so determined we proceed to the flow operating diagram of the pentane-brine system, Fig. 6.13, constructed for 4 mm droplets and a temperature of $75 \,^{\circ}C$. At the ratio of 19 and a holdup 10% below flooding, $\phi = 0.9 \times \phi_f = 0.15$, the superficial velocity of the brine is obtained, $V_c = 0.10$ m/s. The brine throughflow Q_c is specified for the power plant output. Therefore, the diameter of the preheater column is calculated by Eq. (50).

The length of the preheater is calculated by Eq. (34) for the same specified approach temperature of $1.5 \,^{\circ}C$, temperature difference of $9 \,^{\circ}C$ in the boiler, and the mass flow rate ratio of 19. That again is graphically illustrated in Fig. 6.14 constructed for the same system.

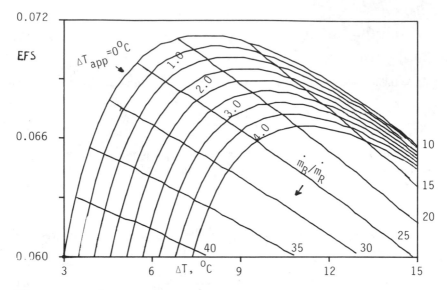

Figure 6.12 System efficiency vs. operating conditions in a power plant [38].

Thus the column proper is sized by specifying the boiler operating conditions.

4.2 Scale-up

The basic scale-up procedures relate to preservation of the temperature gradient along the column. In large columns vortices move along and across the column in a random manner "backmixing" the phases. The scale of the vortices increases with the column diameter. The gradient along the column decreases, and a steep change of temperature or concentration takes place at the inlets. In short columns in which the scale of turbulence approaches the magnitude of the length, the performance of the system resembles a single-stage mixer.

In large columns the preservation of the gradient can be pursued in two ways:

- By extension of length to compensate for local mixing. A large ratio of length to diameter, L/D, makes the vortices decay along the column. The analytical treatment of such problems will progress with the studies of turbulence.

- By installing internal devices in the column, such as baffles or packing, to reduce or eliminate the turbulent eddies. The cross-sectional area may be divided into spaces as small as needed to impose a quiescent regime in the two-phase bulk. In systems in which pressure drop is critical the quiescent flow can be preserved by longitudinal annular baffles [35]. Otherwise, "packing" devices may be used.

The length-diameter ratio has to be experimentally established for a specified performance. The same applies to packing in a spray column. On the

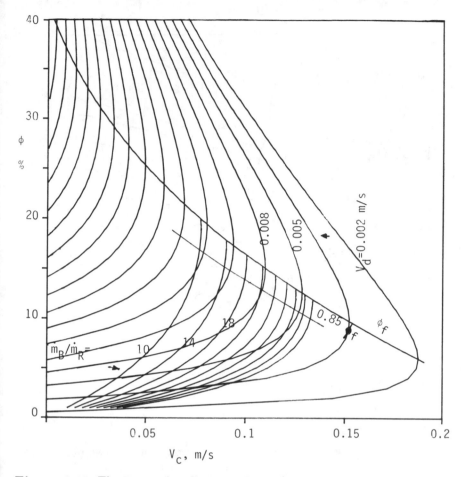

Figure 6.13 Flow operating diagram of a preheater.

other hand, the spacing of annular baffles may be quantitatively estimated for systems in laminar flow. The two-phase bulk flows between the baffles in the same way as in a column of a smaller diameter.

The critical diameter of the column at the onset of turbulence can be estimated in the same way as for one-phase flows. If a larger diameter is needed for higher throughputs, the annular baffles can be spaced accordingly. The critical diameter or spacings are estimated as follows: In a laminar bulk,

$$Re_D \leq (Re_D)_{critical} \tag{65}$$

where the bulk Reynolds number is

$$Re_D = (U_c\, \rho_m D)/\mu_{eff} \tag{66}$$

$$\rho_m = \rho_c(1 - \phi) + \rho_d\phi \tag{67}$$

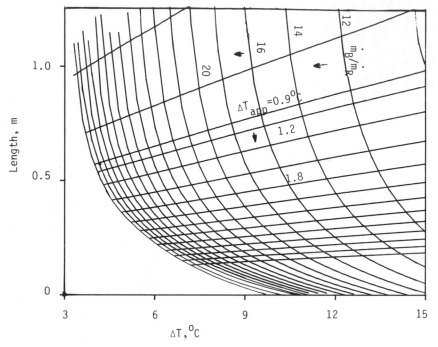

Figure 6.14 Length of a preheater vs. operating conditions.

$$\mu_{eff} = \mu_c \, f(\phi) \tag{68}$$

$$U_c = V_c/(1 - \phi) \tag{9}$$

The density and viscosity are ρ_m and μ_{eff}, respectively, of the two-phase bulk. Combining Eqs. (65)–(68) yields

$$Re_D = \frac{V_c \rho_c D}{\mu_c} \cdot \frac{\left[(1 - \phi) + \dfrac{\rho_d}{\rho_c}\,\phi\right]}{(1 - \phi)\cdot f(\phi)} \tag{69}$$

If the criterion of Eq. (65) is obeyed to preserve a laminar regime of the two-phase bulk, then the column diameter or the spacing between the baffles is determined as

$$D \leq (Re_D)_{crit.} \left(\frac{\nu_c}{V_c}\right) \left[\frac{(1 - \phi)\cdot f(\phi)}{(1 - \phi) + \dfrac{\rho_d}{\rho_c}\,\phi}\right] \tag{70}$$

For the estimates to be made the $(Re_D)_{crit}$ and $f(\phi)$ have to be known. To obtain an order of magnitude of the diameter a one-phase bulk was assumed $((Re_D)_{crit} = 2300)$ and the Roscoe function $f(\phi)$, Eq. (17) was employed [35]. At

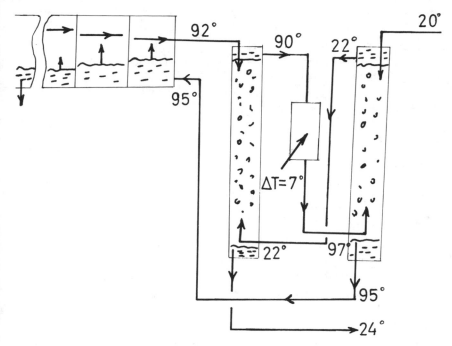

Figure 6.15 Direct-contact desalination process.

lower holdups and with small rigid droplets these assumptions may provide better estimates.

Almost all the experimental data in direct-contact devices were taken from small columns up to 0.15 m in diameter and 1–3 m in height. In the very few larger and taller columns the measurements were conducted within a limited range of conditions.

4.3 Applications

The applications discussed herein relate to processes reported in the literature, such as preheating, crystallization, and extraction. Many other applications can be devised.

Preheating. Preheating of a liquid by another immiscible liquid was first studied in desalination processes [18], and more recently in power plants that utilize low-grade heat sources [32,38,41–44].

In desalination processes water served as the process fluid. An organic liquid like kerosene was circulated for heating and cooling in two separate columns. The process is schematically illustrated in Fig. 6.15. The temperatures are indicated for a numerical example only. Although extensive research was conducted on these topics, the process has not been implemented, mainly because of the apprehension that contact with organics will make the water obsolete for use.

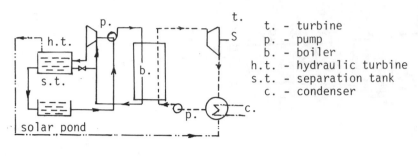

<table>
<tr><td></td><td>t.</td><td>- turbine</td></tr>
</table>

t. - turbine
p. - pump
b. - boiler
h.t. - hydraulic turbine
s.t. - separation tank
c. - condenser

working fluid   —— brine
              --- recovered vapor
cooling water —·—

Figure 6.16 Solar pond power plant (38).

The other process that employs a liquid-liquid preheater relates to preheating of an organic working fluid in power plants operated with geothermal or solar pond brines. A solar pond power plant is schematically illustrated in Fig. 6.16. Experimentation in such systems has been conducted in large and tall columns too [41].

Preheating with geothermal brines has been extensively studied in laboratory columns ($D = 0.15$ m) and in pilot plant columns ($D = 1$ m) too. Urbanek [42] reported data in isobutane and isopentane-brine systems. The volumetric heat transfer coefficients U_V in a separate preheater were about 70 kW/m^3 ° C (4000 Btu/ft^3 hr F) in an 0.15 m in diameter column, 3 m long. The same system experimented with in a combined preheater-boiler, in which the approach temperatures were assumed, yielded heat transfer coefficients of 130 kW/m^3 ° C (7300 Btu/ft^3 hr F) that probably partly corresponded to the boiling section. In an isopentane-brine the preheating proceeded at 35 kW/m^3 ° C (1800 Btu/ft^3 hr F).

Huebner et al. [43] also operated a single vessel for preheating and vaporization of isopentane in brine. The column was 1.2 m in diameter and 2.4 m long. The overall volumetric heat transfer coefficients were 25–40 kW/m^3 ° C (1400–2200 Btu/ft^3 hr F).

Olander [44] reported extensive experimental data obtained in an isobutane-brine system in a column 1 m in diameter and 14 m in total length. The preheater was at first 8 m (27 ft) long, and later was shortened to 5.3 m (16 ft). The volumetric heat transfer coefficients in the longer column were 50 kW/m^3 ° C (2500 Btu/ft^3 hr F) and 130 kW/m^3 ° C (7000 Btu/ft^3 hr F) in the shorter column.

The three works cited above demonstrated performance of the preheaters in a predictable direction. It is apparent that the combination of preheater with boiler has intensified mixing effects in the preheater [43], reducing the heat transfer rate to about half of the previous values (from 70 to 35 kW/m^3 ° C).

The low overall coefficients of 25–40 kW/m^3 ° C, in heating and evaporation together, in the 1.2 m in diameter vessel indicated that the column of 2.4 m as too

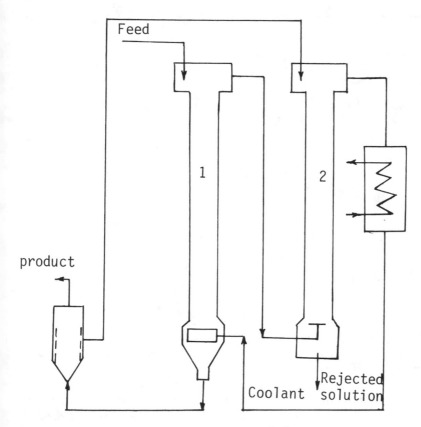

Figure 6.17 Direct-contact crystallization [45].

short to subside the large-scale vortices. Thus the performance of this large column has approached a single mixed stage. However, in a preheater-boiler of about the same diameter [44] but five times taller, the coefficients were three times higher. Any further increase in height brought additional increase in temperature and volume. As these two quantities are related nonlinearly, Eq. (34), the resulting volumetric coefficient in a "subsided" system decreased with the increase in length.

Thus the use of volumetric coefficients for comparison of performance has to be done with caution and comprehension of the physical phenomena in the specific situation.

Crystallization. Crystallization by cooling is applied to solutions of salts which show a decreasing solubility with reduced temperature. The crystallization can be brought about by direct or indirect heat exchange between the crystallizing solution and a colder fluid. In a direct-contact cooled crystallizer the solution contacts an immiscible liquid. Such a process is schematically illustrated in Fig. 6.17. Column 1 serves for crystallization. Column 2 is used for cooling of the coolant.

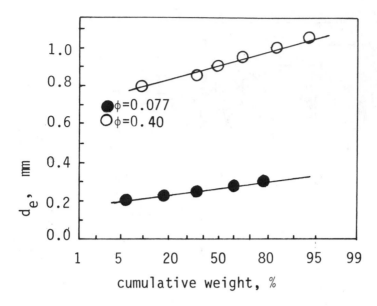

Figure 6.18 Crystal size in a direct-contact crystallizer [45].

In Fig. 6.18 are presented the crystal sizes of magnesium chloride deposited from the aqueous brine cooled by kerosene [45]. The results were obtained at two hold-ups of kerosene: 7.6% and 40%. At the higher holdup larger crystals were formed. In those cases and at any other holdup the crystal size uniformity characterized the direct-contact process. Sizing of the crystallizer is carried out by the same design relations as above and the appropriate physical properties of the contacted phases.

Extraction. Commercial extraction is rarely conducted in spray columns. The main deficiency attributed to this equipment relates to the severe backmixing that prevails in large columns.

The voluminous literature on extraction in spray columns has not provided a reliable correlation to be used for design. The difficulties are due to the poor understanding of the physical mechanism of these processes.

The formulation of the wake model [33] has provided an approach to the problem. Two hundred fifty experimental runs published in the literature were analyzed [34] and compared with values calculated by the wake model. A deviation of less than 20% was obtained for 100 runs. All the data are summarized in full detail in the above-cited work [34]. The design relations and scale-up procedures presented earlier may be applied to extraction processes with limitations and caution.

NOMENCLATURE

A	cross-section area of column
a	surface area of particles per unit volume of column
c	solute concentration
C_D	drag coefficient
c_p	specific heat capacity
D	diameter of column proper
d	diameter of particle or droplet
D_L	longitudinal dispersion coefficient
g	gravity acceleration
h	heat transfer coefficient
K	distribution coefficient
L	length of column proper
M	relative volume of wake to particle
m	relative volume of wake elements shed per unit length of column, also an exponent
n	exponent
Nu	particle Nusselt number, hd/k
P	defined by Eq. (33)
ΔP	pressure drop
Pr	Prandtl number, $\mu c_p/k$
Q	volumetric flow rate
R	flow rate ratio, V_d/V_c
r	Eq. (31)
Re_s	particle Reynolds number, $d\rho_c\,U_S/\mu_c$
Re_d	single particle Reynolds number, $d\rho_c\,U_T/\mu_c$
S	defined by Eqs. (37) and (47)
T	temperature
ΔT_m	logarithmic mean temperature difference, Eq. (21)
t	time
U	linear velocity, also heat transfer coefficient
U_S	slip velocity
U_S^*	slip velocity at constant local holdup
U_T	terminal velocity of a particle
U_V	volumetric heat transfer coefficient
V	superficial velocity
z	height in the column
$\alpha_{1,2}$	defined by Eqs. (36) and (46)
α_E	thermal eddy diffusivity
α_s	angle of separation from rear stagnation point
μ	dynamic viscosity
μ_{eff}	effective viscosity of a dispersion
ν	kinematic viscosity

ρ density
ρ_m density of two-phase bulk
ϕ holdup, volumetric fraction of particles in column
ϕ^l local holdup at height z

Subscripts

c continuous
d dispersed
f flooding
i inlet
l at the top of column proper
o outlet
s slip

REFERENCES

1.   Lapidus, L., and Elgin, J. C., Mechanics of Vertical-Moving Fluidized Systems, *Journal of the American Institute of Chemical Engineers*, vol. 3, no. 1, pp. 63-68, 1957.
2.   Richardson, J. F., and Zaki, W. N., Sedimentation and Fluidization, *Transactions of Institution of Chemical Engineers*, vol. 32, pp. 35-53, 1954.
3.   Zenz, F. A., Calculate Fluidization Rates, *Petroleum Refiner*, vol. 36, no. 8, pp. 147-155, 1957.
4.   Zuber, N., On the Dispersed Two-Phase Flow in the Laminar Flow Regime, *Chemical Engineering Science*, vol. 19, pp. 819-917, 1964.
5.   Beyaert, B. O., Lapidus, L., and Elgin, J. C., The Mechanics of Vertical Moving Liquid-Liquid Fluidized Systems: II Countercurrent Flow, *Journal of the American Institute of Chemical Engineers*, vol. 7, pp. 46-48, 1961.
6.   Letan, R., and Kehat, E., The Mechanics of a Spray Column, *Journal of the American Institute of Chemical Engineers*, vol. 13, pp. 443-449, 1967.
7.   Weaver, R. E. C., Lapidus, L., and Elgin, J. C., The Mechanics of Vertical Moving Liquid-Liquid Fluidized Systems, *Journal of the American Institute of Chemical Engineers*, vol. 5, no. 4, pp. 533-539, 1959.
8.   Price, B. G., Lapidus, L., and Elgin, J. C., The Mechanics of Vertical Moving Liquid-Liquid Fluidized Systems, *Journal of the American Institute of Chemical Engineers*, vol. 5, pp. 93-97, 1959.
9.   Struve, D. L., Lapidus, L., and Elgin, J. C., The Mechanics of Vertical Moving Fluidized Systems, *Canadian Journal of Chemical Engineering*, vol. 36, pp. 141-152, 1958.
10. Finkelstein, E., Letan, R., and Elgin, J. C., Mechanics of Vertical Moving Fluidized Systems with Mixed Particle Sizes, *Journal of the American Institute of Chemical Engineers*, vol. 17, pp. 867-872, 1971.
11. Wilhelm, R. H., and Kwauk, M., Fluidization of Solid Particles, *Chemical Engineering Progress*, vol. 44, pp. 201-218, 1948.
12. Mertes, T. S., and Rhodes, H. B., Liquid Particle Behavior (Part 1), *Chemical Engineering Progress*, vol. 51, pp. 429-437, 1955.
13. Lewis, W. K., Gilliland, E. R., and Bauer, W. C., Characteristics of Fluidized Particles, *Industrial and Engineering Chemistry*, vol. 41, pp. 1104-1117, 1949.
14. Brinkman, H. C., The Viscosity of Concentrated Suspensions and Solutions, *Journal of Chemical Physics*, vol. 20, pp. 571, 1952.
15. Roscoe, R., The Viscosity of Suspensions of Rigid Spheres, *British Applied Physics*, vol. 3, pp. 267-269, 1952.
16. Letan, R., On Vertical Dispersed Two-Phase Flow, *Chemical Engineering Science*, vol. 29, pp. 621-624, 1974.

17.   Bauerle, G. L., and Ahlert, R. C., Heat Transfer and Holdup Phenomena in a Spray Column, *Industrial and Engineering Chemistry Process Design of Development*, vol. 4, pp. 225-230, 1965.

18.   Woodward, T., Heat Transfer in a Spray Column, *Chemical Engineering Progress*, vol. 57, pp. 52-57, 1961.

19.   Letan, R., and Kehat, E., The Mechanism of Heat Transfer in a Spray Column Heat Exchanger: II Dense Packing of Drops, *Journal of American Institute of Chemical Engineers*, vol. 16, pp. 955-963, 1970.

20.   Hiby, J. W., *Longitudinal and Transverse Mixing During Single-Phase Flow Through Granular Beds*, Symposium on Interaction between Fluids and Particles. London: Institution of Chemical Engineers, pp. 313-325, 1962.

21.   Cavers, S. D. and Ewanchyna, J. E., Circulation and End Effects in a Liquid Extraction Spray Column, *Canadian Journal of Chemical Engineering*, vol. 35, pp. 113-128, 1957.

22.   Gier, T. E., and Hougen, J. O., Concentration Gradients in Spray and Packed Extraction Columns, *Industrial and Engineering Chemistry*, vol. 45, no. 6, pp. 1362-1370, 1953.

23.   Kreager, K. M., and Geankoplis, C. J., Effect of Tower Height in a Solvent Extraction Tower, *Industrial and Engineering Chemistry*, vol. 45, no. 10, pp. 2156-2165, 1953.

24.   Letan, R., and Kehat, E., Mixing Effects in a Spray-Column Heat Exchanger, *Journal of the American Institute of Chemical Engineers*, vol. 11, pp. 804-808, 1965.

25.   Kramers, H., Westermann, M. D., de Groote, J. H. and Depont, F. A. A., *The Longitudinal Dispersion of Liquid in a Fluidized Bed*, Symp. on Interaction between Fluids and Particles. London: Institution of Chemical Engineers, pp. 114-119, 1962.

26.   Letan, R., and Elgin, J. C., Fluid Mixing in Particulate Fluidized Beds, *Chemical Engineering Journal*, vol. 3, pp. 136-144, 1972.

27.   Danckwerts, P. V., Continuous Flow Systems, *Chemical Engineering Science*, vol. 2, pp. 1-13, 1953.

28.   Cairns, E. J., and Prausnitz, J. M., Longitudinal Mixing in Fluidization, *Journal of the American Institute of Chemical Engineers, vol. 6, pp. 400-405, 1960.*

29.   Letan, R., and Kehat, E., Residence Time Distribution of the Dispersed Phase in a Spray Column, *Journal of the American Institute of Chemical Engineers*, vol. 15, pp. 4-10, 1969.

30.   Mixon, F. D., Whitaker, D. R., and Orcutt, J. C., Axial Dispersion and Heat Transfer in Liquid-Liquid Spray Towers, *Journal of the American Institute of Chemical Engineers*, vol. 13, pp. 21-28, 1967.

31.   Greskovich, E. J., Ph.D. Dissertation, Pennsylvania State University, 1966.

32.   Plass, S. B., Jacobs, H. R., and Boehm, R. F., *Operational Characteristics of a Spray Column Type Direct Contact Preheater*, AIChE Symposium Series no. 189, vol. 75, pp. 227-234, 1979.

33.   Letan, R., and Kehat, E., The Mechanism of Heat Transfer in a Spray Column Heat Exchanger, *Journal of the American Institute of Chemical Engineers*, vol. 14, pp. 398-405, 1968.

34.   Kehat, E., and Letan, R., The Role of Wakes in the Mechanism of Extraction in Spray Columns, *Journal of the American Institute of Chemical Engineers*, vol. 17, pp. 984-990, 1971.

35.   Letan, R., *Design of a Particulate Direct Contact Heat Exchanger: Uniform Countercurrent Flow*, 16th Natl. Heat Transfer Conf., St. Louis, ASME Paper 76-HT-27, 1976.

36.   Letan, R., A Parametric Study of a Particulate Direct Contact Heat Exchanger, *Journal of Heat Transfer*, vol. 103, pp. 586-590, 1981.

37.   Blanding, F. H., and Elgin, J. C., *Transactions of the American Institute of Chemical Engineers*, vol. 38, p. 305, 1942.

38.   Sonn, A., and Letan, R., *Performance of a Direct Contact Evaporator and its Effect on the Efficiency of a Binary System*, 22nd Natl. Heat Transfer Conf., Niagara Falls, ASME Paper 84-HT-31, 1984.

39.   Zmora, J., and Letan, R., *Direct Contact Cooling of a Crystallizing Brine*, Proc. 6th Int. Heat Transfer Conf., vol. 4, pp. 61-65, 1978.

40.   Taneda, S., *Experimental Investigation of the Wake Behind a Sphere at Low Reynolds Numbers*, Rept. Res. Inst. Appl. Mech., Kyushu Univ., vol. 4, pp. 99-105, 1956.

41.   Goodwin, P., Coban, M., and Boehm, R. F., *Evaluation of the Flooding Limits and Heat Transfer of a Direct Contact Three Phase Spray Column*, 23rd Natl. Heat Transfer Conf., Denver, ASME Paper 85-HT-49, 1985.

42.   Urbanek, M. W., *Experimental Testing of a Direct Contact Heat Exchanger for Geothermal Brine*, Final Rept. ORNL-SUB-79/13564/1 and 79/45736/1 DSS-079, Dec. 1979.

43.   Huebner, A. W., Wall, D. A., and Herlacher, T. L., *Research and Development of a 3 MW Power Plant from the Design Development and Demonstration of a 100 kW Power System Utilizing the Direct Contact Heat Exchanger Concept for Geothermal Brine*, Energy Recovery Program, Final Rept. DOE/ET/28456-T1, Sept. 1980.

44.   Olander, R., *Final Phase Testing and Evaluation of the 500 kW Direct Contact Pilot Plant at East Mesa*, Barber Nichols End., Arvada, Colorado, Dec. 1983.

45.   Shaviv, F., *Performance of a Direct Contact Cooled Crystallizer*, M.Sc. Thesis, Ben-Gurion University, Beer Sheva, 1978.

DISCUSSION OF MASS TRANSFER EFFECTS AND LIQUID-LIQUID TRANSPORT

E. Marschall

1 INTRODUCTION

The explicit purpose of this workshop session was to identify research and development needs in the areas of liquid-liquid direct-contact transport processes and simultaneous direct-contact heat and mass transfer. This task was greatly facilitated by the review papers by Perona [1] and Letan [2], which provided the necessary guidance for the discussion.

Before discussing the various specific areas of needed research and development related to the direct-contact heat- and mass-transfer topics that were presented at this workshop session, a few general observations seem to be in order. Direct-contact heat transfer in its various forms has emerged as a major ingredient in many modern technological processes. Liquid-liquid direct-contact heat transfer is now being used to heat up heavy crude oil for water and gas separation, it has been used successfully in experimental geothermal power plants operating with geothermal brines with high salt contents, and it is still considered a viable alternative in seawater desalination. Other examples include concentration of waste water and potash production. The use of a combination of immiscible liquids, one being close to its freezing point, has shown promising potential for thermal energy storage as well as for transportation of thermal energy. Such mixtures have also been considered for use as cooling fluids in situations where high heat fluxes occur.

Numerous examples of combined direct-contact heat and mass transfer exist. A few recent and promising applications include high pressure solvent extraction with one phase being a supercritical gas, fluidized bed combustion of low-grade coal to minimize No_x and SO_2 production, SO_2 removal in packed beds, and a number of solid-gas and solid-liquid transfer processes in various reactor configurations.

A comprehensive treatment of the various aspects of direct-contact heat transfer does not exist. However, several state-of-the-art reviews of specific subjects in the area of direct-contact heat transfer are available. More than 100 publications on liquid-liquid direct-contact heat transfer up to about 1968 were collected and subsequently discussed by Kehat and Sideman [3]. More recent publications on that subject were reviewed by Marschall, Johnson, and Culbreth [4]. A comprehensive listing of the literature on fluid mechanics, heat and mass transfer of single particles, and drops or bubbles moving in a Newtonian fluid was compiled by Clift, Grace, and Weber [5]. These authors reviewed the relevant literature up to about 1976. A more recent treatment of particle-fluid transport processes was provided by Brauer [6]. Various recommendations for the design of direct-contact heat exchangers were made by Jacobs and Boehm [7] and by Mersmann [8]. A wealth of related information can also be found in reviews of liquid-liquid extraction [9,10] and bubble and drop phenomena [11]. Finally, the handbook of multiphase systems [12] contains discussions of liquid-liquid heat transfer as well as simultaneous direct-contact heat and mass transfer.

The workshop session on mass transfer in heat-transfer processes and on liquid-liquid processes established two major areas of concern. The first area deals with basic understanding and modeling of the heat- and mass-transfer phenomena occurring in these processes, while in the second area emphasis is placed on equipment design and operation. Clearly, these two areas are not entirely separated from each other.

2 RECOMMENDATIONS FOR BASIC RESEARCH

Hydrodynamics, heat transfer and mass transfer in multiphase systems with and without chemical reactions, have been the subject of sometimes very intensive research activities over the past decades. As a result, a fairly good qualitative understanding of these processes for most situations has been obtained. Frequently lacking are reliable quantitative descriptions that allow accurate predictions of momentum, mass, and energy transfer in multiphase systems. This is, of course, exactly what is required for a sound design of direct-contact heat- and mass-transfer systems. The following listing shows research topics identified by the participants of this workshop session as deserving support. The order of listing is not intended to indicate the degree of priority of the topics.

2.1 Hydrodynamics in Liquid-Liquid Systems

Hydrodynamics of single drops in liquid-liquid systems. Single drops rising or falling under gravity in a Newtonian liquid have been studied intensively. Commonly, they are considered to belong to one of three regimes: drops with no internal motion, drops with internal motion, and oscillating drops. Flow characteristics can usually be expressed in terms of nondimensionless groups such as Reynolds numbers, Bond (Eötvös) numbers, Weber numbers, and other groups containing physical properties of both phases. Expressions of that kind generally work well if the liquid-liquid system is not contaminated. If the system is contaminated with surface-active substances, as may very well happen in industrial situations, then the predictive models become frequently very unreliable. Since the understanding of single-drop flow phenomena provides the basis for understanding flow characteristics of drop ensembles, research efforts should be directed toward single-drop flow characteristics in contaminated systems.

Little is known about the hydrodynamics in non-Newtonian liquid-liquid systems. Industrial applications may well involve non-Newtonian fluids. Therefore, studies are recommended that establish quantitatively the behavior of single-drop flow characteristics in liquid-liquid systems in which one or both phases consist of non-Newtonian fluids.

Frequently, solid particles or gases may be present in the liquid-liquid systems. Little is understood of their influence on the hydrodynamics. For instance, it has been observed, although never explained, that one very small gas bubble attached to a liquid drop can change considerably the drop's terminal velocity. Thus research is needed to understand the effect of solid particles and gas bubbles on the flow characteristics of liquid-liquid systems.

Almost all studies of the hydrodynamics in liquid-liquid systems have been made under isothermal conditions or in the absence of mass transfer. However, temperature or concentration gradients are more the rule than the exception in liquid-liquid direct-contact equipment. Therefore, it is recommended to investigate the single-drop flow characteristics in the presence of temperature and concentration gradients.

Oscillating drops may experience some internal turbulence. In addition, the continuous flow may be turbulent either some distance from the drop or in the boundary layer around the drop. Studies are needed to determine quantitatively the effect of turbulence on the flow characteristics of drops moving in a liquid-liquid system.

Drop size can usually be predicted within 10% if the dispersed phase is a Newtonian fluid and does not wet the nozzle rim and if the nozzle is carefully designed and manufactured. Much more difficult is the prediction of flow fields during drop formation and release, especially when non-Newtonian fluids are involved or when the nozzles deviate from the cylindrical, vertical configuration. Given the high heat- and mass-transfer rates experienced during drop formation and release, it is recommended to examine experimentally as well as theoretically the hydrodynamics during drop formation and release in liquid-liquid systems.

Drop size in agitated equipment cannot be predicted with confidence. The same is true for drop size predictions in packed columns. In addition, the description of flow field characteristics even of single drops has not been successfully attempted. Consequently, a necessary research topic appears to be drop formation and flow behavior in agitated liquid-liquid contact equipment and in packed columns.

Hydrodynamics of drop ensembles in liquid-liquid systems. The research topics recommended above are also valid topics for drop ensembles. In addition, a lack of quantitative information on the following items was noted:

Radial and axial drop size and velocity distribution;

Axial and radial mixing depending on drop size distribution and contactor geometry;

Turbulence characteristics of continuous and dispersed flow;

Flooding depending on drop size distribution;

Phase inversion depending on drop size distribution;

Coalescence, and axial and radial variation of the hold-up.

Existing theories on the hydrodynamics of drop ensembles are mostly, though not always, based on a uniform drop size and a constant holdup. These conditions, however, are rarely met in industrial situations.

2.2 Liquid-Liquid Direct-Contact Heat Transfer

For single drops of a pure liquid rising or falling in Newtonian uncontaminated fluid, a number of correlations for area-related internal and external heat-transfer coefficients are available. They can be used with reasonable confidence when physical properties as well as flow characteristics of the fluids are known. Uncertainities exist and research is indicated to determine internal and external heat transfer coefficients in the presence of small solid particles and gas bubbles, and internal and external heat transfer coefficients in the presence of surface-active agents.

Little is known regarding heat transfer to and from single drops for non-Newtonian fluids. Investigations are recommended that establish reliable equations for internal and external heat transfer coefficients in systems with one or both liquids not Newtonian fluids.

Heat transfer equations found for single drops seem to work well for drop ensembles as long as the drops have a fairly uniform size and major backmixing is avoided. If these conditions are not met, then a quantitative prediction of heat transfer rates becomes difficult. For drop ensembles it appears appropriate to study the influence on heat transfer rates due to wide drop size distributions and strong backmixing.

Heat transfer rates during drop formation and release as well as during coalescence cannot be predicted with any reliability. However, it is known that during drop formation and release very high heat transfer rates can occur. It is known that a major portion of the total transferred heat may be transferred in that regime. Consequently, research should address heat transfer during drop formation and release, and heat transfer during drop coalescence.

Predictive methods are based on either area-related or volumetric heat transfer coefficients. Volumetric heat transfer coefficients can easily be obtained in experiments. All that is required is the measurement of flow rates and mixed mean inlet and outlet temperatures. However, the application of experimentally obtained volumetric heat transfer coefficients may lead to considerable errors, as was pointed out by Kehat and Sideman [3]. Volumetric heat transfer coefficients are only useful if they are applied to situations identical with the one in which they were found.

Area-related heat- (and mass-) transfer coefficients depend on local flow, temperature, and concentration fields, and on physical properties of continuous and dispersed phase. Measurement of local area-related heat transfer coefficients is quite involved, and physically sound correlations of these heat transfer coefficients are frequently difficult to obtain. In spite of this it was suggested to put emphasis on predictive methods based on area-related heat- (and mass-) transfer coefficients, since these methods are less restricted than methods based on volumetric heat transfer coefficients.

2.3 Liquid-Liquid Direct-Contact Mass Transfer

Suggestions made for liquid-liquid heat transfer are also valid for liquid-liquid mass transfer. Furthermore, the study of contaminated systems deserves additional attention. While in heat transfer processes surfactants influence the mobility of the liquid-liquid interface, in mass transfer they also appear to create a diffusion barrier. Thus the effect of surfactants on mass transfer differs from the effect on heat transfer. Emphasis should be placed on liquid-liquid mass transfer in contaminated systems.

2.4 Analogy

The concept of analogies between heat and mass transfer is applicable in most situations. However, that concept may yield inaccurate results whenever one mode of transport cannot be mirrored in the other. This may be caused by the reasons listed in [1]. In addition, the nature of packing material in packed or fluidized beds may contribute to the breakdown of the analogy concept. It is recommended to pay attention to the deviation of the analogy concept between heat and mass transfer.

2.5 Simultaneous Direct-Contact Heat and Mass Transfer

No generalized quantitative theory exists for describing simultaneous direct-contact heat and mass transfer, especially when high heat- and mass-transfer rates occur or when chemical reactions occur. This appears to be true not only for liquid-liquid systems but also for gas-liquid, gas-solid, and liquid-solid systems. Since many technical applications involve simultaneous direct-contact heat and mass-transfer, it is recommended to develop a generalized, predictive theory on simultaneous direct-contact heat and mass transfer, with and without chemical reactions.

2.6 Geometric Description of Contact Equipment

A major concern in the prediction of mass-transfer as well as heat transfer rates in direct-contact equipment is obtaining reliable information on the transfer area. Given the complex geometry of packed, fluidized, or spouted beds, or of drop or bubble ensembles in liquid-liquid or liquid-gas systems, this information is not easy to obtain, if at all. However, an accurate knowledge of the transfer area is mandatory information required by most design methods for contact equipment. A high priority should be placed on development or improvement of theoretical and experimental methods for the prediction of transfer area in direct-contact equipment.

3 RECOMMENDATIONS FOR DESIGN AND OPERATION OF DIRECT-CONTACT EQUIPMENT

Understanding qualitatively the mechanism of fluid mechanics, heat transfer, and mass transfer is not sufficient for quantitative predictions. The express goal of the recommended research projects should be obtaining design equations that allow quantitative assessments. If those equations are available, optimal design techniques can then be used to find the best design for a direct-contact heat- or mass-transfer apparatus. The essence of optimal design is to find extreme values for a given set of constraints. Given the competitive nature of the present time, it is recommended to apply and develop further optimal design techniques for direct-contact heat exchangers.

A welcome feedback from applying optimal design techniques should be obtained if these techniques are made the basis of sensitivity analyses. As a result, one should find which of the input parameters in the design process must be known precisely. For instance, such an analysis would tell at what error margin predictive correlations for drop size, holdup, friction factors, heat- and mass-transfer coefficients, flow and mixing effects, etc., are acceptable. Results of this kind could lead to intelligent decisions on the direction in which research efforts should be guided. Thus emphasis should be placed on sensitivity analyses based on optimal design techniques.

In addition, this approach should also provide the tool for judging critically the various heat- and mass-transfer models, e.g., models based on area-related heat- and mass-transfer coefficients and models based on volumetric heat- and mass-transfer coefficients.

Very few studies exist on the dynamic behavior of direct-contact heat exchangers. Given the fact that operational conditions do occasionally change it is recommended to study the dynamic behavior of direct-contact heat exchangers.

Major concerns in operation of direct-contact heat exchangers are the accumulation of scaling products or unwanted products of chemical reactions or of a change of the solubility characteristics. Other concerns include the degradation of loss of working fluids. Design guidelines should be established that address design for ease of cleaning, scale resisting design, design for minimal degradation or loss of working fluid, and design for ease of removal of unwanted byproducts.

Operational guidelines should be formulated that provide recommendations for safe and economical operation of heat transfer equipment.

4 CONCLUSION

The discussion of the survey papers presented by Perona and Letan yielded numerous suggestions for future research and development in the areas of liquid-liquid transport processes and simultaneous heat and mass-transfer. Some of these suggestions concern the qualitative understanding of the involved physical processes. The majority of the suggestions deal with the ability to quantitatively predict heat, mass, and momentum transfer in direct-contact processes. A common theme in the various research proposals is that "real" situations be investigated, as opposed to "ideal" situations. That is, heat, mass, and momentum transfer should be studied in contaminated systems, in systems with nonuniform drop or particle distribution, in systems with nonuniform velocity distribution, etc. In addition, future research should include non-Newtonian fluid behavior as well as the influence of strong temperature and concentration gradients on the transfer processes.

REFERENCES

1. Perona, J., *Mass Transfer Effects in Heat Transfer Processes*, Direct-Contact Heat Transfer Workshop, Solar Research Institute, Golden, Colorado, 1985.
2. Letan, R., *Liquid-Liquid Processes*, Direct-Contact Heat Transfer Workshop, Solar Research Institute, Golden, Colorado, 1985.
3. Kehat, E., and Sideman, S., "Heat Transfer by Direct Liquid-Liquid Contact," *Recent Advances in Liquid-Liquid Extraction*, Chapter 13, Pergamon Press, 1971.
4. Marschall, E., Johnson, G., and Culbreth, W., "Direct-Contact Heat Transfer," *Progress in Chemical Engineering*, VDI-Verlag, Vol. 20, Section A, 1982.
5. Clift, R., Grace, J. R., and Weber, M. E., *Bubbles, Drops, and Particles*, Academic Pres, 1978.
6. Brauer, H., "Particle-Fluid Transport Processes," *Progress in Chemical Engineering*, VDI-Verlag, Vol. 17, Section A, 1979.
7. Jacobs, H. R., and Boehm, R. J., "Direct-Contact Binary Cycles," *Source Book on the Production of Electricity from Geothermal Energy*, U.S. Department of Energy, 1980.
8. Mersmann, A., "Design and Scale Up of Bubble and Spray Columns," *Ger. Chem. Eng.*, Vol. 1, 1978.
9. Hanson, C., *Recent Advances in Liquid-Liquid Extraction*, Pergamon, 1971.
10. Bauer, R. J., "Extraction," *Progress in Chemical Engineering*, VDI-Verlag, Vol. 17, Section C, 1979.
11. Tavlarides, L. L., "Bubble and Drop Phenomena," *I&EC*, Vol. 22, No. 11, 1970.
12. Hetsroni, G., *Handbook of Multiphase System*, McGraw-Hill, 1982.

SOLIDS MOTION AND HEAT TRANSFER
IN GAS FLUIDIZED BEDS

Michael M. Chen

1 INTRODUCTION

When a bed of solid particles is subjected to an upward fluid flow above a critical velocity called the minimum fluidization velocity, the bed becomes fluid-like in a number of respects and is said to be fluidized. Because of the very large surface area, heat and mass exchange between the fluid and the solids are extremely efficient. In the case of typical gas fluidized beds, the solids also possess large mean and fluctuating velocities, thus promoting solids mixing and facilitating the addition and removal of the solids. These conditions favor the use of fluidized beds as chemical reactors, including as a special case combustors for solid, liquid, and gaseous fuels. Furthermore, the intense solids motion enhances heat exchange between the solids and immersed surfaces, permitting direct heat removal at low temperature drop using immersed heat-exchange tubes. Fluidized beds are also used in various configurations as heat exchangers. These and many other applications can be found in standard references (Kunii and Levenspiel, 1969; Davidson and Harrison, 1971; Kunii and Toei, 1983).

This chapter consists of a very personal review of three aspects of fluidization, solids motion, mixing, and heat transfer to immersed surfaces, with a more detailed description of our own work. Before proceeding further, it is useful to first familiarize ourselves with the different regimes of operation of fluidized beds.

The different regimes of operation of fluidized beds are shown in Fig. 8.1. The criteria for the transition from one regime to the next are discussed concisely by Wen and Chen (1984). In brief, for very fine particles there is a bubbleless fluidized regime immediately above minimum fluidization. Above this regime is the bubbling regime with bubbles forming at the bottom, then gradually coalescing and increasing in size as they rise to the top. For larger particles, the bubbleless regime does not exist and the bubbling regime occurs immediately above minimum fluidization. For narrow and tall beds, a slugging regime is encountered when the bubbles grow to fill the bed diameter. At higher velocities, the bubbles become irregular in shape and the flow takes on a very turbulent character, thus forming the turbulent regime. At even higher velocities, the particles are carried out of the bed and are either replaced by new particles or recirculated after separation from the carrier gas. This is the fast fluidization regime. Typically, the higher velocity regimes (bubbling, turbulent, and fast) are of the greatest engineering importance. For example, utility scale fluidized bed combustors tend to operate in the turbulent regime or the higher end of the bubbling regime.

It can be surmised from this presentation that the bubbles play important roles in influencing the heat transfer and dynamics of the bed. This is indeed the case, as can be documented from many studies. Despite the superficial similarity between the bubbles in fluidized beds and the bubbles in liquids, there are important differences. As shown in Fig. 8.2, in beds with fine particles, the gas velocity in the emulsion phase is small relative to the bubble velocity. The bubble thus tends to carry a cloud of circulating gas with it as it ascends. On the other hand, in beds with large particles, the velocity of the gas percolating through the emulsion phase is large relative to the bubble velocity. The bubble thus appears somewhat as a quasi-stationary cavity, permitting a local short-circuit for the gas. In either case, the presence of bubbles represents a mechanism for reducing the contact between the solids and the gas, thus reducing the total heat and mass transfer. It should be recognized, of course, that Fig. 8.2 presents an idealized picture. Actual bubbles, especially those at high fluidization velocities, tend to be more irregular in shape, as shown in Fig. 8.1.

While the gas motion around individual bubbles is of interest, it is the collective effects of the motion of many bubbles that impart the characteristic stochastic nature of the fluidized bed dynamics. First of all, the bubbles influence each other's motion, often drawing closer as they rise. Second, adjacent bubbles tend to coalesce, forming larger bubbles as they ascend to the top. The laterally nonuniform distribution of the bubbles is a major driving force for the solids circulation in the bed, as will be discussed below. In addition, the motion of the bubbles also serves as the random forcing function for exciting the turbulent fluctuations in bed. This includes the rather large amplitude sloshing motion of the bed surface. These three types of motion—large scale circulation and random fluctuation of the solids and solids motion in response to the passing bubble—combine to

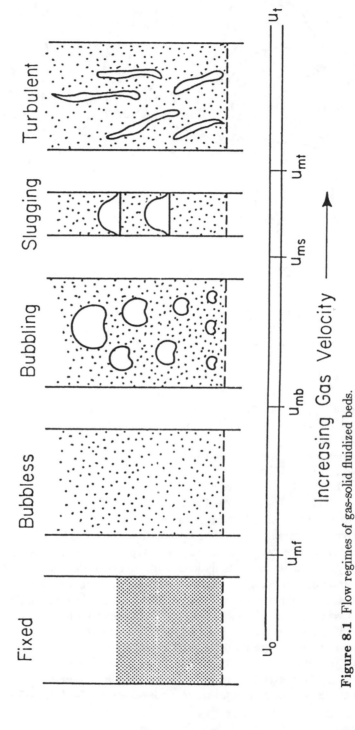

Figure 8.1 Flow regimes of gas-solid fluidized beds.

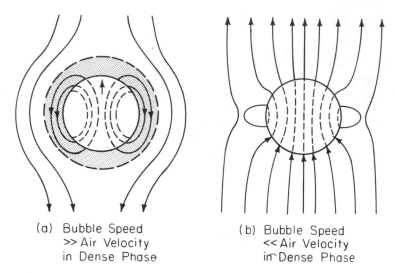

(a) Bubble Speed
>> Air Velocity
in Dense Phase

(b) Bubble Speed
<< Air Velocity
in Dense Phase

Figure 8.2 Differences in gas streamlines between a fast bubble and a slow bubble (adapted from Catipovic et al., 1978).

promote mixing processes in the bed. Here we speak not only of species mixing, of obvious importance to chemical reactor operation, but also of thermal mixing, the approach to a more uniform temperature. A uniform temperature is clearly important in highly exothermic or endothermic reactions. Furthermore, temperature uniformity also promotes the total heat transfer between the gas and solid.

This short discussion thus serves as the motivation for selecting the three topics of presentation in this review. Clearly an understanding of the mean and random solids motion is of vital importance to understanding fluidized bed behavior. This is the first topic to be discussed below. Thermal mixing and species mixing are analogous processes and will be discussed together, forming the second topic. Ultimately, the easiest means of extracting heat from a fluidized bed are to use internal heat-exchange tubes. This is the third topic of our discussion.

This review is somewhat prejudiced in favor of experimental studies rather than theory or modeling. It is therefore appropriate to say a few words about the latter. Research on rigorous formulation of multiphase problems is progressing, but at a pace that may be trying to practical engineers. Short of a rigorous theory, models, or formulations with heuristically motivated assumptions and thoroughly tested against representative experiments, offer the best temporary relief. At the present time, modeling of fluidized bed dynamics can be roughly put in two categories. The first category consists of ad hoc models based on the many specific observations of fluidized bed behavior accumulated over many decades. These models make no pretense of presenting a general "equation of motion" suitable for all occasions but concentrate on predicting only the few specific quantities at hand. An excellent review of such models is found in Wen and Chen (1984). The second category consists of models cast in the form of equations of motion, derived with various degrees of rigor and thoroughness, coupled with a numerical solution scheme based on finite difference or finite element formulation. Early examples of this type of modeling have focused on flows around single bubbles (Jackson, 1963a,b; Murray, 1965a,b; Anderson and Jackson, 1967,1969).

In recent years the models have been applied to computing global flow patterns. An excellent review is presented by Crowe in this workshop. Other examples include the formulations by Soo (1983) and Gidaspow (see Gidaspow et al., 1984). It is clear that the task facing these models is formidable. For example, the interaction of the bubbles with the solids would require not only the volume fraction but also the bubble size and indeed the bubble size distribution. The interaction of the bubbles themselves also depends on the local and global distribution of all bubble sizes in complex and possibly unknown ways. The selection of appropriate simplifying assumptions is thus crucial. This author is therefore convinced that more definitive experimental data are required to motivate and validate the simplifying assumptions to be used with these models.

2 SOLIDS MOTION IN THE FLUIDIZED BED

2.1 Brief Review

For many years, the mainstay of the techniques for observing solids motion in fluidized beds was the "two-dimensional bed." This is fluidized bed confined between two vertical surfaces, at least one of which is transparent. The device is particularly useful for observing bubble dynamics. This is shown in Fig. 8.3, which is taken from Rowe (1971). Clearly this type of observation can only be considered qualitative because the behavior of two-dimensional bed cannot be taken to be identical to a three-dimensional bed. An interesting development of the two-dimensional bed is the so-called "two-and-a-half-dimensional bed" or the semi-cylindrical bed. This is half a cylindrical bed with a transparent wall at the diagonal plane. Amazingly, some of the measurements obtained in a two-and-a-half-dimensional bed have closely paralleled those obtained in a three-dimensional bed.

A better technique of making observations in a three-dimensional bed is with X-rays (Rowe 1971). This has been very useful for the study of large bubbles in smaller beds. For large beds or for very small bubbles, the technique would be quite expensive because of large amounts of attenuation of the X-rays and because of the poor resolution that accompanies time-dependent measurements.

The techniques for measuring the local velocity of solid particles can essentially be classified into five categories. The first uses a small obstacle to detect the drag force exerted on it by the flow of solid particles. The second is based on the measurement of the rate of heat transfer by solid particles. The third is based on a laser technique. The fourth is based on the statistical cross-correlation of the optically observed movement of particles. The fifth is a tracer method based on following the track of a tracer.

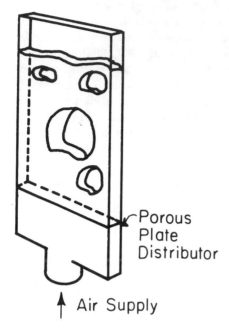

Porous
Plate
Distributor

Air Supply

Figure 8.3 Arrangements for two-dimensional bed experiments.

Drage force method. The drag force method is based on measuring the force exerted on a small obstacle inserted in the flowfield. A small needle probe (0.2 mm diameter) developed by Heertjes et al. (1970/71) was used to detect the force resulting from the collision of particles with the probe. The strain of the needle probe was converted into an electric signal by means of a piezoelectric device. This method gives an output signal that can be directly related to the particle velocity, provided that the physical parameters such as the mass of particles or porosity of the bed are known. Unfortunately, it is not an easy task to measure these physical parameters as well as the signal output of the probe simultaneously. For example, when the particles have a size distribution, it is difficult to measure the particle velocity based on the formulation of this technique because the masses of individual particles colliding with the needle tip must be simultaneously known, along with the output of the probe.

Heat transfer method. The heat transfer method developed and employed by Marscheck and Gomezplata (1965) is a fairly sophisticated technique that uses a principle similar to that of the hot wire anemometer, and the device is called a thermistor probe. They applied the probe to measure the local mass-flow rate of solid particles along with their direction in a gas fluidized bed. The probe consists of two thermistors: one is used for heating the particles and the other for measuring the temperature of the particles. If the two thermistors are aligned in the flow direction with the heating thermistor upstream, the downstream thermistor exhibits a maximum temperature. This maximum temperature depends on the local mass-flow rate. Since a high mass-flow rate causes a high heat-transfer rate by the particles, the mass velocity can be related to the temperature of the thermistor. This probe has to be calibrated to determine the particle velocities each time a different type of particle is used.

Laser method. A laser technique has been developed by Yong and coworkers (1980) to study the particle movement in a two-dimensional bed. A laser beam from a nitrogen laser source was projected through the transparent bed wall intermittently onto the particles, which emit fluorescence for a definite duration after each bombardment, and thus the particle movement was observed and the velocity of a particle was determined by the distance between two successive bright points divided by the period of the pulsating emission of the laser. Very important qualitative conclusions were drawn in regard to the motion of particles and the role of rising bubbles through the application of this method.

Cross-correlation method. The cross-correlation technique is based on individual detection of moving particles at two locations aligned in the direction of flow using a fiber optic probe. The probe consists of a pair of bundles of two small optical fibers. One bundle of fibers is used to illuminate individual particles and the other to detect the light reflected by the particles. The velocity is determined by computing the following cross-correlation function $C(\tau)$:

$$C(\tau) = \lim_{\tau \to \infty} \frac{1}{T} \int_{o}^{T} x(t - \tau) y(t) \, dt \tag{1}$$

where $x(t)$ and $y(t)$ are, respectively, fluctuating signals detected at one location upstream and another location downstream. The time required for the particles to travel between these two locations is the transit time τ_m, where the cross-correlation function is maximum. The particle velocity v_p, therefore, is calculated as

$$v_p = \ell/\tau_m \tag{2}$$

where ℓ is the known distance between the two detecting positions. The most recent probe by Oki et al. (1980) consisted of a cluster of six light receiving fibers surrounding a central light projector. The optical-fiber array probes, useful as they are, suffer from the following serious disadvantages: (a) the very presence of the probe in the bed interferes with the particle motion, particularly when a large array (10×10 sensors used in the above, for imaging void phase) is employed and (b) its inability to distinguish particles of different density, shape, and size.

Tracer method. The tracer method is based on the motion of a single tracer particle put into the flowfield, whose velocity is determined by following the position of the particle. Perhaps this is the most quantitative technique used to date to measure the local velocity of solids in fluidized beds; it has the advantage over the other methods that the velocity is measured without obstructing the flow of particles.

Kondukov et al. (1964) measured particle flow patterns in a fluidized bed (17.2 cm ID, 50 cm height) using a tracer particle (2.8 mm diameter) marked by radioactive isotope Co^{60}. The position of the tracer particle was continuously determined by means of radiation detectors. The detectors were fixed in pairs along three mutually perpendicular axes to allow three-dimensional measurement. Due to the difficulty in processing the very large volume of data necessary for statistically meaningful results, few quantitative results were given on the flow pattern of particles over the entire volume of the bed. A number of other investigators have employed similar concepts in studying solids motion. Among these are Masson et al. (1981), who examined the circulation of large and light spheres in a fluid bed of glass beads and (Van Velzen et al., 1974) who measured the flow pattern of particles in a spouted bed.

The radio pill tracer method, developed by Handley and Perry (1965), uses a radio pill as a tracer to obtain the particle flow pattern, and the position of the tracer is observed externally by an aerial antenna and a receiver. With a similar method, Merry and Davidson (1973) measured the particle circulation rate on a two-dimensional fluidized bed, where the gas was unevenly supplied from the air distributor. The radio pill was contained in a ping-pong ball of 38 mm diameter so as to be neutrally buoyant in an incipiently fluidized bed. The velocity of the particles was measured by determining the time required for the pill to travel the distance between the two antennas fixed at two points in the bed for detection of passing radio pill.

2.2 CAPTF

The computer-aided particle tracking facility (CAPTF) at the University of Illinois in Urbana-Champaign is a facility for the study of fluidized bed dynamics developed under the sponsorship of the National Science Foundation and the Department of Energy. Its basic principle is similar to the method of Kondukov (1964), except that the tracer particle is dynamically identical to the bed particles under study and full advantage is taken of the ability of modern laboratory computers to acquire and process data. This has permitted the use of multiple detectors and long test runs to improve the quality of the statistically processed data. As a consequence, a number of interesting quantitative results on solids motion in the bed, hitherto not available in the literature, have been obtained.

The tracer particle was made of Scandium-46 and was closely matched to the size and density of the glass beads used as fluidized particles. Scandium-46 was selected for the following reasons: (a) It has a specific gravity of 2.80, which is only slightly greater than that of the glass beads. Hence, only a very small amount of nonradioactive material needs to be added to match the density. This results in a tracer of high radiactivity. (b) It emits gamma rays at 0.89 Mev and 1.12 Mev. The energy range is such that the mass absorption coefficient is relatively independent of most absorbing materials. (c) It has a relatively high specific activity and moderate half-life (84 days)), making the activation of the tracer in a reactor reasonably convenient. Two different sizes of Scandium balls coated with epoxy were used in the experiments, the finished diameters being 0.7 mm and 0.5 mm, respectively. Experiments indicated that radioactivity of 100 to 150

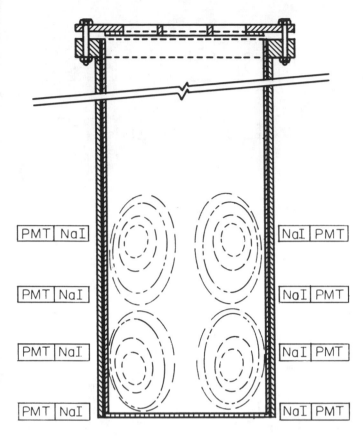

Figure 8.4 Detector arrangement for single particle tracking.

microcuries was adequate, yielding approximately 50,000 count/sec when the tracer was at the center of the empty bed of 14 cm diameter.

Sixteen photomultiplier tubes (PMT) incorporating Bicron 2 in × 2 in sodium-iodide crystals were used to continuously monitor the gamma-ray emission from the tracer. The 16 detectors were arranged in a staggered configuration at four different heights, with four in each level as shown in Fig. 8.4. The arrangement offers the advantage that, wherever the tracer is, there are several detectors nearby, thus providing accurate distance measurements. As the tracer moves around the bed, 16 intensity measurements are made by the detectors at any time giving the "instantaneous" position of the tracer. A data reduction scheme for determining the tracer location from the intensity measurements will be discussed later. Additional detectors could be installed to improve resolution or to study solids motion in deeper beds.

The photon counting scheme and the data acquisition system are shown schematically in Fig. 8.5. Details of the electronic circuitry can be found in Lin (1981). The raw signals have a noisy background mainly originating from

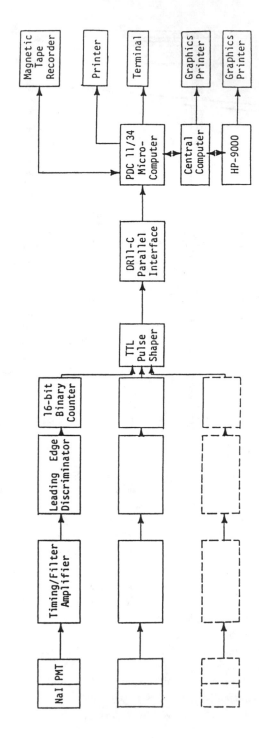

Figure 8.5 Schematic diagram of the improved data acquisition and processing systems for UIUC's CAPTF.

secondary emissions due to the interaction of gamma-ray with bed materials. Since the secondary emissions are essentially of fairly low energies, their effect can be effectively removed by employing a Schmitt trigger with an adjustable threshold.

The measured count rates are then fed into a microcomputer-based data acquisition system via a multiplexer. The count rates are then converted to particle to detector distances on the basis of a previously established calibration curve.

Calculation of tracer position. The tracer position is calculated from the 16 distances using a weighted least square scheme, after using a clever linearization scheme developed by Lin (1981). It should be noted that for the position calculations, a minimum of three distances would be adequate. The use of 16 distances provides redundancy that can be utilized to improve the accuracy. The weighted least square gives more weight to the measurements which are more accurate. The details of the method can be found in Lin (1981) and Lin et al. (1985).

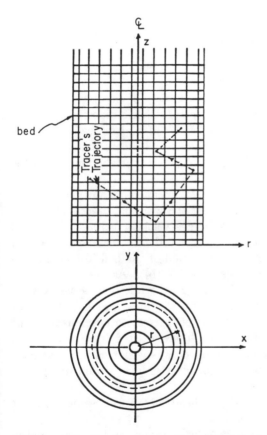

Figure 8.6 Configuration of compartments.

Calculation of mean velocity distribution. Instantaneous solids velocities were obtained by differentiating two successive locations at a known data sampling rate. For the purpose of determining the mean velocity distribution, the bed was divided into 154 (i.e., 7×22) sampling compartments as shown in Fig. 8.6. It should be noted that the azimuthal dependence of the solids velocity was found to be small, and thus the average was taken over the entire 360 degrees despite the capability of the present techniques for measuring three-dimensional motion. By running experiments for sufficiently long durations, the repeated appearance of the tracer in a given compartment enables the ensemble-average of the instantaneous velocities to be calculated. For a typical running time of two hours using a sampling interval of 30 milliseconds, a total of 240,000 data points were obtained for an experiment. Since the probability of the tracer's occurrence in each compartment varies from compartment to compartment, the confidence level of ensemble-averaged data varies accordingly.

3 RESULTS AND DISCUSSION

3.1 Effects of Fluidizing Velocity on Particle Circulation Pattern

A number of experiments were conducted with the superficial air velocity, U_o, ranging from 32 cm/s to 80 cm/s. The corresponding ratio U_o/U_{mf} range from 1.65 to 4.60. It was found that the bed exhibited a variety of circulation flow patterns, being strongly dependent on the gas flow rate. Distributions of the solids mean velocity are summarily presented in Figs. 8.7(a–d) for glass beads of diameters in the range of 0.42 to 0.6 mm. In these figures and others that follow, the starting point of the vector denotes the location in question; it also represents the center of the sampling compartment.

At the lowest fluidization velocity, the basic mean circulatory pattern of solids was that of a toroidal vortex ascending near the wall and descending at the center (AWDC) as shown in Fig. 8.7(a). As the gas velocity increased, a second toroidal vortex in the reverse direction, ascending at the center and descending near the wall (ACDW), appeared in the upper portion of the bed although solids at the center remained flowing downward (see Fig. 8.7(b)). As the gas flow was further increased, the extent of the ACDW vortex grew while that of the AWDC diminished, as illustrated in Fig. 8.7(c), and eventually only the ACDW vortex existed in the bed as shown in Fig. 8.7(d). It is of interest to note that for the experimental conditions of Fig. 8.7(c) and (d), there existed a narrow region between the two vortices where the solids movement was very weak or near stagnation. This narrow region or layer was occasionally visible as a stagnant ring on the cylindrical wall of the bed.

Similar trends were also observed for glass beads of 0.6 to 0.8 mm diameter (Lin, 1981).

The phenomena described in the foregoing paragraphs can be interpreted in terms of the bubble behavior. Werther and Molerus (1973) reported that close to the distributor plate intensified bubble activity exists in an annular region near the wall. As the bubbles detach and rise, they tend to move toward the center with

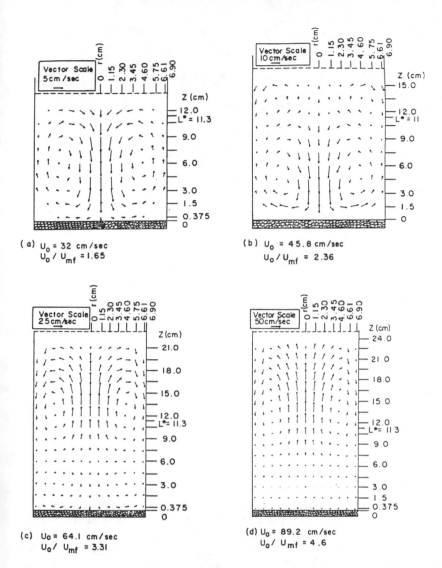

Figure 8.7 Particle circulation patterns of various fluidizing velocities for a gas fluidized bed of 0.42–0.4 mm dia. glass beads, L^* denotes static bed height.

increasing height. If the bed is sufficiently deep, they would eventually merge at the center. Whitehead et al. (1976) also observed that close to a uniform distributor, bubbles formed preferentially near the wall of a large bed (1.2 m × 1.2 m vessel) because of the inherent pressure maldistribution near the distributor. Since the solids are carried upwards in the wake of the bubbles, besides the drift of particles in the proximity of bubbles, they basically move along the bubble tracks. The downward motion of the solids ensues to maintain continuity.

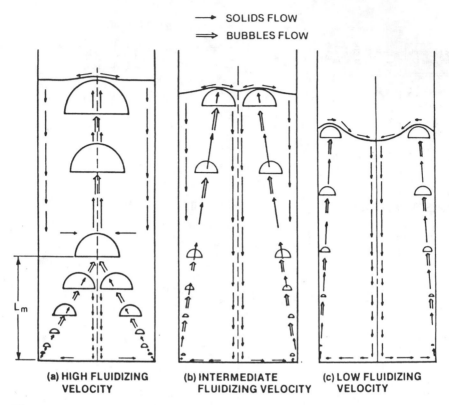

Figure 8.8 Development of solids circulation versus bubbles motion at various fluidizing velocities.

In light of the presently measured solids recirculation patterns and the known bubble flow behavior, a schematic representation of their relationship is proposed and is illustrated in Figs. 8.8(a–c). The figures are to a large extent self-explanatory. At high fluidizing velocities, the bubbles would merge in the central region of the bed. This suggests that the phenomenon can be characterized by a length L_m, the bubble merging height. In general, L_m is a function of the superficial gas velocity, particle size, particle density, bed diameter, etc. It may be reasoned that the simultaneous existence of both upward and downward solids flow in the central region would occur if $(L_m/L_f) < 1$, L_f being the expanded bed height. Werther and Molerus proposed pictorial presentation similar to that shown in Figs. 8.8(a–c), except that attention was directed to the influence of bed height instead of fluidizing velocity.

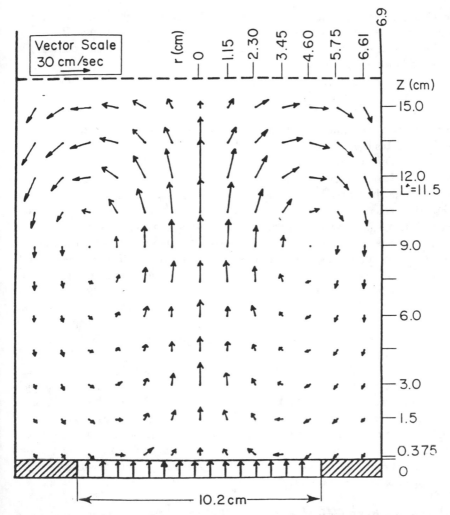

Figure 8.9 Velocity distribution for a fluidized bed of 0.6-0.8 mm glass beads. $L = 11.5$ cm; $D = 13.8$ cm; $d = 10.2$ cm. The scale of the velocity vectors is 26.7 cm/s/cm. $U_o = 63.12$ cm/s.

3.2 Solids Flow as Influenced by Nonuniform Distributor Plate

The configuration of gas distributor plate plays a key role in bubble development, which in turn determines the flow patters of solids. Experiments were thus conducted with a distributor plate that restricted the air flow to its central region, as illustrated in Fig. 8.9. The ACDW vortex was intensified to such an extent that the AWDC vortex disappeared completely. Figure 8.9 also indicates the existence of a low velocity region near the corners adjacent to the distributor plate.

3.3 Lagrangian Autocorrelation of Fluctuating Velocities

To develop some feeling about the statistical behavior of the particle motion, Lagrangian autocorrelation of the fluctuating velocities has been evaluated.

The Lagrangian autocorrelation coefficient $R_{\alpha\alpha}(\mathbf{x},\tau)$ at a given position $\mathbf{x}$ is defined by Tennekes and Lumley (1972) as

$$R_{\alpha\alpha}(\mathbf{x},\tau) = \frac{<v_\alpha'(\mathbf{a},t)\, v_\alpha'\,(\mathbf{a},t+\tau)>}{<v_\alpha'^2\,(\mathbf{a},t)>} \tag{3}$$

where v'_α is the fluctuating velocity in either axial $(\alpha = z)$ or radial $(\alpha = r)$ direction and $\mathbf{a}$ denotes the initial position of the particle, which in the present instance is the tracer. The ensemble average (denoted by $<>$ in Eq. (3)) for $R_{\alpha\alpha}$ at any given location is determined in the following way. The event of interest is initiated each time the tracer is found in the sampling compartment at that location. Its fluctuating velocity and that at later times are computed. The event is considered to be reinitiated when the tracer is again found in the same compartment. To ensure that the two sequences of data are statistically independent, the interval between the two succeeding starting times must be greater than a preset limit. This limit is equal to or greater than the time lag for which the Lagrangian autocorrelation becomes negligible.

Figure 8.10 shows some sample results for the axial autocorrelation coefficient R_{zz} at four axial locations not far from the centerline. The operating conditions were the same as those given in the caption of Fig. 8.7(c). It is seen that the axial motions near the distributor plate are essentially uncorrelated after a short time lag of a few hundredths of a second (corresponding to a distance of the order of 0.3 mm). At $z = 6$ cm, which roughly corresponds to the region separating the ACDW and AWDC vortices, teh particles exhibit the longest memory. For greater values of z, the axial autocorrelation curves cross the zero axis again at shorter times. It is estimated that the frequencies of particle random motion ranges from 1.6 Hz in the relatively stagnant layer to 16 Hz near the distributor plate, while the frequencies in teh ACDW vortex varied within a much narrow range (about 3.8 to 5 Hz).

3.4 Effects of Internal Heat-Exchange Tubes on Bed Circulation

It is known that the packing density of internal heat exchange tubes has a strong influence on bed circulation. Two types of qualitative evidence are available. It has been reported that when the density of tubes is too high, severe temperature nonuniformities may exist in fluidized bed combustors. Furthermore, increasing the tube packing density has been found to drastically reduce the tube erosion rate. Unfortunately, quantitative data are lacking. CAPTF is ideally suited to providing this type of data. This work is described in detail in Chen et al. (1984).

Four rod bundles were tested as internals in the present study. The first consists of rods of 6.35 mm (0.25 in) diameter, arranged in horizontal banks with center-to-center distances of 19.05 mm (0.75 in). These banks are stacked

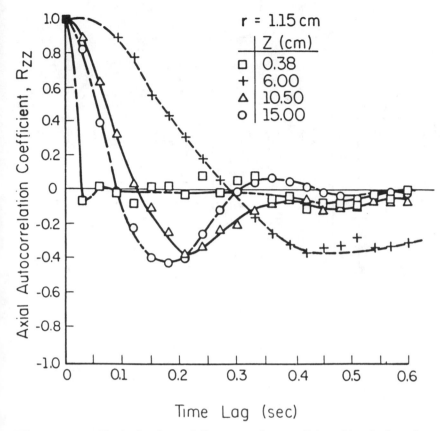

Figure 8.10 Typical values of R_{zz}, operation conditions identical to those in Fig. 8.7c.

vertically, with center-to-center distances of 25.4 mm. The orientation of the rods in the horizontal plane is rotated by 90 degrees for each succeeding bank, to minimize overall rotational asymmetry. This rod bundle, with horizontal pitch ratios of 3:1, represents a densely packed rod bundle.

The second rod bundle consists of the same rods, with horizontal center-to-center spacing of 38.1 mm (1.5 in), and vertical center-to-center spacing of 50.8 mm (2 in). This represents a sparsely packed rod bundle.

Two other rod bundles of slightly different rod diameters and pitch ratios are also tested. The results are consistent with those of the above two rod bundles, but are not presented because of space limitations.

The height of the rod bundles corresponds to the expanded bed height when no internals are present.

The results presented are for two superficial velocities corresponding to $u_o/u_{mf} = 4.0$ and 6.0, respectively. The minimum fluidization velocity for the bed without internal structures is 22.3 cm/s.

Figures 8.11 and 8.12 show the solids circulation patterns of $u/u_m = 4$ and 6, respectively. In each figure, the left margin represents the centerline of the

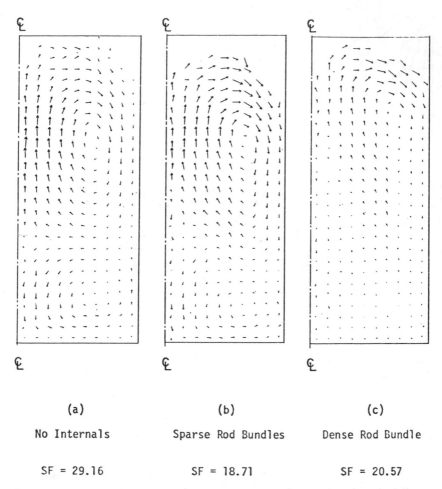

(a)

No Internals

SF = 29.16

(b)

Sparse Rod Bundles

SF = 18.71

(c)

Dense Rod Bundle

SF = 20.57

Figure 8.11 Solids velocity distributions for a fluidized bed with different internal rod bundles compared to the same bed without rod bundles, $u_o/u_{mf} = 4.0$.

circular bed. To enhance the legibility of velocity vectors in the figures, the range of amplitudes of the vectors have been "compressed" somewhat by the use of a nonlinear scale. The velocities are related to the length of the vectors by the following relationships: Velocity = (Arrow Length $\times$ SF)$^{1.5}$, where SF is the scale factor in units of (cm/s)$^{2/3}$cm shown under each figure.

The figures show that the recirculation pattern for the conditions shown consists of two counter-rotating torodial vortices, the lower vortex having a descending core and the upper vortex an ascending core. As the superficial velocity increases, the size of the upper vortex increases at the expense of the lower vortex. This phenomenon is consistent with previously reported observations by Lin, 1981,1985. It is interesting to note that the general circulation pattern is

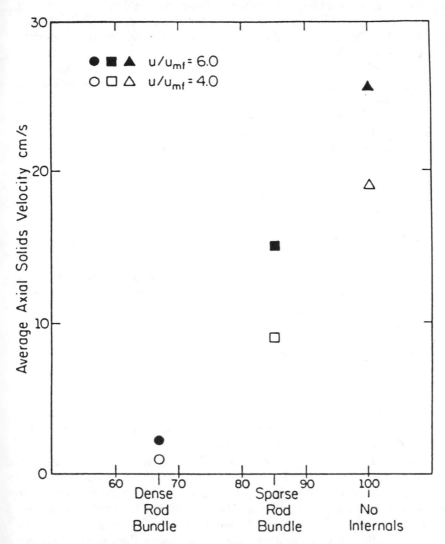

Figure 8.12 Effect of internal rod bundles on the magnitude of solids circulation (the ordinate is the vertical velocity of the core region at an elevation corresponding to the center of the upper vortex; the abscissa is the percentage of unblocked cross-section area).

essentially unchanged by the presence of the internals, aside from the concentration of higher solids velocity near the top of the bed. The latter phenomenon is apparently caused by the fact that some solids in the expanded bed are above the top of the rod bundle and hence are unhampered by the rods. Even the overall dimensions of the vortices remain the same for the three cases at the same superficial velocities. In view of the current understanding that the solids circulation is driven by bubbles, our finding is consistent with the observation of bubble motion by Glass (1967), who found that bubbles are only slightly affected by small internals. The present results, however, represent the first direct, quantitative demonstration of this phenomenon.

An entirely different picture emerges when intensity of solids circulation is examined. An indicator of the intensity of solids circulation is the average vertical velocity at the center of the bed at an elevation corresponding to the core of the upper vortex. This velocity is shown in Fig. 8.12 for the six cases. It is shown that the presence of the internals dramatically reduces the intensity of solids circulation. This type of quantitative data has not previously been available in the literature.

The results reported above reveal two remarkable facts concerning the influence of internals on solids circulation in the fluidized bed. Both facts have been suspected in the past but never demonstrated quantitatively by actual measurement.

The first fact is that even when an internal structure occupies a very small fraction of the total volume in the bed, it could lead to significant reduction of the average velocities involved. The second remarkable fact is that even in the presence of a relatively densely packed rod bundle, the basic circulation patterns in the bed remain essentially unchanged. It is hoped that the present measurements may be useful in furthering the understanding of mixing processes in the bed.

4   MIXING PROCESS IN THE BED

A prerequisite to the understanding of heat transfer processes in fluidized bed heat exchangers and combustors is the prediction of the temperature distribution in the bed. Because of the efficient heat exchange between the solids and the gas, and because of the fact that all but a tiny fraction of the heat capacity of the mixture is vested in the solids, the temperature distribution in the dense phase is dominated by the thermal mixing behavior of the solid particles. In addition, it is not difficult to demonstrate that the effective Peclet number $\rho c u L / k_m$, the ratio of the convective flux $\rho c u \Delta T$ to the conductive flux $k_m \Delta T / L$, is extremely large de to the low mixture conductivity k_m. Hence thermal mixing is dominated by turbulent eddy mixing processes, analogous to species mixing processes. Since species mixing can be studied in a fluidized bed much more easily than thermal mixing, considerable insight can be gained on the latter by studying the former. Species mixing is also important to reaction kinetics, which affects the temperature distribution through the release of chemical energy.

Solids mixing considerations, of course, do not address the topic of nonequilibrium gas temperatures in large bubbles, which can potentially influence heat transfer between the gas phase and the immersed surfaces.

The study of solids mixing in fluidized beds has had a long history (Kunii and Levenspiel, 1969; Davidson and Harrison, 1971; Wen and Chen, 1984). The recent development of the UIUC-CAPTF have further contributed to the understanding of the mixing process by providing a means of correlating the mixing data with in situ measurements of solids dynamics, including mean and turbulent quantities.

The measurements of the mean and turbulent dynmic quantities, including the Lagrangian auto-correlations, have already been described in previous sections. Their importance to mixing is due to the fact that mixing is the combined consequence of two effects: convection associated with the mean circulation pattern, and eddy diffusion or local dispersion due to the accumulated effects of random turbulent motion. A theory by G. I. Taylor relates the turbulent diffusion coefficient directly to the Lagrangian integral time scale, which is the integral of the normalized Lagrangian auto-correlation function, already discussed in the previous section. Thus the coefficents for both the convective and diffusive contributions of mixing can be determined from bed dynamics measurements employing the UIUC-CAPTF. With these data, a mathematical model of the mixing process can be constructed, using the convective-diffusive mass transfer equation.

Independently, a direct measurement of mixing can also be performed in the CAPTF. This is based on the swarm tracking scheme, in contrast to the single particle tracking scheme described previously. A swarm, or collection of radioactive particles, of total volume of approximately several milliliters, is released at a selected location in the bed. Its subsequent migration due to mean motion, and dispersion due to random motion, are tracked by the radiation detectors, until the radioactive particles are thoroughly mixed with the non-radioactive bed particles. This technique can therefore be viewed as the modern counterpart of the "stratified bed" experiments using dyed or otherwise marked particles. The present method is however free from the artifacts frequently introduced when the bed must be stopped and contents carefully removed to determine the composition distribution.

A sample comparison of the mixing data with the results from the model calculations is shown in Fig. 8.13. Here the detector output for the numerical simulation was computed by integrating the contributions of the tracers according to computed concentration distributions. In view of the fact that no adjustable parameters were used in the model, the agreement is seen to be excellent. Similarly good agreement was also found for several different operating conditions, though the agreement became poorer at velocities just above the minimum fluidization velocity. These results indicate that the role of convection due to mean solids circulation is important to the mixing process, and that if the fluidization velocity is sufficiently high, mixing can be accurately predicted from data on solids dynamics. For further details on these studies, see Moslemian (1986).

In turbulent flow, the temperature distribution is governed by a convective-diffusive equation identical to that which governs mixing, with most likely the same eddy diffusivity. Therefore the procedure described above for the prediction of species mixing can also be used to predict the temperature distribution. This procedure, however, has not yet been tested. Investigations along these lines should prove valuable.

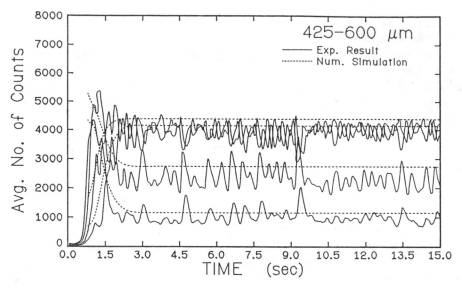

Figure 8.13 Comparison of the numerical and experimental mixing results for 425-600 μm glass beads, $u_{mf} = 21.9$ cm/s, $\mu_o = 170.3$ cm/s.

5 HEAT TRANSFER TO IMMERSED SURFACES

5.1 Literature Survey

To gain a better physical understanding of the processes involved, we shall consider the various potential mechanisms of heat transfer in disperse multiphase flows and then examine the current literature specifically devoted to this mechanism. We present the important mechanisms as well as the literature. The literature that is not mechanistic in nature shall be presented separately.

Continuum descriptions. In this category we include all mechanisms that can be treated by essentially using continuum description. The bulk properties involved, however, may be strongly influenced by microscopic particulate behavior unique to a disperse two-phase system. Such a continuum description is valid if the relevant scale of the temperature distribution is large compared with the particle diameter. For fully developed laminar flows or for slowly moving or slowly varying flows without thin boundary layers, this length scale is of the order of the dimension of the device, such as tube diameter. For rapidly moving fluids or for rapidly varying fluids (see packet renewal theory discussion below), thermal boundary layers of the order of $(\alpha D/u)^{1/2}$ or $(\alpha t)^{1/2}$ exist with t the characteristic time scale of the fluctuating conditions. If these boundary layers are too thin in comparison with the particle diameter, then the mechanisms based on continuum description of the heat transfer processes are not valid.

The continuum description for the radiation heat transfer must consider the relevant scale of the radiation field as compared with the particle scale. This continuum criterion could be the same as discussed above but will generally differ. This is due to radiation absorption, scattering properties, and the spectral variation, which are not tied to the temperature field. The bulk radiation properties of a continuum description are strongly influenced by the geometric and optical properties of the particulate and fluid phase.

In the following, we first examine mechanisms that alter the effective bulk properties of the mixture. Since extensive thermodynamic and mechanical properties such as enthalpy and mass are additive, simple mixture rules exist for their intensive counterparts, such as specific enthalpy and density, as long as the sample volume is sufficiently large to include a large number of particles. These mixture rules are not fundamentally different from similar mixture rules for single phase, multicomponent mixtures. On the other hand, transport properties in multiphase mixtures are known to depend on flow parameters, unlike their single-phase counterparts, which are all material properties. A well-known example is the viscosity of suspensions, which is known to depend on the local shear rate and frequently on other flow parameters as well. Such a flow dependence also exists in thermoconductivity.

Bulk properties of near-equilibrium conditions. The first step in a continuum description is to arrive at the bulk properties—the effective thermodynamic, optical, and transport properties to be used with continuum equations. The latter are equations governing either the temporally or the spatially averaged dependent variables such as temperature, velocity, and intensity. There are two common approaches. The first approach is phenomenological. In this case, the bulk properties are evaluated empirically in systems large enough to contain many particles. In the case of transport properties, they are defined as the coefficients in postulated constitutive relationships. The second approach is to attempt to compute the bulk properties from the properties of the constituents. Examples of this approach include the review of transport properties by Batchelor (1974) and the careful considerations of volume averaging by Whitaker (1967). It should be pointed out, however, that such careful analysis is often only possible when the gradients are small. Otherwise the question of local thermodynamic equilibrium will have to be examined. The situation is not unlike the situation with macromolecular solutions.

A mechanism that belongs to this class is the consideration of gas phase heat transfer, which is shown to dominate at high gas velocities in beds with large particles (Adams and Welty, 1979).

Microconvective effects. Microconvective effects arise from two different contributions. One contribution results from the shear-induced particle-scale convection while the other contribution results from the difference in the velocities of each phase. These two contributions are discussed here.

In shear flow the particles may rotate, collide, and execute random migration relative to the fluid. The exact nature of the motion depends on the Reynolds number and other properties of the mixture, but even in the lowest Reynolds number ranges the rotation and collision would still lead to considerable particle-scale convection. This phenomenon should lead to enhanced thermal conductivity and mass diffusivity for the mixture in comparison with the stationary values. Such enhancements were suspected in the early studies on transport processes in blood and were confirmed qualitatively by the experiments of Singh (1968) and Collingham (1968). More recently, Sohn and Chen (1981) quantitatively showed that the effect was quite significant when the particle Peclet number ($Pe = ed^2/\alpha$) based on the shear rate e, particle diameter d, and fluid heat diffusivity α was high. At the moderate Peclet numbers of the order of 1,000 reached in the experiments, a five-fold enhancement of conductivity was observed, and conductivity was seen to be proportional to the one-half power of Pe. Unfortunately, theoretical understanding of the phenomenon is poor, and theories exist only for low solid concentrations (Nir and Acrivos, 1973; Leal, 1973). Experimental measurements for dilute suspensions at low shear rates (Chung and Leal, 1982) appear to agree with the theory of Leal (1973). However, there appears to be no study of the phenomenon for the high solids loading and high shear rate conditions that are of the greatest engineering interest. Because the effect has been shown to be quite significant at moderate shear rates (Sohn and Chen, 1981), it can be concluded that it should be even more important at higher shear rates.

The contribution of particle diffusion to the effective conductivity, especially at high Reynolds number flows, was investigated by Soo (1967). However, the microconvective contribution due to the stirring of the fluid by the migrating particles was not taken into consideration. Soo's theory is thus expected to be more applicable to gas solid systems, for which the fluid phase has negligible heat capacity, than for liquid-solid systems, for which fluid convection plays a greater relative role. Experimental verification for Soo's theory (1967) on both the diffusion coefficient and conductivity is also needed.

When the two phases in a mixture are not traveling at the same velocity, the microscopic flowfields of the fluid around each particle should also lead to microconvective contributions to the apparent macroscopic conductivity. Unfortunately, the phenomenon is not widely investigated, the only studies appear to be those of Gelperin (1940) and Aerov (1951) (both cited in Gelperin and Einstein, 1971) and Yagi and Kunii (1957), who concluded that such an incremental conductivity for packed and fluidized beds should be about 0.1. Their results should be applicable to flowing two-phase mixtures if the velocity is replaced by the relative velocity. Even in fluidized beds critical and independent experimental verifications of this formula have been few. The first-order dependence on Peclet number implied by this expression is also somewhat at variance with other convection results, which generally have a fractional power dependence. Further investigation of this phenomenon is clearly warranted.

Transient conduction of "packets" of particles. This is the "packet renewal model" originally proposed by Mickley and Fairbanks (1955) for fluidized bed heat transfer. The mechanism invoked is essentially a macroscopic one. During each "residence," there is a heat transfer to a thin conduction boundary layer of thickness $(\alpha t)^{1/2}$. Thus the mean heat transfer coefficient is proportional to $(\alpha t_m)^{1/2}$, with t_m a suitably defined mean residence time. The conductivity used in this model is an effective conductivity for the mixture that often must be determined from other assumptions. Later modifications have added equivalent contact resistances to improve agreement with short residence-time data (Wunschmann and Schlunder, 1975; Saxena, 1978). An alternate explanation of the observed disagreement, however, may be the failure of the continuum calculations to properly account for particle-scale conduction problems. Attempts to treat the solids and fluids separately have also been made (Gabor, 1970; Antonishin et al., 1974). Recently, heat transfer measurements were compared with predictions of packet renewal theory using residence time distributions carefully measured from capacitance probe measurements (Chandran et al., 1980). A composite model based in part on this consideration was recently published by Chandran and Chen (1985).

Another extension of the packet renewal model is to consider the response of the solid and fluid separately (Gabor, 1970; Antonishin et al., 1974), corresponding to two-temperature, continuum modeling.

It would seem that the packet renewal model is most suitable for slowly bubbling beds with little or no motion of the dense phase in the absence of bubbles. For high velocity fluidized beds or for flows of suspensions, it is doubtful that the mixture can be legitimately assumed to be stationary at any time. In the presence of significant motion, the thermal boundary layer should be related to $(\alpha D/u)^{1/2}$ rather than $(\alpha t)^{1/2}$. In either case, the model is of questionable merit if the boundary layer thickness is of the order of particle diameter or smaller.

The packet renewal concept has been utilized in the analysis of the radiation heat transfer in beds. The radiation heat transfer at the surface is composed of a contribution during the time the surface sees the bubble and the time the surface is exposed to the packet. The bubble is treated as transparent with a bubble wall emissivity of the bed (Bock, 1983; Thring, 1977). Alternately, the bed emittence is calculated from bed radiation properties (Chen, 1981). This analysis is subject to the same questions posed directly above for high velocity fluidized beds.

Discrete particle layer models. This is, in essence, the steady convection version of the discretized packet renewal model described above. In short, the model considers discrete particle layers with the major heat transfer resistance residing in the particle-particle conductances sliding past each other. This model has also been utilized for radiation transport (Borodulya et al., 1983).

Macroscopic modeling. This type of modeling is based on continuum formulation of convection, with the conductivity based on the microconvective enhanced values discussed previously in the subsection entitled "Microconvective effects." For laminar pipe flows this was done recently by Sohn and Chen (1984). Similarly, Gabor (1970b) used the enhanced conductivity of Yagi and Kunii (1957) to calculate gas convective heat transfer in fluidized beds.

The modeling of radiation heat transfer from basic principles is complicated by the high volume fraction of particulates in the bed. Until recently, the continuum approach utilizing the equation of transfer was felt to be inappropriate due to significant interference effects from the close spacing of the particulate.

Experimental measurements (Brewster and Tien, 1982) now indicate that the continuum approach is applicable with correctly evaluated radiation properties.

Nonmechanistic continuum models. In addition to model considerations based on specific hypothetical mechanisms, a number of theoretical studies attempt to by-pass the mechanisms and compute the the turbulent temperature field and the heat transfer by extensions of single-phase turbulence models (Tien, 1961; Azad and Modest, 1981). To a large extent these seem to be intended for dilute suspensions. In view of the complex particle-surface interactions and their importance to the heat transfer process, the profitability of this approach to dense multiphase flows remains to be demonstrated.

Fundamental microscopic treatment. For external flows or for rapidly fluctuating flows, the thermal boundary layer thickness can be quite thin. For example, for fluids with thermal diffusivity of 10^{-7} m^2/s flowing at velocity greater than 1 m/s, the thermal boundary layer thickness around an object of a few centimeters in diameter is of the order of 0.1 mm or thinner. Accordingly, under these conditions, the macroscopic description of the heat transfer process would be inadequate, and a microscopic consideration of the mechanisms would be required.

Heat transfer to stationary particles. This is essentially the microscopic version of the packet renewal model. Instead of packets of mixtures assumed to be a continuum, one or more particles in a string or array are considered to be in temporary but stationary contact with the heat transfer surface (Botterill et al., 1962,1967; Gabor, 1970; Kobayashi et al., 1970). By considering particles instead of a packet, the model is free of the criticisms of a continuum description, though the highly regular arrangements of the particles have been somewhat contrived. On the other hand, the fact that the collection of particles is assumed to be essentially motionless during each residency clearly makes the mechanism not suitable for high speed fluidized beds or for flowing suspensions.

The radiation analysis of stationary packed beds has been considered in insulation applications as well as fluidized bed applications. The insulation application considered regular arrays of spherical particles and addressed the radiation transport in specified cells (Chan and Tien, 1974). A review of this type of analysis has been presented by Vortmeyer (1978). Fluidized bed applications are similar to those works discussed in the packet renewal modeling. In light of the experimental results of Brewster and Tien (1983), the radiation analysis based on individual particles does not appear to be needed or appropriate.

Heat transfer due to particle impact. Here we consider the heat exchange during the contact between the particle and the surface during impact, neglecting the mediating influence of the continuous phase fluid.

For solid-surface impact, the problem has been considered by Soo (1962,1969,1983). In short, the duration and area of contact are determined from the theory of elasticity, allowing an estimate of heat exchange for each impact. The results appear to show that this mechanism accounts for only a small fraction of the actual heat transfer in solid fluid systems. However, Soo used the particle diameter as a scale for the estimate of heat conduction, instead of the transient thermal layer thickness $(\alpha t)^{1/2}$ and did not consider damping due to the fluid, which should be much smaller. Hence the result is probably an underestimate. Furthermore, the possibility of multiple impacts and of sustained rolling along the surface have not been adequately assessed. The problem thus merits further critical examination.

A related mechanism exists in dispersed droplet flows. In this case, the heat exchange between the wall and the droplet, which deforms completely during impact, is extremely high. Hence particle-surface impact is an important heat transfer mechanism (Hersroni, 1982; Mastanaian and Ganic, 1981).

In either case, there is considerable interest to estimate the rate of particle impact on the surface. Treatment for both the external flow, such as tubes in crossflow, and the internal flow, such as turbulent pipe flow, can be found in the literature (Soo, 1962,1983; Hetsroni, 1982). One problem of interest is the cushioning effect of the fluid in reducing the velocity of the solid particles near the surface.

Fluid-mediated particle-surface heat exchange. Under this heading are two physically distinct but inseparable mechanisms. During particle-surface impact, a significant fraction of the surface area of the particle lies in close proximity to the large body. If the small gap between the two surfaces is filled with a fluid, the latter may act as a conduction intermediary and contribute significantly to the heat transfer through direct solid contact along. This mechanism could be especially important if the particle and the surface both have high moduli of elasticity, leading to small contact areas. However, there appears to be no study in the literature concerning this effect.

A mechanism distinct from the above is heat exchange between the particle and the thin fluid thermal boundary layer while the particle is on its way to and from the surface. Unless the thermal boundary layer is extremely thin, it is likely that the contact time between the particle and the thermal boundary layer is longer than the contact time between the solid and surface. Hence this could be a more important mechanism than the purely conductive heat exchange between the particle and the surface.

Both these mechanisms are concerned with heat transfer to the particle, and hence are distinct from the surface-fluid heat transfer described in the mechanism in the next subsection.

Enhanced fluid convection due to particle stirring. This mechanism is physically related to the microconvection mechanisms described in previous sections. If the scale length for heat transfer is small relative to the particle diameter, then the effect can no longer be treated in the form of effective thermal conductivity. This condition is encountered when the thermal boundary layer or laminar sublayer is thin due to high velocities. For the sake of classification, we shall include in this category only fluid convection, with the particle playing only a mechanical stirring role. The case when the particle, with finite heat capacity and conductivity, removes a significant amount of energy from the boundary layer is already included in the mechanism directly above. Collective convection of the solids is described in the next mechanism.

The mechanism has been investigated most intensively in connection with fluidized beds (Leva, 1959; Levenspiel and Walton, 1954; Dow and Jakob, 1951; Wasan and Ahluwalia, 1969). Other investigators, however, point out that the mechanism cannot be too important for gas fluidized beds, where the heat capacity of the particles should play a major role (van Heerden et al., 1953; Ziegler and Brazelton, 1964). Their conclusion, however, does not apply to liquid-solid suspensions where the fluid heat capacity is not only important but tends to dominate. The mechanism thus deserves closer scrutiny for both liquid fluidized beds and liquid-solid flows.

Collective solids convection. For gas-solid systems, the solids constitute the overwhelming fraction of the total heat capacity. Even in liquid-solid systems, in dense suspensions solids generally contribute a fraction of the total heat capacity equal to those of the liquid. Thus convection based on the heat capacity of the particulates would play a major role in the heat transfer of any disperse two-phase system. On the other hand, since the particulates constitute the discontinuous phase, the heat transfer mechanisms are extremely numerous and complex. Many of these mechanisms take place simultaneously in an overlapping way. For this reason, systematic study appears to be essentially nonexistent.

Experimental studies. A large number of experimental studies exist, usually concentrated in a few special applications. For example, the literature on fluidized bed heat transfer is quite large, as summarized in several recent reviews and symposium volumes (Gelperin and Einstein, 1971; Botterill, 1975; Saxena et al., 1978; Kunii and Toei, 1984). There is also some literature on heat transfer in gas solid suspensions in pipe flow (Danziger, 1963; Farbar and Depew, 1963; Depew and Farbar, 1963; Gorbis and Bakhtiozin, 1962; Kim and Seader, 1983; Mamaev et al., 1976; Pfeffer et al., 1966; Quader and Wilkinson, 1981; Schlunderberg et al., 1961; Tien, 1961). Although the solids loading by weight can be quite high, the volume fraction of these suspensions tend to be fairly dilute.

5.2 Heat Transfer Measurements at the University of Illinois

Due to the lack of local velocity and density data, most previous experimental studies on heat transfer to immersed surfaces in fluidized beds have correlated the results with total bed parameters, such as fluidization velocity. Such an approach does not easily yield mechanistic understanding of the processes involved. The UIUC-CAPTF can yield mean and fluctuating components of local dynamic parameters. A series of heat transfer measurements was carried out to investigate the relationship between the local heat transfer coefficient and the local velocity and density.

Because of space limitations, only one set of such experiments will be discussed here. More comprehensive results can be found in Iwashko (1985) and Moslemian (1986). For comparison, the velocity distribution and the conditions for this set of experiments are shown in Fig. 8.14, with the two positions of the

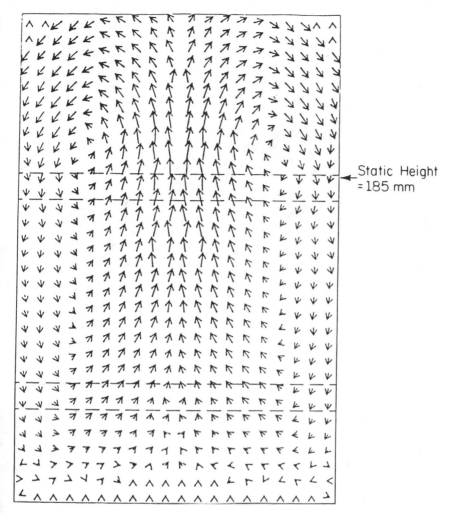

Static Height = 185 mm

Figure 8.14 Vector velocity distribution of circumferentially averaged mean solids velocities in a fluidized bed with single immersed rod; $u_o/u_{mf} = 2.5$.

simulated heat exchange tube indicated by dashed lines. The heat transfer measurements were made with a guarded-heater probe assembly, shown in Fig. 8.15, which formed one segment of the simulated heat exchange tube. Both the probe and the guard heater were made of copper, separated from each other by an air gap with the aid of nylon spacers. They were heated with small resistors and the temperatures were measured with small thermistors. External power supplies were adjusted until the temperatures were equal, at which time the power to the probe represented the heat flux. The probe had a surface area of 70.17 mm. The diameters of both the assembly and of the simulated tube, were 15.88 mm.

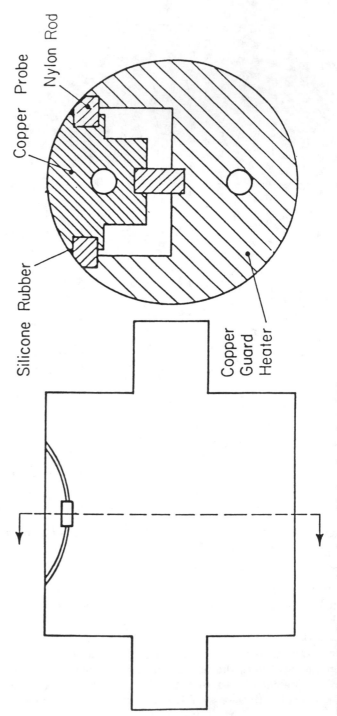

Figure 8.15 Cross-section of assembled probe and guard heater.

Data were obtained for two rod heights ($z/2R = 0.34$ and 1.03 and two radial positions) at three values of fluidization velocity ($\mu/\mu_{mf} = 2.0$, 2.5 and 3.0).

The measured average local heat transfer coefficients are presented in polar form in Figs. 8.16 through 8.19 with $\theta = 0^o$ corresponding to the bottom of the rod, upstream with respect to the direction of air flow. Ten measurements were taken at each probe location and angle with the mean values being presented of the plots. Standard deviations of the mean were usually found to be less than 5%.

In most of the cases (with exceptions to be described later), the heat transfer coefficient is found to be lowest on the bottom side of the rod ($\theta = 0^o$) and to reach a maximum on the top side ($\theta = 180^o$). This tendency was observed by Noack (1970) and Gelperin et al. (1966,1968) for relatively high fluidizing velocities. At lower fluidizing velocities, the former investigators found peak values at the equatorial lateral zones. This behavior was not observed in the present study because of the high fluidization rates.

Careful comparison of these observations suggests that the magnitude and direction of the mean flow have strong influences on the local heat transfer coefficient. In particular, an additional heat transfer contribution appears to be present at the upstream side of the rod. Since this phenomenon is not predicted by existing theories of fluid bed heat transfer, and for lack of better terminology, the following discussion will associate the upstream contribution with the "impact" of the solids on the tube. This terminology should be interpreted as a concise description of an experimental observation and not as the author's commitment to a mechanistic explanation of the phenomenon. Impact as a heat transfer mechanism in multiphase flow has previously been considered by Soo (1983).

The highest value of the heat-transfer coefficient on the top side of the rod for a given fluidizing velocity is obtained near the wall ($r/R = 0.84$) for the high rod position ($z/2R = 1.03$). At this location, the impact of the solids is greatest because of the downward flow of the solids in the upper vortex near the wall. The lowest top side value is generally found in the center of the bed ($r/R = 0$) at the upper rod height. There, the flow of particles is upward and of the highest velocity. The particles are forced to flow around the rod with little or no impact occurring on the top side. These trends are amplified with an increase in the fluidizing velocity.

On the bottom side of the rod, however, the heat transfer coefficient is greatest at the center of the bed at the high rod position. Here, the solids impact is the most vigorous. The lowest bottom side value is found near the wall, as expected, where the flow of solids is downward. Again, this behavior becomes more evident as the fluidizing velocity is increased.

Some interesting conclusions may be drawn by looking at the results, location by location, particularly in the case of the upper rod positions presented in Figs. 8.17 and 8.18. At the center ($r/R = 0$), the heat transfer coefficient on the bottom side of the rod increases with an increase in the air fluidizing velocity while the top side coefficient decreases. At the high fluidization rate ($U_o/U_{mf} = 3.0$), in fact, the bottom side coefficient overtakes the top side value.

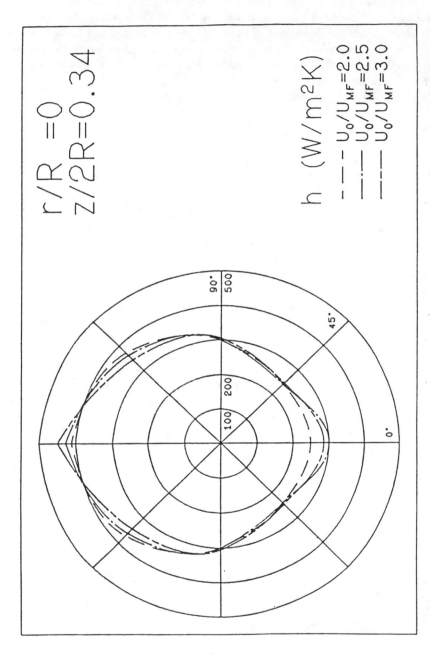

Figure 8.16 Azimuthal variation of local heat transfer coefficient around the circumference of an immersed horizontal rod.

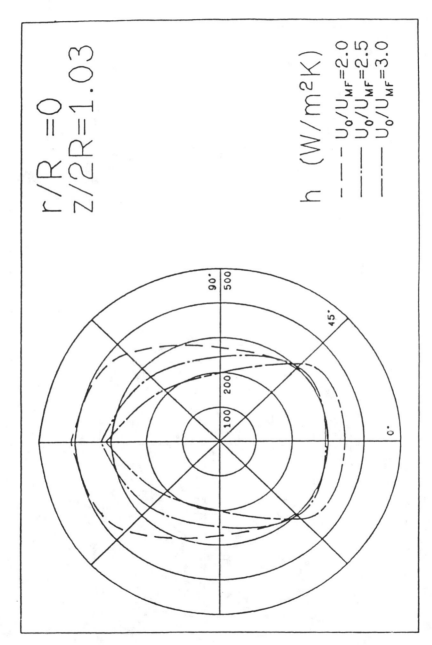

Figure 8.17 Azimuthal variation of local heat transfer coefficient around the circumference of an immersed horizontal rod.

159

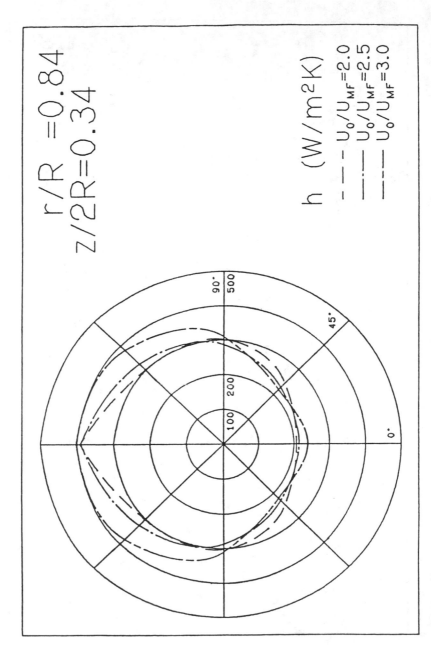

r/R =0.84
z/2R=0.34

h (W/m²K)

- — U₀/U$_{MF}$=2.0
- —·— U₀/U$_{MF}$=2.5
- —·· U₀/U$_{MF}$=3.0

Figure 8.18 Azimuthal variation of local heat transfer coefficient around the circumference of an immersed horizontal rod.

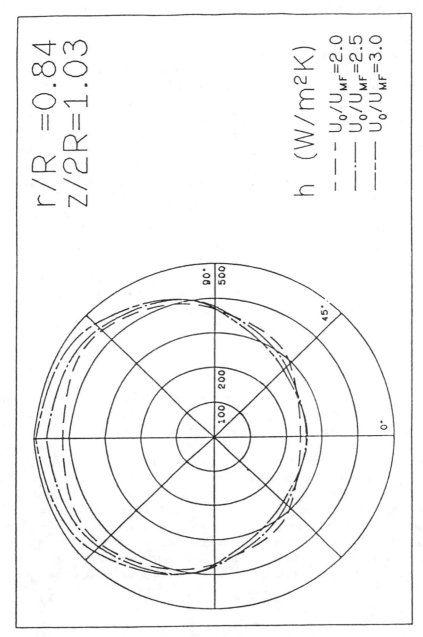

r/R = 0.84
z/2R = 1.03

h (W/m²K)

– – – U_0/U_{MF}=2.0
–··– U_0/U_{MF}=2.5
––– U_0/U_{MF}=3.0

90°
500
45°
200
100
0°

Figure 8.19 Azimuthal variation of local heat transfer coefficient around the circumference of an immersed horizontal rod.

This is the only position for which this occurs, indicting the importance of the high velocity solids impact on heat transfer. The increase in fluidizing velocity raises the velocity of the upward flowing solid particulates, which in turn increases the solids impact velocity and thus raises the heat transfer coefficient on the bottom side of the rod. Because of the high velocity of the solids in the upper central locations of the bed, the upward flowing particles tend to bypass the top side of the rod. An increase in the fluidization rate causes a more complete particle wake to form on the top side thus lowering the rate of heat transfer.

Near the wall, meanwhile, an increase in the fluidization rate raises the value of the heat transfer coefficient on the top side of the rod. Again, this can be attributed to the increase in the solids impact rate, this time with the downward flowing particulates of the upper vortex. Fluidizing velocity, however, has little effect on the bottom side of the rod at this location as shown in Fig. 8.18. This may be because the overall solids velocities are too low to clearly reveal the effect. The magnitudes of the solids velocities there are only about 25% of those found in the center of the bed at the same rod height. Similarly, no trends can be established at the lower probe position as shown in Fig. 8.19, where the magnitudes of the solids velocities are still lower.

Experimental results presented in this report show that there is an incremental contribution to heat transfer at the upstream side of the cylinder with respect to the mean solid particle motion. The data presented suggest that this additional heat transfer is associated with solid particle impact on the tube. This new observation was made possible by a direct comparison of its measured local heat transfer rates with solids velocity data that were heretofore unavailable. It is hoped that these results will stimulate further experimental and theoretical research to clarify the role of solids motion on heat transfer to immersed surfaces in a fluidized bed.

6 CONCLUDING REMARKS

In this chapter we have reviewed three aspects of the fluidized bed behavior of relevance to current and future energy applications. It is seen that in all cases the current level of understanding is quite poor. Recent efforts at the University of Illinois toward gaining a better understanding of these problems has also been described briefly. The key measurements have been the mean and fluctuating velocity distributions of the solids in the bed. Because the solids constitute the overwhelming fraction of the mass and heat capacity in the bed, their importance to the bed dynamics, mixing, and heat transfer cannot be overemphasized. At present only a limited number of data on mixing and heat transfer have been obtained in conjunction with solids motion data. It is hoped that as more data become available, they can be useful in furthering the understanding of the mixing, heat transfer, and in turn the reaction kinetics in the fluidized bed.

ACKNOWLEDGMENT

The research described herein is the collaborative work of the author with his colleague, Professor B. T. Chao, and his graduate students, J. S. Lin, J. Liljegren, D. Moslemian, M. Iwashko, and J. G. Sun. Their contributions and successive sponsorship of the National Science Foundation, the Department of Energy, and the Illinois Coal Research Board through the Center for Research on Sulfur in Coal are hereby gratefully acknowledged. The author especially wishes to thank D. Moslemian for his invaluable editorial help and the Department of Mechanical and Industrial Engineering Publications Office for typing this manuscript on very short notice.

REFERENCES

Adams R. L., and J. R. Welty (1979), "A Gas Convective Model of Heat Transfer Aerov, M. E., 1951, *Diss., Inst. Khim. Machinostr.*, Moscow Antonishin, N. V. Geller, M. A. and A. L. Parnas, 1974, *J. Engng. Phys.*, (Engl. trans.), 23:353.

Anderson, T. B., and R. Jackson (1969), "A Fluid Mechanical Description of Fluidized Beds: Comparison of Theory and Experimental," *Ind. Eng. Chem. Fundamentals*, 8:1, 137-144.

Anderson, T. B., and R. Jackson (1967), "A Fluid Mechanical Description of Fluidized Beds: Equations of Motion," *Ind. and Eng. Fundamentals*, 6:4, 527-539.

Azad, F. H., and M. F. Modest (1981), "Combined Radiation and Convection in Absorbing, Emitting and Anisotropically Scattering Gas-particle Tube Flow," *Int. J. Heat Mass Transfer*, 24:1681-98.

Batchelor, G. K. (1974), "Transport Properties of Two Phase Materials with Random Structure," *Ann. Rev. Fluid Mech.*, 6:227-255. Bock, H. J. (1983), "Heat Transfer in Fluidized Beds," *Fluidization VI*, 5-2:1-8.

Borodulya, V. A. and V. I. Korensky (1983), "Radiative Heat Transfer Between a Fluidized Bed and a Surface," *Int. J. Heat and Mass Transfer*, 26:2;277-287.

Botterill, J. S. M. (1975), *Fluid-Bed Heat Transfer*, Academic Press.

Botterill, J. S. M., K. A. Redish, D. K. Ross, and J. R. Williams (1962), *Proc. Symp. Interact. Fluids Particles*, 183.

Botterill, J. S. M., M. H. D. Butt, G. L. Cain, and K. A. Redish (1967), *Proc. Int. Symp. Fluidization*, 442.

Brewster, M. Q., and C.-L. Tien (1982), "Radiative Transfer in Packed Fluidized Beds: Dependent versus Independent Scattering," *J. of H. Transfer*, 104:0;573-579.

Chan, C. K., and C.-L. Tien (1974), "Radiative Transfer in Packed Spheres," *J. Heat Transfer*, 96:52-58.

Chandran, R. L. (1980), *Local Heat Transfer and Fluidization Dynamics around Horizontal Tubes in Fluidized Beds*," Ph.D. Dis., Lehigh University, Bethlehem, Pa.

Chandran, R., and J. Chen, "A Heat Transfer Model for Tubes Immersed in Gas Fluidized Beds," *AIChE J.*, 31-244-252.

Chang, H-C (1983), "Effective Diffusion and Conduction in Two Phase Media," *AIChE J.*, 29:846-53.

Chen, M. M., J. Liljegren, and B. T. Chao (1984), "The Effects of Bed Internals on the Solids Velocity Distribution in Gas Fluidized Beds," *Fluidization*, D. Kunii and R. Toei, Eds., Engineering Foundation, New York.

Chen, M. M., J. S. Lin, and B. T. Chao (1981), *Computer-Aided Particle Tracking: A Technique for Studying Solid Particle Dynamics in Gas or Liquid Fluidized Beds*, presented at AIChE Annual Meeting, New Orleans, La.

Chung, Y. C. and L. G. Leal (1982), "An Experimental Study of the Effective Thermal Conductivity of a Sheared Suspension of Rigid Spheres," *Int. J. Multiphase Flow*, 8:605-22.

Collingham, R. E. (1968), Ph.D. Thesis, University of Minnesota.

Danziger, W. J. (1963), "Heat Transfer to Fluidized Gas Solid Mixtures in Vertical Transport," *Ind. Eng. Chem.*, 2:269-76.

Davidson, J. F. and D. Harrison (1971), *Fluidization*, Academic Press, New York.

Depew, C. A. and L. Farbar (1963), "Heat Transfer to Pneumatically Conveyed Glass Particles of Fixed Size," *J. Heat Transfer*, C85:164-72.

Dow, W. M., and M. Jakob (1951), *Chem. Eng. Prog.*, 47:537.

Farbar, L., and M. J. Morley (1957), "Heat Transfer to Flowing Gas-solid Mixtures in a Circular Tube," *Ind. Eng. Chem.*, 49:1143-50.

Farbar, L., and C. A. Depew (1963), "Heat transfer effects to gas-solid mixtures using solid spherical particles of uniform size," *Ind. Engr. Chem. Fundam.*, 2:130-5.

Gabor, J. D. (1970a), *Chem. Eng. Prog., Symp. Ser.*, 66/105:76.

Gabor, J. D. (1970b), *Chem. Eng. Sci.*, 25:959.

Gelperin, N. I. (1940), Khim. Mahinostr. No. 3, 1 Gelperin, N. I., and V. G. Einstein, 1971 "Heat Transfer in Fluidized Beds," in *Fluidization*, J. F. Davidson and D. Harrison, Eds., Academic Press. 471-536.

Gelperin, N. I., V. G. Einshtein, L. A. Korotyanskaya, and J. P. Peierozchikova (1968), *TOKHT*, 2:430.

Gelperin, N. I., V. G. Einshtein, and A. V. Zaikovski (1966), "Variation of Heat-Transfer around the Perimeter of a Horizontal Tube in a Fluidized Bed," *J. Eng. Phys*, 10-473-475.

Gidaspow, D. B., B. Ettahadieh, and R. W. Lyczkowski (1984), *Hydrodynamics of Fluidization in a Semicircular Bed with a Jet, AIChE J.*, 30:4, 529-536.

Glass, D. H. (1967), Ph.D dissertation, University of Cambridge.

Gorbis, Z. R., and R. A. Bakhtiozin (1962), "Investigation of Convection Heat Transfer to a Gas Graphite Suspension in Vertical Channels," *Sov. J. At. Energy, 12:402-9*.

Handly, M. F., and M. G. Perry (1965), *Rheol. Acta.*, 4:225.

Heertjes, P. M., J. Verloop, and R. Williams (1970/71), "The Measurement of Local Mass Flow Rates and Particle Velocities in Fluid-Solid Flow," *Powder Technology*, 4:38.

Hetsroni, G. (Ed.) (1982), *Handbook of Multiphase Systems*, McGraw-Hill.

Iwashko, M. A. (1985), "Effect of Solids Circulation on Heat Transfer from an Immersed Horizontal Rod in a Gas Fluidized Bed." MS Thesis, University of Illinois at Urbana-Champaign.

Jackson, R. (1963a) "The Mechanics of Fluidized Beds: Part I. The Stability of the State of Uniform Fluidization," *Trans. Inst. Chem. Engrs.*, 41:13-21.

Jackson, R. (1963b), "The Mechanics of Fluidized Beds: Part II. The Motion of Fully Developed Bubbles," *Trans. Inst. Chem. Engrs.*, 41:22-28.

Kim, J. M., and J. D. Seader (1983), "Heat Transfer to Gas-solids Suspensions Flowing Cocurrently Downward in a Circular Tube," *AIChE J.*, 29:306-12.

Kobayashi, M., D. Ramaswami, and W. T. Brazelton (1970), *Chem. Eng. Prog., Symp. Ser.*, 66/105-58.

Kondukov, N. B., A. N. Kornilaev, I. M. Skachko, A. A. Akromenkov, and A. S. Kruglov (1964), "An Investigation of the Parameters of Moving Particles in a Fluidized Bed by a Radiosotropic Method," *Int. Chem. Eng.*, 4:1, 43-47.

Kunii, D., and O. Levenspiel (1969), *Fluidization Engineering*, John Wiley, New York.

Kunii, D. and R. Toei (1984), *Fluidization*, Engineering Foundation, New York.

Leal, L. G. (1973), "On the Effective Conductivity of a Dilute Suspension of Spherical Drops in the Limit of Low Particle Peclet Number," *Chem. Engng Commun.*, 1:21-31.

Levenspiel, O., and J. S. Walton (1954), *Chem. Eng. Prog., Symp.*, Ser. 50-9:1.

Liljegren, J. C. (1984), M.S. Thesis, University of Illinois at Urbana-Champaign.

Lin, J. S. (1981), "*Particle Tracking Studies for Solids Motion in a Gas Fluidized Bed,*" Ph.D. Thesis, Department of Mechanical and Industrial Engineering, University of Illinois at Urbana-Champaign.

Lin, J. S., M. M. Chen, and B. T. Chao (1985), "A Novel Radioactive Particle Tracking Facility for Measurement of Solids Motion in Fluidized Beds," *AIChE J.*, 31:3, 465-473.

Mamaev, V. V., V. S. Nosov, N. I. Syromyatnikov, and V. S. Barbolin (1976), "Heat Transfer of Gas-suspension Flow in Horizontal and Vertical Tubes," *J. Eng. Phys.*, 31:1146-9.

Marscheck, R. M., and A. Gomezplata (1965), "Particle Flow Patterns in a Fluidized Bed," *AIChE J.*, 11:167.

Masson, H., K. Dan Tran, and G. Rios (1981), "Circulation of a Large Isolated Sphere in a Gas-Solid Fluid Bed," *Int. Chem. Eng. Symposium*, Series No. 65.

Mastanaiah, K., and E. N. Ganici (1981), "Heat Transfer in Two Component Dispersed Flow," *J. Heat Trans.*, 103:300-306.

Merry, J. M., and J. F. Davidson (1973), "Gulf Stream Circulation in Shallow Fluidized Beds," *Trans. Inst. Chem. Engrs.*, 51:351-368.

Mickley, H. S., and F. Fairbanks (1955), "Mechanism of Heat Transfer to Fluidized Beds, *AIChE J.*, 1:3, 374-386.

Moslemian, D. (1986), "Study of Solids Motion, Mixing and Heat Transfer in Gas Fluidized Beds," Ph.D. Thesis, University of Illinois at Urbana-Champaign.

Murray, J. D. (1965a), "On the Mathematics of Fluidization: Part I. Fundamental Equations and Wave Propagation," *J. Fluid Mech.*, 21:3, 465-493.

Murray, J. D. (1965b), "On the Mathematics of Fluidization: Part II. Steady Motion of Fully Developed Bubbles," *J. Fluid Mech.*, 22:1, 57-80.

Nir, A., and A. Acrivos (1973), "The Effective Thermal Conductivity of Sheared Suspensions," *J. Fluid Mech.*, 78:33-40.

Noack, R. (1970), *Chem. Eng. Tech.*, 42:371.

Oki, K., M. Ishida, and T. Shirai (1980), "The Behavior of Jets and Particles near the Gas Distributor Grid in a Three-Dimensional Fluidized Bed," *Proc. Intl., Conf. on Fluidization*, Henniker, NH, pp. 421-428.

Pfeffer, R., S. Rossetti, and S. Licklein (1966), "Analysis and Correlation of Heat Transfer Coefficient and Friction Factor Data for Dilute Gas Solid Suspensions," NASA TN D-3603. Rowe, P. N. (1971), "Experimental properties of bubbles," in *Fluidization*, J. F. Davidson and D. Harrison, Eds., Academic Press.

Saxena, S. C., N. S. Grewal, J. D. Gabor, S. S. Zabrodsky, and D. M. Galershtein (1978), "Heat Transfer Between a Gas Fluidized Bed and Immersed Tubes," *Advances in Heat Transfer*, 14:149-247.

Schlunderberg, D. C., R. L. Whitelaw, and R. W. Carlson (1961), "Gaseous suspension-a New Reactor Coolant," *Nucleonics*, 19:67-8,70-2,74,76.

Singh, A. (1968), Ph.D. Thesis, University of Minnesota.

Sohn, C. W., and M. M. Chen (1984), "Heat Transfer Enhancement in Laminar Slurry Pipe Flow with Power Law Thermal Conductivities," accepted for publication, *J. of Heat Transfer*.

Sohn, C. W., and M. M. Chen (1981), "Microconvective Thermal Conductivity in Disperse Two Phase Mixtures as Observed in a Low Velocity Couette Flow Experiment," *J. Heat Transfer*, 103-47-51.

Soo, S. L. (1962), *Proc. Symp. on Interaction between Fluids and Particles*, Inst. of Chem. Engrs., London, p. 50.

Soo, S. L. (1967), *Fluid Dynamics of Multiphase Systems*, Blaisdell, Waltham, Mass.

Soo, S. L. (1969), *Advanced Heat Transfer*, B. T. Chao, Ed., University of Illinois Press, Urbana, IL.

Soo, S. L. (1983), *Multiphase Fluid Dynamics*, S. L. Soo Associates, 2020 Curaton Dr., Urbana, IL 61801.

Soo, S. L. (1962), *Proc., Symp. on Interaction between Fluids and Particles*, Inst. of Chem. Engrs., London, p. 50; see also Soo, S. L. (1983), *Multiphase Fluid Dynamics*, S. L. Soo, Associates, Pub. (2020 Cureton Drive, Urbana, IL 61801).

Tennekes, H., and J. L. Lumley (1972), *A First Course in Turbulence*, The MIT Press, Cambridge.

Thring, R. H. (1977), "Fluidized Bed Combustion for the Stirling Engine," *Int. J. Heat Mass Transfer*, 20:911-918.

Tien C. L. (1961), "Heat Transfer by Turbulently Flowing Fluid-solids Mixtures in a Pipe," *J. Heat Transfer*, C83:183-8.

Van Heerden, L., P. Nobel, and D. W. Van Krerilen (1953), *Ind. Eng. Chem.*, 45:1237.

Van Velzen, D., H. J. Flamm, H. Langenkamp, and E. Casile (1974), "Motion of Solids in Spouted Beds," *Can. J. Chem. Eng.*, 52:156-161.

Vortmeyer, D. (1978), "Radiation in Packed Solids," *VI Int. Heat Transfer Conf. Proc.*, Toronto, 525-539.

Wasan, D. T., and M. S. Ahluwalia (1969), *Chem. Eng. Sci.*, 24:1535.

Wen, C. Y., and L. H. Chen (1984), "Flow Modeling Concepts of Fluidized Beds," in *Handbook of Fluids in Motion*, N. Cheremisinoff and R. Gupta, Eds., 665-691, Butterworths, Boston.

Werther, J., and O. Molerus (1973), "The Local Structure of Gas Fluidized Beds--II. The Spatial Distribution of Bubbles," *Int. J. Multiphase Flow*, 1:123-138.

Whitaker, S. (1967), "Diffusion and Dispersion in Porous Media," *AIChE J.*, 13:420-427.

Whitehead, A. B., G. Gartside, and D. C. Dent (1976), "Fluidization Studies in Large Gas-Solid Systems, Part III. The Effect of Bed Depth and Fluidizing Velocity on Solids Circulation Patterns," *Powder Technology*, 14:61-70.

Wunschmann, J., and E. V. Schlunder (1975), *Trans. Int. Conf. Heat Trans.*, CT2.1, 49.

Yagi, S., and D. Kunii (1957), "Studies on Effective Conductivity in Packed Beds," *AIChE J.*, 3:373-81.

Yong, J., Y. Zheging, Li Zhang, and W. Zhanwan (1980, "A Study of Particle Movement in a Gas Fluidized Bed," *Proc. Intl. Conf. on Fluidization, Henniker, NJ, 365-372.*

Ziegler, E. N., and W. T. Brazelton (1964), "Mechanism of Heat Transfer to a Surface," *Ind. Eng. Chem. Fund.*, 3:94-8.

HIGH-TEMPERATURE
SOLIDS-GAS INTERACTIONS

M. Q. Brewster

ABSTRACT

This paper reviews recent studies of high-temperature, solid-particle, direct-contact heat transfer devices. Three different particle-gas flow configurations are covered: fluidized bed, entrained flow, and free-falling particle flow. Several preliminary experimental studies have been conducted using each of these flow configurations. These are discussed and comparisons between them are made. Some theoretical modeling of the radiative transport and gas-particle heat transfer has also been done. These various models are discussed and a review of pertinent techniques for modeling radiative transport in particulate media is given. Based on several modeling efforts some recommendations for improving solids-gas direct-contact heat transfer are given. These include promotion of lateral particle mixing in entrained and free-falling flows to reduce infrared emission losses and investigating "windowless" means of containment for fluidized bed solar receivers.

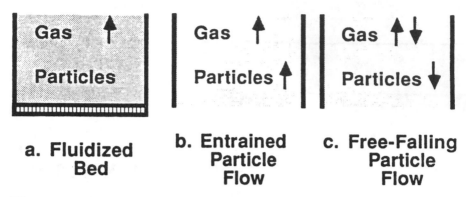

Figure 9.1 Solid particle direct-contact heat transfer.

1 SCOPE

The aim of this review paper is to catalogue recent findings in the area of high-temperature (greater than 1000 K) gas-solid particle direct-contact heat-transfer devices. These devices include fluidized beds, free-falling particle films, and entrained particle flows. In addition, an attempt will be made to indicate which areas of future study promise the greatest returns.

Primary emphasis is given to the radiative mode of heat transfer, as opposed to convective/conductive heat transfer issues. As a result most of the examples and studies cited here involve direct-contact, solid particle solar receiver devices, since by definition radiative transfer plays an important role in these systems. The main emphasis of this paper, however, is not the solar application but the fundamentals of radiative transfer in flowing gas-solid particulate media.

No specific attempt is made to limit or categorize discussion according to the application of the device. Thus various configurations are considered. However, many of the systems discussed have as their end-use application thermal electric power generation or chemical processing.

Finally, while there are many practical design and materials considerations that present themselves, those will not be the main focus of this paper. Martin [1,2] has given a thorough discussion of many of the important design and production considerations associated with solid particle direct-contact heat transfer devices as well as possible applications.

2 OVERVIEW

Figure 9.1 shows a schematic representation of three possible configurations under consideration for use as solid particle direct-contact heat transfer devices. The fluidized bed (Fig. 9.1a) uses upwardly flowing gas through a bed of particles. When the velocity of the gas is sufficient, the pressure drop through the particles multiplied by the cross-sectional area of the bed equals the weight of the bed, and the particles become "fluidized." Their motion is random and chaotic with a net velocity of zero. For greater gas velocities, the particles will move with a net velocity in the direction of the gas flow, and an entrained flow results (Fig. 9.1b). Both fluidized and entrained flows require a forced flow of gas.

If no forced gas flow is utilized and the particles are allowed to fall under gravity, a free-falling particle flow results (Fig. 9.1c). In this situation the particle motion is downward, and the gas flow can be either downward or upward depending on the relative strength of gas entrainment and buoyant convection effects. If gas entrainment (momentum transfer from particles to gas, which tends to drag the gas downward) dominates, the gas flows downward. If buoyant convection (due to radiative particle-to-gas heating) dominates, the gas flows upward.

In any of these three flow configurations, radiative heating of the particles can conceivably be accomplished by one of two means. The first is to make the containment vessel walls entirely or partly from high-temperature transparent material (preferably abrasion-resistant material) and direct concentrated radiation through the transparent walls on to the particles. The other method is to direct the radiation through an open, "windowless" surface on the particles. This method has not received as much attention as the first.

This paper will proceed in the following manner. First a review of actual laboratory or field test experiences involving high-temperature solid particle direct-contact devices will be given. Then pertinent radiative modeling techniques will be discussed. Drawing upon that discussion, the influence of key parameters on radiative transport among particles will then be considered. Finally, recommendations for future efforts will be made.

3 REVIEW OF RECENT EXPERIMENTAL STUDIES

Significant experimental investigations have been conducted using all three types of particle flow configurations mentioned in the overview. Fluidized bed solar receivers have been built and tested both at the Laboratoire d'Energetique Solaire in Odeillo, France and, under sponsorship of the Solar Energy Research Institute (SERI), at the Georgia Tech Research Institute (GTRI). Entrained flow devices have been constructed and tested at GTRI and at Lawrence Berkeley Laboratories. And, an experimental free-falling particle chute has been investigated at Sandia National Laboratories (Albuquerque, N.M.).

3.1 French Fluidized Bed

Flamant and coworkers [3–5] at the Laboratoire d'Energetique Solaire in France have done extensive experimental testing of both fluidized and packed bed solar receivers. Their work has also included a substantial modeling effort that will be discussed later. In their experiments a 6.5 kW (2200 kW/m^2) solar concentrator has been used to heat packed and fluidized beds of refractory particles contained in transparent quartz tubes (Fig. 9.2). The dimensions of the fluidized bed were 6.5 cm diameter by 15 cm height. Receiver efficiency was defined as the increase in gas internal energy divided by the incident solar energy. As the gas flow rate

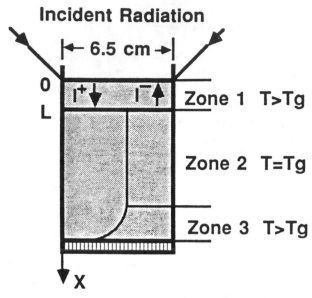

Figure 9.2 Flamant solar fluidized bed.

was increased, efficiency also increased. Typical values for receiver efficiency of the fluidized bed ranged from 0.4 to 0.7 for darker SiC particles and from 0.2 to 0.4 for lighter ZrO_2 particles. In the packed bed the efficiencies were consistently lower than in the fluidized bed, the difference being due to greater emission loss in the highly nonisothermal packed bed than in the nearly isothermal fluidized bed.

In addition to the receiver efficiency, Flamant and Olalde [4] also considered the product of receiver and thermal cycle (e.g., Brayton) efficiency. Since receiver efficiency is a decreasing function of outlet fluidized gas temprature, and cycle efficiency is an increasing function of that temperature, the product of those two efficiencies exhibited a maximum at intermediate temperatures. For the more efficient SiC particles this maximum overall efficiency was 0.27 and occurred at outlet gas temperatures between 750 and 950 K.

One potentially serious drawback to solar fluidized bed receivers is long-term degradation of the transparent tube material. The French studies do not mention this problem. They do recognize the significant loss in their open experimental design due to thermal emission and propose a cavity configuration to overcome this deficiency. This proposed design would have an annular fluidized bed contained presumably in quartz. The center region inside the annulus would form the cavity, with an opening at the bottom for the incident solar flux. While the estimated cavity efficiency for this design is reported as 83% [4], the quartz degradation appears to be a major technical problem that must be overcome.

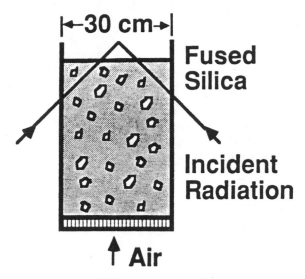

Figure 9.3 GTRI solar fluidized bed.

3.2 GTRI Fluidized Bed

A fluidized bed experiment, somewhat larger in scale than the French studies, was conducted at GTRI [6,7]. In that work a 30 cm diameter fluidized bed received concentrated solar irradiation, through fused silica walls from below, as pictured in Fig. 9.3. Various bed materials were tested including SiC, Al_2O_3, sand, copper, and translucent crushed fused silica. Due to a limitation in the available air velocities, fluidization was unsatisfactory, and the reported efficiencies ranged between approximately 30% and 40%. The mean particle sizes in the GTRI study were larger than those in the French studies, 1000–3000 μm compared with 250–700 μm. Larger particles require higher gas velocities for effective fluidization, and this may have contributed to the fluidization difficulties in the GTRI study.

Another significant finding of the GTRI study was that the fused silica tube was discolored by operating the bed with nonoxide particles (e.g., SiC and copper). This discoloration took place within only a few hours of operation. The GTRI fluidized bed was operated for more than 70 hours. The most serious equipment problem encountered was the window discoloration. It is still not clear, however, how serious the problem of optical degradation of the window due to mechanical abrasion will be over longer periods of time.

3.3 GTRI Entrained Flow

An entrained flow solar receiver/reactor has also been tested at GTRI [8,9]. This device, pictured in Fig. 9.4, also used a fused quartz tube for flow containment and as a window for solar irradiation. Both inert and reactive particles were added to the steam carrier gas through a concentric inner tube. The particles, fed

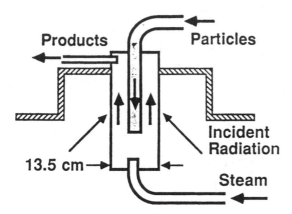

Figure 9.4 GTRI entrained flow reactor.

by a screw feeder, would fall under gravity down the inside of the inner (alumina) tube, exit the bottom of the feed tube, and be entrained in the steam flowing upward through the annulus. To avoid gravitational settling, 60–90 μm inert particles (Al_2O_3, SiC, quartz, and glass) and 40 μm carbon particles were used. Theoretical predictions for spherical particles in Stoke's flow indicated that 76 μm particles with densities up to that of iron would be able to be entrained by the steam. In actual tests, however, only 50% of the inert particles were entrained.

In the inert particle tests, it was noted that convective heating of the flow was relatively important with respect to the radiative heating. This conclusion was based on the observation that the steam temperature would increase by 450–550 C upon passing through the tube, even without any particles added and that with the addition of particles the exit temperature only increased by an additional 50–100 C. It was also noted, however, that even at the highest particle loadings that could be successfully entrained, the flow was not opaque and the absorption efficiency could have been improved substantially if the opacity of the flow could have been increased. It was recommended that ways of increasing the opacity of entrained flows, within the particle entrainment limitations, be further studied.

Devitrification and discoloration of the quartz tube were also noticed in the entrained study at GTRI, as in the fluidized bed. This occurred within four hours of operation. However, it did not appear that this had a significant impact on the performance of the device. Apparently, convective heating of the steam improved as the tube darkened to offset the loss of direct radiant heating.

3.4 LBL Entrained Flow

Hunt and coworkers at Lawrence Berkeley Laboratory (LBL) demonstrated the feasibility of a small particle heat exchanger receiver (SPHER), which used micron-sized carbon particles as absorbers [10–14]. The particles were generated

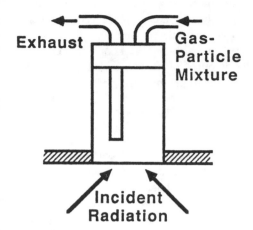

Figure 9.5 LBL small particle entrained flow absorber.

upsteam of the receiver by pyrolysis of acetylene in Argon. The particles were then mixed with air and introduced as an air-particle flow into the receiver (Fig. 9.5). The particles were forced to flow through the region of maximum radiant flux before the flow was exhausted from the receiver through a quartz exit tube. Upon passing through the intense radiation the particles would initially heat the surrounding gas and eventually oxidize. Thus the particles would be consumed by the process. It was noted that the amount of carbon necessary was small relative to the amount that would have been necessary to produce equivalent heating by direct combustion of the carbon.

The peak measured exhaust temperature for this configuration was reported as 1200 K with an incident flux of 4000 kW/m^2 and particle volume fraction less than 5×10^{-6}. Measured efficiencies were not reported, although theoretical considerations indicated that very high efficiencies should be attainable (85%–90%).

Small particle receivers are inherently capable of high efficiencies due to the excellent absorptive properties of small particles. Rayleigh scattering theory indicates that as particle diameter decreases, particles that are inherently absorbing (i.e., not transparent) will become absorption-dominated and scattering becomes insignificant. In the limit as diameter goes to zero the scattering coefficient goes to zero, while the absorption coefficient remains constant. The loss due to incident radiation being scattered back out of the receiver is therefore very small for a small particle receiver.

In comparing the GTRI entrained flow receiver with the LBL device, advantages and disadvantages with each are evident. The GTRI approach relies on entrainment of commercially available power (40 μm) while the LBL SPHER approach uses much smaller (approximately 0.1 μm) carbon generated from hydrocarbon pyrolysis. While commercial powders have the advantage that the experimenter is able to control the particle size, there is a lower limit on the size of powder that can be economically dispersed in an entrained flow. Fine powders have a greater tendency to agglomerate than coarse powders. On the other hand, small particles generated from thermal decomposition have the disadvantage that it is more difficult for the experimenter to control the particle size distribution. Since smaller particles are optically more favorable (greater opacity and less scattering) the ideal combination would be a source of micron or submicronparticles that could be controlled by the experimenter to produce the desired size distribution.

3.5 Sandia Free-Falling Flow

A test involving a free-falling configuration of radiantly heated particles was conducted at Sandia National Laboratories [15]. Particles of SiC and silica sand were dropped 10 m through a 15 by 30 cm refractory-lined sheet metal chute. The particles were heated radiantly by infrared lamps mounted on one side of the chute behind fused silica plates (Fig. 9.6). The particles were caught at the bottom of the chute in a bin, where the temperature was measured. Particle velocity was measured at various heights using laser Doppler velocimetry.

The maximum particle temperature reached in this apparatus was 1300 K for 500 μm SiC particles with 500 kW/m^2 incident flux. The efficiencies reported (based on particle heating) were rather low, less than 25%, but a cavity geometry could improve that by reducing the emission loss.

Significant trends noted were the influence of particle mass flow rate and radiant heat flux on receiver efficiency and final particle temperature. As mass flow increased, the efficiency increased while particle temperature decreased. The opposite trend was observed when the incident flux was varied. As heat flux increased, the efficiency decreased due to emission loss while particle temperature increased.

Several interesting flow and heat transfer effects were also noticed in this study. It was found that buoyancy-induced convection currents due to particle heating were substantial. In fact, 300 μm particles could not be successfully used in the test because convective currents carried more than half of the particles out of the chute. Buoyant convective flow also influenced the velocity of the larger SiC particles (500 and 1000 μm). Initially the particles would accelerate due to gravity. But at some intermediate height the particle velocity would begin to decrease due to convective heating of the air. As a result, the residence time of the darker SiC particles was larger than that of the sand particles. The less absorptive sand particles did not show a decrease in velocity due to convection but accelerated through the entire chute. These findings indicated that wall convective heating of the air was small relative to particle convective heating of the air. If the convection currents had been caused by wall heating, the sand particle velocity would have also been retarded, similar to the SiC particles.

The studies outlined above have all been preliminary in nature and mainly useful for demonstrating the various concepts. One thing evident is that the fluid mechanics and heat transfer are coupled. Therefore, the motion of the gas and particles cannot be predicted for a cold flow with heat transfer effects added afterward. Energy and momentum transfer are strongly coupled in these flows.

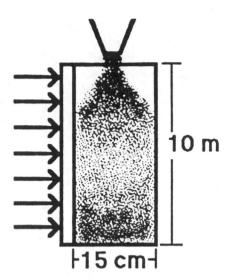

10 m

├15 cm┤

Figure 9.6 Sandia free-fall radiant heating test.

4 PERTINENT RADIATIVE MODELING TECHNIQUES

Because of the strong coupling between heat transfer and fluid-solids motion, accurate modeling of the radiative transport is particularly important in solid particle direct-contact heat transfer devices. This modeling can take place at various levels of complexity. At the simplest level there is the single particle model. This model does not address the multiple scattering problem but only looks at how a single particle interacts with whatever radiation is incident upon it. This approach is of limited usefulness for solid particle absorbers, receivers, and heat exchangers since the radiative transfer in these devices will be dominated by multiple scattering. At the more complex level there is the radiative transfer equation, which includes the effects of multiple scattering of radiation among the particles. Two solutions of the transfer equation will be presented: the two-flux model, which results in simple, approximate closed-form solutions, and the discrete ordinate method, which is used when greater accuracy is required.

4.1 Single Particle Model

The simplest model of radiant heating of particles is not a model of radiative transfer (i.e., multiple scattering) at all. It is simply a Lagrangian energy balance on a single particle (Fig. 9.7). Such an energy balance is given in Equation (1).

$$\rho_s C_s \, V \, DT/Dt = \alpha \int_A q^- \, dA - A\epsilon \, \sigma T^4 + hA \, (T_\infty - T) \tag{1}$$

In this equation, A represents the surface area of the particle, T the (lumped) temperature of the particle, and D/Dt is the derivative following the particle. This energy balance accounts for the increase in internal energy of the

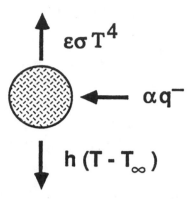

Figure 9.7 Single particle model.

particle on the left side and for absorbed radiation, emitted radiation and convective gain from the surrounding gas, respectively, on the right side. The difficulty in solving Equation (1) comes in specifying the heat transfer coefficient h, the ambient temperature T_∞, the particle velocity (contained in the D/Dt term), and the radiant flux incident on the particle, q^-. Of necessity the radiative transport problem must be solved independent of Equation (1) in order to specify q^-. However, since the radiative field is influenced by emission from the particles (and therefore by the particle temperature) the radiative transport equation and the particle energy equation are coupled. The usefulness of Equation (1), then, comes in assuming what the particle irradiation is and solving (1) for the particle temperature approximately.

In making a preliminary feasibility study of a free-falling, windowless, solid particle solar central receiver, Martin and Vitko [1] used the single particle approach to estimate that it would take 1.8 s (approximately 9 m of free fall) to heat a 460 μm SiC particle from 298 K to 1223 K, subject to 1366 K cavity wall radiant heating. The temperature of 1366 K is approximately what a concentrated solar flux heating apparatus could be expected to achieve. More generally, the time required to heat a particle to a given temperature was found to be proportional to the particle diameter, thus placing an upper limit on the size of particles that could be used. Falcone [16] estimated that particles subjected to 1000 kW/m^2 irradiation would reach 1260 K after 3.7 s of heating. It must be remembered, however, that for the reasons already mentioned, and also because convection with the surrounding gas was neglected, these predictions are subject to considerable variation and uncertainty.

4.2 Radiative Transfer Equation

This section will give an introductory description of the radiative transfer equation in a one-dimensional slab geometry. Such descriptions and derivations exist in plentitude in the literature [17–20] so this portion will necessarily omit some of the details.

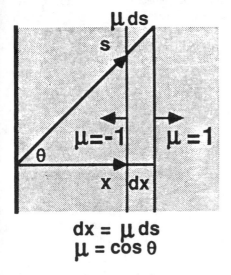

$$dx = \mu\,ds$$
$$\mu = \cos\theta$$

Figure 9.8 Slab coordinates for radiative transfer equation.

The radiative transfer equation in a one-dimensional slab geometry is a spectral, radiant intensity balance on a differential element of the slab (Fig. 9.8). Furthermore, it is an intensity balance along only one path. In Fig. 9.8 the intensity balance is taken along the optical path labeled s. Equation (2) gives the transfer equation for this configuration:

$$\frac{dI(x,\theta,\psi)}{ds} = -(\sigma + a)\,I(x,\theta,\psi) + aI_b(x)$$
$$+ \sigma/4\pi \int_0^{2\pi}\int_0^{\pi} I(x,\theta',\psi')\,p(\theta,\psi;\theta',\psi')\sin\theta'\,d\theta'\,d\psi' \qquad (2)$$

The left side of Equation (2) represents the change in intensity with respect to distance in traversing the differential element along a path that lies in the direction of the slab polar and azimuthal angles, θ and ψ, respectively. (For simplicity the azimuthal angle is not pictured in Fig. 9.8.) The differential element is located at a perpendicular distance from the origin of magnitude x. The three terms on the right side represent the decrease in intensity due to scattering and absorption primarily by solid particles; the increase in intensity due to thermal emission, again, primarily by solid particles; and the increase in intensity due to scattering of radiation out of the θ'/ψ' directions into the θ/ψ direction.

For azimuthally symmetric boundary conditions and randomly-oriented scattering particles the intensity field will also be azimuthally symmetric and the transfer equation reduces to (3):

$$\mu\,\frac{dI(x,\mu)}{dx} = -(\sigma + a)\,I(x,\mu) + aI_b(x) + \sigma/2 \int_{-1}^{1} I(x,\mu')p(\mu,\mu')d\mu' \qquad (3)$$

Assuming the medium to be composed of uniformly distributed, monodisperse, independently scattering, spherical particles, the scattering and absorption coefficients, σ and a, can be related to the particle volume fraction f_v, diameter d, and scattering and absorption efficiencies $Q_{s,a}$, by Equation (4):

$$\sigma, a = 1.5 \; Q_{s,a} \; f_v/d \qquad (4)$$

This relation holds even for fluidized beds ($f_v \approx 0.5$) as long as the interparticle clearance divided by characteristic wavelength c/λ is greater than about 0.3 [21].

In general $Q_{s,a}$ are complicated functions of the particle size parameter $(\pi d/\lambda)$ and the particle complex refractice index $(n-ik)$. However, for values of $\pi d/\lambda > 5$ (approximately) the simple results of geometric optics can be used [20]. These results are expressed by Equation (5):

$$Q_a = 1 - Q_s = \epsilon \qquad (5)$$

In Equation (5), ϵ is the spectral hemispherical emissivity of the particle material. Several assumptions are associated with Equation (5). First, the particles are assumed to be much larger than the characteristic wavelength of radiation ($d \gg \lambda$). Second, the particles are assumed to be spherical, opaque, and either diffuse- or specular-reflecting. Third, diffraction by the particles is neglected. This is justified for heat transfer calculations since the diffracted component is all concentrated in the forward directions and can be treated as transmitted (unscattered) radiation. Fourth, the spectral-directional reflectivity has been assumed to be independent of incidence angle. Given the uncertainty associated with most radiative property data, these assumptions are appropriate for the present treatment.

The phase function $p(\theta,\psi,\theta',\psi')$ appearing in Equation (2) is the normalized function that gives the directional distribution of the scattered intensity from the primed direction into the unprimed direction (Fig. 9.9). It is related to the scattering phase function of the individual particles that make up the medium $P(\phi)$, by Equations (6) and (7):

$$p(\theta,\psi,\theta',\psi') = P[\phi(\theta,\psi;\theta',\psi')] \qquad (6)$$

$$\cos \phi = \mu\mu' + \left(1 - \mu^2\right)^{1/2} \left(1 - \mu'^2\right)^{1/2} \cos (\psi - \psi') \qquad (7)$$

Here, ϕ is the angle between the direction of incidence and the direction of scattering for a single particle. Equation (7) is a result from analytic geometry relating the polar angle in the particle geometry (ϕ) to the polar and azimuthal angles in the slab geometry (θ,ψ) [19].

Making use of symmetry in the ψ-direction, the phase function for scattering from the μ' directions into the μ directions can be obtained by integrating Equation (8), or

$$p(\mu,\mu') = 1/\pi \int_{0}^{\pi} p(\mu,\psi = 0; \mu',\psi') \; d\psi' \qquad (8)$$

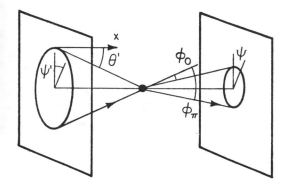

Figure 9.9 Relation between slab and particle scattering geometry.

To carry out the integration, Equation (7) is differentiated implicitly and substituted into (8) giving (9):

$$p(\mu,\mu') = 1/\pi \int_{\phi_o}^{\phi_\pi} \frac{P(\phi)\sin\phi d\phi}{[(1 - \mu^2)(1 - \mu'^2) - (\cos \phi - \mu\mu')^2]^{1/2}} \qquad (9)$$

The limiting angles on ϕ for a given μ and μ' are given by Equations (10) and (11). These are obtained by substituting 0 and π for the quantity $(\psi - \psi')$ in (7):

$$\cos \phi_0 = \mu\mu' + (1 - \mu^2)^{1/2} (1 - \mu'^2)^{1/2} \qquad (10)$$

$$\cos \phi_\pi = \mu\mu' - (1 - \mu^2)^{1/2} (1 - \mu'^2)^{1/2} \qquad (11)$$

To restate in words the meaning of Equation (9), $p(\mu,\mu')$ is the phase function for scattering from the direction μ' (or θ') into the direction μ (or θ), that is from the μ' "cone" into the μ "cone." For a given μ and μ' there are many values of ϕ (single particle scattering angle) possible. Equation (9) averages over all possible values of ϕ.

For the same assumptions stated earlier for Equation (5), the expressions for $P(\phi)$ are given in Equations (12) and (13) [20]:

$$\text{Specular}: \quad P(\phi) = 1 \qquad (12)$$

$$\text{Diffuse}: \quad P(\phi) = 8 \left(\sin \phi - \phi \cos\phi\right)/3\pi \qquad (13)$$

Equations (12) and (13) and Figure 9.10 indicate that specular-reflecting particles are isotropic scatterers (within the assumption of directional reflectivity independent of incidence angle), and diffuse-reflecting particles are somewhat backscattering in nature.

As a conclusion to this section, a formulation of the transfer equation will be presented that is preferable when the irradiation on the particle slab is collimated (Figure 9.11). First the transfer equation (2) for azimuthal dependence will be rewritten in the optical depth notation as Equation (14):

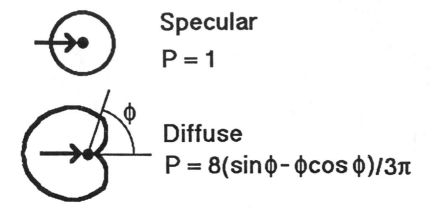

Figure 9.10 Phase function for opaque geometric spheres.

$$\mu \, \frac{dI(t,\mu,\psi)}{dt} = - \, I(t,\mu,\psi) + (1 - \omega) \, I_b(t)$$

$$+ \frac{\omega}{4\pi} \int_0^{2\pi} \int_{-1}^{1} I(t,\mu',\psi')p(\mu,\psi;\mu',\psi') \, d\mu' \, d\psi' \qquad (14)$$

The optical depth t, and albedo ω, are defined by Equations (15) and (16):

$$t = (\sigma + a) \, x \qquad (15)$$

$$\omega = \sigma/(\sigma + a) \qquad (16)$$

The total intensity, previously written as I and now written as I_{tot}, is expressed as the sum of a scattered and an unscattered contribution:

$$I_{tot}(t,\mu,\psi) = I(t,\mu,\psi) + I_o \, \exp \, [-t/\mu_o] \, \delta(\mu - \mu_o) \, \delta(\psi - \psi_o) \qquad (17)$$

Here, $I(t,\mu,\psi)$ is the intensity of radiation that has suffered at least one scattering event, I_o is the collimated flux that is incident in the direction μ_o, and δ is the Dirac function. Substituting this expression into Equation (14) gives the transfer equation suitable for collimated incident flux, Equation (18):

$$\mu \, \frac{dI(t,\mu,\psi)}{dt} = I(t,\mu,\psi) + (1 - \omega) \, I_b(t)$$

$$+ \frac{\omega}{4\pi} \int_0^{2\pi} \int_{-1}^{1} I(t,\mu',\psi')p(\mu,\psi;\mu',\psi')d\mu' \, d\psi'$$

$$+ \frac{\omega}{4\pi} \, I_o \, p(\mu,\psi;\mu_o,\psi_o) \exp \, [-t/\mu_o] \qquad (18)$$

Two-flux model. The two-flux model [22–26] is an approximate solution of the transfer equation. It is obtained by assuming that the intensity distribution is semi-isotropic, that is, constant with a value of I^+ in the forward hemisphere

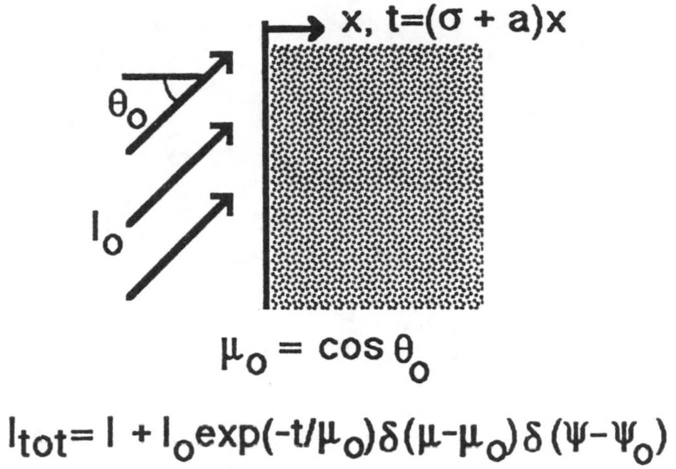

$$x, t=(\sigma + a)x$$

$$\mu_0 = \cos \theta_0$$

$$I_{tot} = I + I_0 \exp(-t/\mu_0)\delta(\mu-\mu_0)\delta(\psi-\psi_0)$$

Figure 9.11 Skewed collimated incidence.

($\mu > 0$) and constant with a value of I^- in the backward hemisphere ($\mu < 0$). This assumption is illustrated in Fig. 9.12.

The governing equations are obtained by integrating the transfer Equation (3) over the forward and backward hemispheres:

$$\frac{1}{2}\frac{dI^+}{dx} = -(\sigma B + a)\,I^+ + a\,I_b + \sigma B\,I^- \tag{19}$$

$$-\frac{1}{2}\frac{dI^-}{dx} = -(\sigma B + a)\,I^- + a\,I_b + \sigma B\,I^+ \tag{20}$$

Here the definition of the two-flux back-scatter fraction B has been used:

$$B = \frac{1}{2}\int\limits_0^1\int\limits_{-1}^0 p(\mu,\mu')\,d\mu'\,d\mu \tag{21}$$

This definition of the back-scatter fraction is recommended for use over the single, back-scatter fraction b, which is often quoted [18] in the form:

$$b = \frac{1}{2}\int\limits_{-1}^0 P(\phi)\,d(\cos\,\phi) \tag{22}$$

The two-flux back-scatter fraction B is consistent with the formulation of the two-flux model. The single, back-scatter fraction b is not. Figure 9.13 illustrates the difference between these two parameters. While the definition of B accounts for the various directions of incidence upon the particles, the definition of b assumes that all the radiation is traveling in either the direction forward ($\mu = 1$) or direct backward ($\mu = -1$) direction, as far as scattering is concerned.

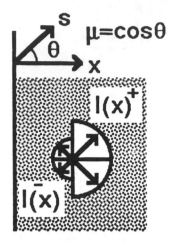

Figure 9.12 Two-flux model.

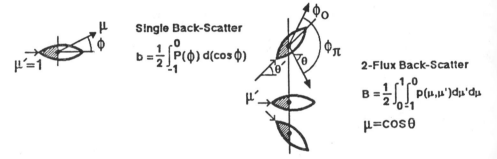

Figure 9.13 Single vs. two-flux back-scatter fraction.

Equations (9) and (21) have been integrated for the phase functions given in Equations (12) and (13) to yield the two-flux back-scatter fractions for specular and diffuse, geometric, spherical particles.

Specular: $B = 0.5$ (23)

Diffuse: $B = 0.667$ (24)

A word of caution is in order regarding the integration of Equation (9). The integrande is singular at the limits given in (10) and (11). This leads to significant errors if ordinary numerical integration is attempted. One way of avoiding this problem is to assume $P(\phi)$ is constant over small intervals of ϕ and to perform the integration of (9) analytically over each interval.

There are numerous applications of the two-flux model to be found in the literature. Flamant [3,4] used the two-flux model to predict radiant flux as a function of depth in a fluidized bed heated by concentrated solar radiation directed at

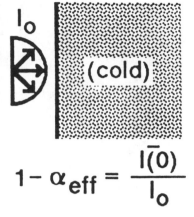

$$1 - \alpha_{\text{eff}} = \frac{\overline{I(0)}}{I_o}$$

Figure 9.14 Model for effective absorptivity.

the top of the bed (Fig. 9.2). While the experimental measurements and theoretical predictions for flux were in good agreement for highly absorbing materials (e.g., SiC) the agreement was poor for highly scattering materials (e.g., ZrO_2). This result is likely due to the assumption made of diffuse incident flux and the semi-isotropic assumption of the two-flux model when in fact the experimental incident flux the intensity distribution in the first few particle layers of the fluidized bed would be highly anisotropic. The tendency would be for this deficiency in the model to become more pronounced as the albedo or reflectivity of the particles increased. Therefore, the observations of good agreement at low albedos and poor agreement at high albedos certainly seem consistent. The Flamant studies confirm what has been observed by others [25,26], that the usefulness of the two-flux model is rather limited in situations involving collimated or highly directional incident radiation, highly anisotropic scattering particles, and/or high scattering albedos. The two-flux model has also been used to determine the effective absorptivity and emissivity of a semi-infinite slab composed of monodisperse, spherical particles [27]. The effective absorptivity was obtained by solving equations (19) and (20), omitting the emission source term, subject to the "cold-medium" boundary conditions, (25) and (26) (see Fig. 9.14):

$$I^+ (x = 0) = I_o \tag{25}$$

$$dI^+/dx \ (x \rightarrow \infty) = 0 \tag{26}$$

The effective absorptivity, defined by Equation (27), is given by Equation (28):

$$1 - \alpha_{e\!f\!f} = I^- (x = 0)/I_o \tag{27}$$

$$\alpha_{e\!f\!f}(\epsilon,B) = \left[\frac{\epsilon}{(1 - \epsilon)B} \left[\frac{\epsilon}{(1 - \epsilon)B} + 2 \right] \right]^{1/2} - \frac{\epsilon}{(1 - \epsilon)B} \tag{28}$$

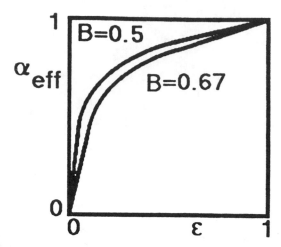

Figure 9.15 Effective absorptivity.

Figure 9.15 shows a plot of the effective absorptivity as a function of particle emissivity/absorptivity for $B = 0.5$ (specular particles) and $B = 0.667$ (diffuse particles). It can be seen from Fig. 9.15 that the effective absorptivity is always greater than the particle absorptivity, due to multiple scattering. Also, the effective absorptivity increases as back-scattering decreases. Finally, for any value of back-scattering fraction B, the effective absorptivity can be increased by increasing the absorptivity of the particles.

The effective emissivity was obtained by solving Equations (19) and (20), including the emission term, with I_o set equal to zero in Equation (25). To account for the particle temperature variation near the boundary, $x = 0$, an exponential temperature profile was assumed (Fig. 9.16):

$$\frac{T - T_o}{T_b - T_o} = 1 - \exp\left(-x/\delta\right) \tag{29}$$

In Equation (29) T_o is the particle temperature at the surface and T_b is the bulk temperature of the particles deep below the nonisothermal layer. Note from Fig. 9.16 and Equation (29) that T_o can be either less than or greater that T_b, to accommodate either heating or cooling of the particles. For these conditions, the effective emissivity, defined by Equation (30), is given in Equations (31–33) and in Fig. 9.17:

$$\epsilon_{eff} = I^-(x = 0)/I_b(T_b) . \tag{30}$$

$$\epsilon_{eff}(\epsilon, B, f_v\delta/d, T_o/T_b)$$

$$= \frac{\epsilon}{(1 - \epsilon)B}\left\{\sum_{n=0}^{4} \frac{4!}{n!(4-n)!}\frac{(\eta-1)^n}{(\varsigma + n)}\right.$$

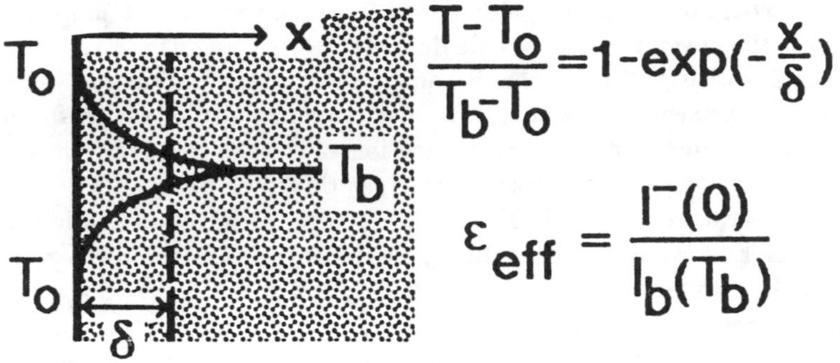

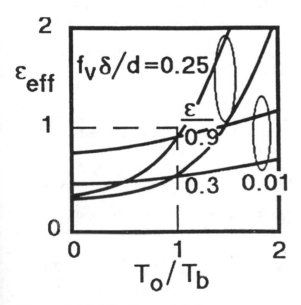

$$\frac{T-T_o}{T_b-T_o}=1-\exp\left(-\frac{x}{\delta}\right)$$

$$\varepsilon_{eff} = \frac{I^-(0)}{I_b(T_b)}$$

Figure 9.16 Model for effective emissivity.

Figure 9.17 Effective emissivity.

$$\left[\varsigma\left(1 + \frac{2B(1-\epsilon)^{1/2}}{\epsilon} + \right) + n\right] - \eta^4\right\} \tag{31}$$

where

$$\eta \equiv T_o/T_b \tag{32}$$

$$\varsigma \equiv 3(f_v\ \delta/d)\left[\epsilon(2(1-\epsilon)\ B + \epsilon)\right]^{1/2} \tag{33}$$

Four dimensionless parameters influence the effective emissivity. They are B

(back-scatter fraction), ϵ (particle emissivity), T_o/T_b (particle temperature at the surface over bulk particle temperature), and $f_v \, \delta/d$ (particle volume fraction times nonisothermal layer thickness divided particle diameter).

From inspection of Fig. 9.17 it can it can be seen which parameter variations are favorable in the sense of reducing the effective emissivity of the medium. If $T_o/T_b > 1$, as for example in the heating section of a solar receiver, then it would be desirable to reduce the nonisothermal layer thickness parameter $f_v \, \delta/d$ and the particle temperature at $x = 0$, T_o. If $T_o/T_b < 1$, as in an adiabatic section of a solar receiver, it would be desirable to increase $f_v \, \delta/d$, since for a given particle emissivity the curves in Fig. 9.17 cross at $T_o/T_b = 1$.

Hruby and Falcone [16,28] have used the two-flux solution of the transfer equation in connection with the two-dimensional particle energy and momentum equations to predict the temperature of a sand-size, free-falling particle medium subject to concentrated solar irradiation. It is difficult, however, to assess the role of the two-flux model in that study since the radiative properties were obtained from experimental measurements of transmission in packed beds and not from the fundamental particle properties.

Finally, Chen and Chen [29] have developed an analytical model for coupled radiation and conduction in fluidized beds. They employed the two-flux model for radiative transport and a transient energy equation with a statistical treatment to account for the alternate presence of bubbles and emulsion packets at the surface. Their study pointed out that while the conduction mechanism dominates near the wall, radiation dominates in the region farther away from the wall and needs to be included for proper heat transfer modeling.

Discrete ordinate method. In the method of discrete ordinates [17] the radiation field is divided into more than just two discrete streams in order to obtain improved accuracy over the two-flux model. They key to the discrete ordinate method is to replace the in-scattering integral term in the transfer equation with a numerical quadrature formula. Chandrasekhar [17] suggests the use of the Gauss quadrature formula

$$\int_{-1}^{1} f(\mu) \, d\mu \cong \sum_{i=1}^{N} w_i \, f(\mu_i) \tag{34}$$

where the weights are given by

$$w_i = \frac{1}{P_N(\mu_i)} \int_{-1}^{1} \frac{P_N(\mu)}{\mu - \mu_i} \, d\mu \tag{35}$$

and the divisions μ_i correspond to the zeroes of the Legendre polynomials $P_N(\mu)$. In (34) and (35) N is the total number of discrete streams or ordinates. The advantage of the Gauss formula is that the quadrature approximation is exact if $f(\mu)$ is a polynomial of order less than $2N$. Although the discrete ordinate method does not actually make the assumption of constant intensity over finite solid angles, as the two-flux model does, the intensity at the points of division $I(\mu_i)$, can be interpreted that way for visualization purposes (Fig. 9.18). For collimated

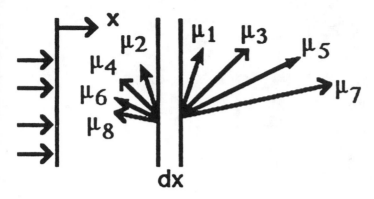

Figure 9.18 Method of discrete ordinates.

radiation incident normally on the particle slab the intensity will be independent of ψ. Using the discrete ordinate approximation in the ψ-independent form of the transfer equation for scattered intensity (18) results in Equation (36):

$$\mu_i \, \frac{dI(t,\mu_i)}{dt}$$

$$= -\, I(t,\mu_i) + (1 - \omega) \, I_b(t) + \frac{\omega}{2} \sum_{j=1}^{N} I(t,\mu_j) \, p(\mu_i,\mu_j) w_j$$

$$+ \frac{\omega}{4\pi} \, I_o \, p(\mu_i,1) \, \exp\,[-t] \tag{36}$$

Equation (36) is a system of simultaneous linear differential equations. The solution can be obtained as follows. A solution to the homogeneous system of equations is first found by assuming an exponential form of the solution. This leads to an eigenvalue problem. Once the eigenvalues and eigenvectors have been solved for, the nonhomogeneous solution can be found. This can be done relatively easily for the nonemitting case. For this case the only nonhomogeneous term in (36) is the source term due to collimated incidence. The particular solution can then be found by the method of undetermined coefficients by assuming an exponential solution. For wavelengths where the emission source term must also be included (e.g., near infrared and infrared) the particular solution can be found using the method of variation of parameters. The final step is to solve for the undetermined coefficient vector in the general solution by applying the boundary conditions. Further details of recent studies using the discrete ordinate method may be found in references [21, 26, and 30].

Workers at Sandia have systematically explored radiative transfer effects in a free-falling film of sand particles in a central cavity receiver using the discrete ordinate method. Houf and Greif [31] have performed a detailed study of the effect of radiative properties on the radiant transport in the particle curtain, neglecting momentum considerations. Hrudy and Falcone [16] have included two-dimensional particle momentum considerations to predict the effects of particle

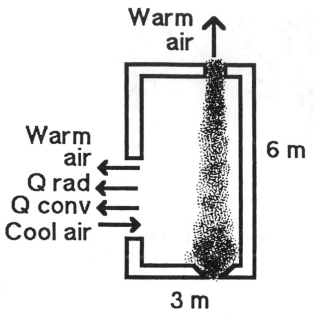

Figure 9.19 Solar cavity receiver (Evans et al. [32]).

mass flow rate and incident flux on cavity efficiency. And Evans et al. [32] have included the radiative interaction between the particle curtain and the cavity to create a detailed model of a solar central receiver with a free-falling particle curtain. Some specific findings of these studies will be discussed in the following paragraphs.

In [31] the effect of several radiative properties on local volumetric absorption by the particles was investigated. The analysis was carried out at the peak of the solar spectrum (0.5 μm) and hence emission by the particles was neglected. The incident radiation was assumed to be diffuse. The radiative properties varied were single scatter albedo, back-wall reflectivity, optical thickness of the particle curtain, and the scattering distribution (phase function). The principal finding was that darker particles (e.g., SiC with an albedo of 0.1) produce greater overall absorption of solar energy than lighter particles (e.g., Al_2O_3 with an albedo of 0.7). However, the local rate of absorption is less uniform for darker particles. This would have the effect of producing less uniform temperature profiles for darker particles, which would be accompanied by greater emission losses. A similar trend was also noticed for the optical thickness. For greater optical thicknesses the overall absorption would be greater but the heating would be less uniform. Rear wall reflectivity and scattering distribution appeared to play relatively minor roles compared with optical depth and albedo.

Evans et al. [32] modeled the flow of air and particles and the heat transfer inside a solar-heated, open cavity containing a falling curtain of 100–1000 μm solid particles. The cavity with the associated particle, air and heat fluxes, is depicted in Fig. 9.19. Detailed attention was given to two-phase flow and direction-dependent radiative transport effects. The efficiency for the cavity (increase in particle internal energy divided by incident energy) was predicted to be around 35% for an incident flux of 250 kW/m^2 and a particle mass flow rate of 1.0 kg/s. By increasing the incident flux to 500 kW/m^2 and the particle flow to 1.5 kg/s the efficiency was predicted to increase to 42% with the particle exit temperature remaining near 1150 K. A further increase of incident flux and particle flow to 1000 kW/m^2 and 2.5 kg/s, coupled with a smaller cavity, resulted in an efficiency of 72%. It appears that the losses are split approximately equally between the particle-to-air convective loss and radiative emission loss. Both are substantial and both must be reduced to realize acceptable solid particle cavity receiver performance. Increasing the particle mass flow rate appears to decrease the convective loss by forcing the particles to remain closer together during free-fall. However, the fluid mechanics dictate an upper limit to the particle volume fraction that can be achieved during free-fall. The particles always accelerate after being released and therefore spread out in the streamwise and spanwise directions. This makes it difficult to reduce the convection loss beyond a certain limit without going to another flow configuration (e.g., fluidized bed). Decreasing the radiative loss is also an important consideration and will be discussed later in the section on the effect of important variables on radiative transport.

4.3 Other Models

Many radiative models other than the transfer equation have been applied to particulate media, particularly fluidized beds. For various reasons they are not as appropriate as the transfer equation, however, and will be mentioned only briefly here.

The alternate slab model [33] consists of alternate slabs of gas and solid, with empirically determined spacing. The pile model [34] is very similar with an alternating assembly of reflecting, absorbing, and emitting plates. These, together with various packet and cell models [35] that have also been developed, all suffer from the drawback that the properties of the cell, packet, slab, or pile must be empirically related to the fundamental particle properties. This requires extensive experimentation.

The Monte Carlo technique has also been applied to analyze radiative transfer in particulate media [36,37]. While this technique seems to be suitable for packed beds that have predictable, albeit complicated, geometries, it does not seem suitable for entrained, fluidized, or free-falling particle flows that do not have predictable, fixed particle orientations.

5 EFFECT OF IMPORTANT VARIABLES ON RADIATIVE TRANSPORT

In this section the influence of various parameters on the radiative heating of the particles will be discussed. This topic can be approached from a pure radiative point of view and from an overall receiver efficiency point of view. From a purely radiative point of view the primary concern would be how to vary the parameters to maximize the receiver efficiency (which itself may be defined in various ways). The second point of view includes not only the radiative transfer considerations but also the particle-gas fluid mechanics and convective heat transfer. Obviously the second point of view is inclusive of the first and is the more important. Yet it is instructive to consider first radiative transfer alone, as much as possible, and add convective effects afterward.

As stated above, the primary concern of the purely radiative point of view is to maximize absorption of solar energy and minimize thermal infrared emission losses by the particles. This is equivalent to maximizing the effective absorptivity in Fig. 9.15 and minimizing the effective emissivity in Fig. 9.17. Although the results in Fig. 9.15 and 9.17 were obtained for a semi-infinite medium with an assumed exponential temperature profile, the trends would also apply to a slab of finite optical thickness with a temperature profile determined by energy conservation.

To maximize absorptivity of the particle medium, Fig. 9.15 indicates that it would be desirable to use particles with a high emissivity and a small back-scatter fraction. Thus specular SiC would be preferable to diffuse Al_2O_3 or SiO_2. To minimize effective emissivity of the particle medium, Fig. 9.17 indicates that it would be desirable to use particles with a low infrared emissivity. This trend is opposite that which is favorable for maximum absorption. Assuming the particle emissivity is independent of wavelength, both effective emissivity and absorptivity cannot be optimized with respect to particle emissivity. It is conceivable that a spectrally selective particle would be of value for these purposes, but this point will be taken up later. Given that relatively gray particles were selected for use it would be better to use a high emissivity particle and maximize absorption. Although this would tend to increase emission loss from the particles the net energy gain by the particles would still be greater.

Figure 9.17 also indicates two other nondimensional parameters that can greatly influence the effective emissivity. These are the particle volume fraction times nonisothermal thickness over particle diameter, $f_v \delta / d$, and the ratio of the particle temperature at the surface of the cloud to the (assumed) constant temperature deep in the cloud T_o / T_b. Reducing T_o / T_b always lowers the effective emissivity. For $T_o / T_b > 1$ reducing $f_v \delta / d$ lowers the effective emissivity whereas for $T_o / T_b < 1$ increasing $f_v \delta / d$ lowers ϵ_{eff}.

Referring to Equations (4), (5), and (15), the parameter $f_v \delta / d$ can be interpreted approximately as the optical depth based on nonisothermal layer thickness t_δ. It would therefore seem desirable (for $T_o / T_b > 1$) to simply reduce t_δ from say 1 to 0.1 or 0.01 in order to reduce emission losses. The difficulty with doing this is that for entrained and free-falling particle flows, due to inherently poor lateral particle mixing, the nonisothermal layer thickness δ, and the slab thickness L are of the same order, i.e., $t_\delta \approx t_L$. To maintain efficient solar absorption t_l must be at least of the order of 1. Reducing $t_\delta (\approx t_L)$ from 1 to 0.1 or 0.01 would reduce the absorptivity of the particle curtain to unacceptably low levels. The key to reducing t_δ without reducing t_L is to somehow introduce sufficient lateral particle mixing so that the nonisothermal layer is confined to a narrow region at the surface of the particle curtain ($\delta \ll L$). The fluidized bed, with vigorous bubble-induced lateral mixing, seems to be well-suited for this purpose. The superiority of fluidized beds over packed beds for maintaining a more uniform temperature profile and reducing emission loss has already been established by Flamant and Olalde [4]. Other schemes that induce lateral particle mixing in entrained and free-falling particle flows are also conceivable and should be investigated. Finally, as a result of enhanced lateral particle movement, T_o/T_b would also be decreased, which would further help in reducing the effective emissivity of the particle curtain.

From the viewpoint of overall receiver performance, the main objective is to maximize the efficiency of the receiver. Depending on whether gas heating is included, the efficiency can be defined at least two ways. If the objective of the process is to heat the particles only, the gas heating should not be included in the efficiency. If the heated gas can be recovered and used profitably or if the objective of the process includes heating of the gas, it would be logical to include that energy term in the efficiency.

Evans et al. [32] considered particle heating only to be the objective of the process. Their comprehensive numerical study of particle and air flow and heat transfer in an open cavity receiver point out some interesting effects. They point out that since smaller particle diameters mean larger optical thicknesses and larger particle residence times in the cavity, greater cavity efficiencies might also be expected. However, their results show that these favorable trends are offset by the increase in convective heat loss to the air, which is very sensitive to particle diameter. Overall, cavity receiver efficiency for free-falling sand particles appears to be relatively insensitive to particle size, for the range of sizes studied (100–1000 μm).

Evans et al. [32] also point out an interesting effect associated with particle single-scattering albedo. It was suggested earlier in this section that spectrally selective particles might offer some advantage over gray particles. That is, particles with a low solar albedo (high solar emissivity) and a high infrared albedo (low infrared emissivity) would be both efficient absorbers of solar energy and poor emitters of their own infrared thermal energy. The results of [32] indicate, however, that this is not true. As the infrared albedo was increased, with solar albedo held constant, the cavity efficiency actually decreased, instead of increasing. This was attributed to a decrease in the ability of the particle curtain to absorb the significant amount of infrared radiation being emitted by the inside cavity walls. It appears then that, at least in a cavity receiver configuration, particles that are black over the entire wavelength spectrum are the most effective.

6 SUMMARY AND RECOMMENDATIONS

To summarize, the high-temperature (radiative) heat transfer characteristics of three types of direct-contact gas-particle flow configurations have been examined:

entrained flow, free-falling flow, and fluidized bed. Each configuration has its peculiar strengths and weaknesses.

The entrained flow configuration makes use of very small particles, typically micron-size or less. This configuration is therefore suitable for applications where heating of the carrier gas is the objective of the process (e.g., chemical process or Brayton cycle). In general, the particles would be used in a once through fashion. From a radiative standpoint, the particle sizes used in entrained flows can range from the Rayleigh limit $(d < 0.1\ \mu m)$ to the geometric optics limit $(d > 2\text{-}3\ \mu m)$. The smaller Rayleigh particles are more favorable than larger particles because back-scattering loss of incident radiation is reduced. However, lateral mixing is generally poor in entrained flows, which results in less uniform heating. Less uniform heating means higher surface to bulk particle temperature ratios and larger nonisothermal layer thicknesses, both of which mean higher infrared emission loss from the medium.

The free-fall configuration makes use of relatively large particles, 500–1000 μm. This configuration is therefore suitable for applications where heating of the particles is the objective. To be cost-effective this method would have to overcome significant dusting and sintering problems as well as develop efficient methods for utilizing the convectively heated gas. Poor lateral mixing of particles also means that this flow configuration is prone to high infrared emission losses.

The fluidized bed configuration uses intermediate to large particles (100–1000 μm). This type of flow is suited for applications where either particle or gas heating or both are the objective. Vigorous lateral particle mixing in fluidized beds means emission losses are lower. Also, high particle loadings occur naturally, so achieving sufficient optical depth for optimum efficiency is not problem. The main limiting feature of fluidized beds as solar receivers is the material limitation of the transparent containing wall. Not only is the window material apt to undergo severe optical degradation over the life of the device, but it must also withstand the maximum temperature of the system. (The window would be subject to higher temperature than the refractory particles.) A "windowless" fluidized bed with fluidic or other means of particle ·containment is an attractive possibility for overcoming this limitation.

The solid particle solar receiver is one application of high-temperature direct-contact solid-gas heat transfer in which radiative transport would play a key role. Efficiencies of these devices are still lower than those for the more-developed molten salt and liquid metal technologies. However, significant improvement could be made in relatively short time. The experimental studies with solid particle receivers done to date have been at the proof-of-concept level. Measured efficiencies have been expectedly lower than predicted efficiencies. In general, entrained and fluidized efficiencies, which are based on gas heating, have been higher than free-fall efficiencies, which are based on particle heating. As the fundamental mechanisms of fluid-particle interaction are better understood, measured efficiencies are certain to increase significantly. At this stage, however, there is at least one significant feasibility problem with each of the methods proposed. Rather than focus on a particular configuration at this point it would be more beneficial to do exploratory research, investigating widely different fluid-particle flow arrangements that have the potential for dramatic improvements in performance.

Modeling of radiative transport in gas-solid direct-contact devices is, of itself, fairly well understood. Given sufficient computer time even complicated directional and spectral variations can be accounted for. The biggest lack of understanding right now is in the area of coupled momentum and energy effects, which certainly will play a key role in many of these devices at high temperature.

In conclusion, some specific recommendations for future work in this area are given:

1.   Emphasize exploratory research which makes use of innovative, untried particle-gas flow configurations.

2.   Introduce lateral mixing in entrained and free-falling particle flows to reduce infrared emission losses.

3.   Look at "windowless" means for containing fluidized beds, e.g., fluidic, acoustic, electrostatic, etc.

4.   Develop efficient methods for recovering and using the heated gas in fluidized beds which do not interfere with solar irradiation.

5.   Concentrate particle selection efforts on materials as close to black across the entire spectrum as possible. Selective absorption seems unjustifiably costly when infrared emission loss can be reduced through lateral flow mixing.

6.   Modify existing solutions of the radiative transfer equation to include azimthally non-symmetric incident radiation so that they will better represent concentrated solar receiver towers, etc.

7.   Emphasize research which improves the understanding of coupled momentum and energy effects at high temperature.

8.   Develop experimental techniques for independent particle and gas temperature measurement such as an infrared fiber optic probe.

REFERENCES

1.   Martin, J. and Vitko, J., Jr., *ASCUAS: A Solar Central Receiver Utilizing a Solid Thermal Carrier*, Sandia Report, SAND82-8203, January 1982.

2.   Martin, J., *Solid Thermal Carriers for High Temperature Solar Applications*, International Seminar on Solar Thermal Heat Production and Solar Fuels and Chemicals, W. Hoyer, Editor, German Aerospace Research Establishment, Oct. 13-14, 1983, Stuttgart, Germany.

3.   Flamant, G., *Theoretical and Experimental Study of Radiant Heat Transfer in a Solar Fluidized-Bed Receiver*, AIChE J., Vol. 28, No. 4, July 1982, pp. 529-535.

4.   Flamant, G. and Olalde, G., *High Temperature Solar Gas Heating Comparison Between Packed and Fluidized Bed Receivers-I*, Solar Energy, Vol. 31, No. 5, 1983, pp. 463-471.

5.   Flamant, G., Olalde, G., and Gauthier, D., High Temperature Solar Gas-Solid Receivers, *Alternative Energy Sources V. Part B: Solar Applications*, editor by T. N. Veziroglu, Elsevier Science Publishers, B. V. Amsterdam, 1983.

6.   Bachovchin, D. M., Archer, D. H., Keairns, L. M., and Thomas, L. M., *Design and Testing of a Fluidized-Bed Solar Thermal Receiver*, Final Report by Westinghouse R & D Center and Georgia Institute of Technology to the Solar Energy Research Institute, August 1980 (Subcontract No. XP-9-8321-1).

7. Bachovchin, D. M., Archer, D. H., Neale, D. H. Brown, C. T., and Lefferdo, J. M., *Development and Testing of a Fluidized Bed Solar Thermal Receiver*, Proc. 1981 Annual Meeting - Amer. Section International Solar Energy Society, TJ810-T56-81.

8. Neale, D. H. and Cassanova, R. A., *Solar Thermal Hydrogen Production with a Direct Flux Chemical Reactor*, presented at 6th Miami International Conference on Alternative Energy Sources, Dec. 1983.

9. Neale, D. H. and Cassanova, R. A., *Water Gas Production with a Solar Thermal Direct Flux Chemical Reactor*, presented at 22nd ASME/AIChE National Heat Transfer Conference, Aug. 1984.

10. Hunt, A. J., Ayer, P. H., Miller, F., Russo, R., and Yuen, W., *Solar Radiant Processing of Gas-Particle Systems for Producing Useful Fuels and Chemicals*, presented at 23rd National Heat Transfer Conference ASME/AIChE, Denver, CO., Aug. 4-7, 1985.

11. Hunt, A. J., *A New Solar Thermal Receiver Utilizing a Small Particle Heat Exchanger* LBL Report LBL-8520, presented at 14th Intersociety Energy Conversion Engineering Conference, Boston, MA, Aug. 5-19, 1979.

12. Fisk, W. J., Wroblewski, D. E., Jr., and Hunt, A. J., *Performance Analysis of a Windowed High Temperature Gas Receiver Using a Suspension of Ultrafine Carbon Particles as the Absorber*, LBL Report LBL-10100, presented at American Session of International Solar Energy Society Annual Meeting, Phoenix, AX, June 2-6, 1980.

13. Hunt, A. J. and Brown, C. T., *Solar Test Results of an Advanced Direct Absorption High Temperature Gas Receiver (SPHER)*, LBL Report LBL-16497, Proceedings of the 1983 Solar World Congress, Perth, Australia, Aug. 15-19, 1983.

14. Hunt, A. J., *Solar Radiant Heating of Small Particle Suspensions*, LBL Report LBL-14077, Symposium Series *Fundamentals of Solar Energy*, Vol. 3, 1982.

15. Hruby, J. M. and Steele, B. R., *Examination of a Solid Particle Central Receiver: Radiant Heat Experiment*, presented at the Solar Energy Conference, Knoxville, Tennessee, March 1985.

16. Hruby, J. M. and Falcone, P. K., *Momentum and Energy Exchange in a Solid Particle Solar Central Receiver*, to be presented at 1985 ASME/AIChE National Heat Transfer Conference, Denver, Colorado, August 5-7, 1985.

17. Chandrasekhar, S., *Radiative Transfer*, Dover, New York, 1960.

18. Hottel, H. C. and Sarofim, A. F., *Radiative Transfer*, McGraw-Hill, New York, 1967.

19. Ozisik, M. N., *Radiative Transfer and Interactions with Conduction and Convection*, Wiley-Interscience, New York, 1973.

20. Siegel, R. and Howell, J. R., *Thermal Radiation Heat Transfer*, 2nd ed., McGraw-Hill, New York, 1981.

21. Brewster, M. Q. and Tien, C. L., Radiative Transfer in Packed/Fluidized Beds: Dependent vs. Independent Scattering, *J. Heat Transfer*, Vol. 104, No. 4, Nov. 1982, pp. 573-579.

22. Schuster, A., Radiation Through a Foggy Atmosphere, *Astroph. J.*, Vol. 21, pp. 1-22, 1905.

23. Hamaker, H. C., *Phillips Research Reports*, Vol. 2, pp. 55, 103, 112, 420; 1947.

24. Chu, C. M. and Churchill, S. W., *J. Phys. Chem.*, Vol. 59, pp. 855-863, 1955.

25. Brewster, M. Q. and Tien, C. L., Examination of the Two-Flux Model for Radiative transfer in Particulate Systems, *Int. J. of Heat and Mass Transfer*, Vol. 25, No. 12, Dec. 1982, pp. 1905-6.

26. Daniel, K. J., Laurendeau, N. M., and Incropera, F. P., Prediction of Radiation Absorption and Scattering in Turbid Water Bodies, *J. of Heat Transfer*, Vol. 101, Feb. 1979, pp. 63-67.

27. Brewster, M. Q., *Effective Emissivity of a Fluidized Bed*, presented at ASME Winter Annual Meeting, New Orleans, LA, Dec. 9-14, 1984, HTD-Vol. 40, pp. 7-13.

28. Falcone, P. K., Noring, J. E., and Hruby, J. M., *Assessment of a Solid Particle Receiver for a High Temperature Solar Central Receiver System*, Sandia National Laboratories, SAND85-8208, 1985.

29. Chen, J. C. and Chen, K. L., Analysis of Simultaneous Radiative and Conductive Heat Transfer in Fluidized Beds, *Chem. Eng. Commun.*, Vol. 9, 1981, pp. 255-271.

30. Hottel, H. C., Sarofim, A. F., Vasalos, I. A., and Dalzell, W. H., Multiple Scatter: Comparison of Theory with Experiment, *J. Heat Transfer*, Vol. 92, 1970, pp. 285-291.

31. Houf, W. G. and Greif, R., *Radiative Transfer in a Solar Absorbing Particle Laden Flow*, presented at ASME/AIChE Heat Transfer Conference, Denver, Colorado, August 5-7, 1985.

32. Evans, G., Houf, W., Greif R., and Crowe, C., *Particle Flow within a High Temperature Solar Cavity Receiver Including Radiation Heat Transfer*, presented at ASME/AIChE Heat Transfer Conference, Denver, Colorado, Aug. 5-7, 1985.

33. Kolar, A. K., Grewal, N. S., and Saxena, S. C., Investigation of Radiative Contribution in a High Temperature Fluidized-Bed Using the Alternate-Salb Model, *Int. J. Heat Mass Transfer*, Vol. 22, 1979, pp. 1695-1703.

34. Borodulya, V. A. and Kovensky, V. I., Radiative Heat Transfer Between a Fluidized Bed and a Surface, *Int. J. Heat Mass Transfer*, Vol. 26, No. 2, 1983, pp. 277-287.

35. Baskakov, A. P., Berg, B. V., Vitt, O. K., Filippovsky, N. F., Kirakosyan, V. A., Goldobin, J. M., and Maskaev, V. K., Heat Transfer to Objects Immersed in Fluidized Beds, *Power Technology*, Vol. 8, 1973, pp. 273-282.

36. Yang, Y. S., Howell, J. R., and Klein, D. E., Radiative Heat Transfer Through a Randomly Packed Bed of Spheres by the Monte Carlo Method, *J. Heat Transfer*, Vol. 105, May 1983, pp. 325-332.

37. Abbasi, M. H. and Evans, J. W., Monte Carlo Simulation of Radiant Transport Through an Adiabatic Packed Bed or Porous Solid, *AIChE J.*, Vol. 28, No. 5, Sept. 1982, pp. 853-854.

DIRECT-CONTACT HEAT TRANSFER
IN SOLID-GAS SYSTEMS

James R. Welty

Heat transfer between solid particles and a gas in direct contact is a subject as old as when solid fuel combustion was first observed. The combustion phenomenon was the principal application of this direct-contact process for many years and may remain so today. The thrust of the current workshop has been the examination of fundamental heat transfer phenomena, however, so the situation with chemically reacting species will not be considered here in depth. We are essentially considering the state of knowledge and continuing needs for describing the exchange between a gas and solids—either particles, solid boundaries, or both—when there are temperature differences between the media.

All of the basic heat transfer modes are present in gas-solid systems. Conduction will occur at points of contact between particle surfaces and other boundaries whether they be container walls or internal heat-exchange surfaces. Convection will naturally always be present since the gas will have some sort of motion and this motion will affect the energy transport directly. A "gas" will generally refer to a single phase, however, with quite small particles present, a more-or-less homogeneous emulsion may be treated as a dense gas phase. Radiation will be a significant heat transfer mode if relatively large temperature differences exist across a heat-transfer path. Such effects normally occur between particles that are at or close to the bed temperature and solid boundaries such as internal surfaces.

Heat transfer processes in gas-solid systems are intimately associated with the relative motions of the phases; this is the major challenge in this area—that of understanding and describing gas and solid-particle motions. When one observes, visually, a bed of particles in motion as a result of fluid interaction the complexity of this process is readily apparent. The process is chaotic and is affected by numerous variables such as particle size and distribution, fluid characteristics—principally as a function of temperature, particle properties, bed geometry, the presence and geometric arrangement of bed internals, and the manner in which the bed particles are confined and fluidized. A phenomenon of extreme importance in this regard is that involving "bubbles" of particle-free gas that exist in fluidized beds and may be fundamental in affecting the heat transfer.

Following the acceptance of the notion that this is a complex business, we are left with a range of needs to satisfy different audiences. Academicians and others whose approaches are basic wish to understand gas-solid systems sufficiently well that heat transfer and, obviously, the motions of the phases can be described from first principles, given a few system parameters. Practitioners are interested in operational information adequate for describing a process already in existence or for designing a system to satisfy a definite operational need. Unfortunately, the state of knowledge at present leaves us quite a distance from satisfying any of these needs in complete fashion.

1 FLUID FLOW MECHANISTIC CONSIDERATIONS

Some introductory remarks have already been made on this subject. Chen describes the process whereby a bed of particles becomes fluidized by the upward flow of a gas. Beyond the velocity at which minimum fluidization occurs bubbles begin to form at the distributor plate and their upward motion, whether "fast" or "slow," will influence heat transfer in a major way. "Bubbles," in the fluidized-bed sense, are regions where the gas is free of particles over a distance that is large compared with the size of particles. In contrast to "bubbles" in the usual gas-liquid sense, ours are regions through which gas is flowing. In a case where gas convection is significant in a heat transfer sense, the effective heat transfer coefficient will be much different when the gas is flowing rapidly between the interstices of adjacent particles and/or internal surfaces than when flow is relatively slow within a bubble. Bubbles themselves are subject to some complex effects. Chen describes the relative motion of bed particles when bubble flow is "fast" or "slow." This motion is, as yet, not fully predictable. Certainly, the resulting particle motion is of interest for heat transfer purposes.

It is well known that a horizontal tube immersed in a fluidized bed will, at relatively low superficial velocities, experience variable effects around its periphery. At the bottom a pseudo-stagnation point effect will exist. Around the sides the gas-solid-boundary motion is quite dynamic. Near the top a "stack" of stationary particles will form and remain in place until being displaced by a passing bubble. This stack region is one of low heat transfer; thus if the stack is not displaced relatively frequently, the average heat transfer for the cylinder will be reduced markedly. This stacking phenomenon will exist for arrays of horizontal tubes as well as single tubes; however, bubble motion will naturally be quite different as the tube arrangement becomes more involved and concentrated.

The mechanism of flow of each phase is also complex and directly related to particle size. When particles are small enough they may be suspended in the gas to the extent that the resulting emulsion is described adequately as a single phase. Even when the gas and solid phases are quite distinct, particles that are relative small, and thus of nominal thermal mass, will experience a relatively rapid temperature change when in the vicinity of a surface that is at a temperature appreciably different from that of the bed. Large particles with considerable thermal mass will remain nearly isothermal under similar conditions unless their residence time is quite long. Clearly, these different cases will be associated with appreciably different heat transfer rates. Currently, there is no complete model to connect residence time, particle size, and other parameters in a way that can be used to predict performance. The terms "large" and "small" particles are, likewise, somewhat ambiguous terms.

2 THEORETICAL MODELS

Some phenomenological background has been described already concerning gas-solid interactions that affect heat transfer. It might be appropriate here to separate radiation concerns from those relating to conduction and convection.

Brewster's paper is an excellent overview of the ideas and approaches associated with radiant energy exchange in these systems. The major complexity in the fluidized bed case concerns again fluid flow effects and complicated geometries. For example, a complete radiation analysis between an immersed surface that is cooler than the surrounding fluidized bed and the bed itself must consider the following:

1)   Is the bed "heat source" composed effectively of hot particles and gas at a uniform temperature or are the particles, nearer to the cool surface, at a temperature significantly below that of the bed region?

2)   Does the surface see an emulsion of gas and particles that has a relatively constant fraction of particles per unit volume or does the surface see a particle string that is the effective boundary of a passing bubble?

3)   Do the interacting heat transfer mechanisms result in a process that is time-variant over an interval that is comparable to particle residence times in the vicinity of an immersed surface?

4)   What properties of the particles, gas, and solid boundaries are significant in this process, and how are they affected by the temperature changes involved?

5)   At what range of ΔT does radiation become a significant effect?

Other relevant questions might also be asked on this same subject. Radiation is obviously an effect that must be evaluated when high-temperature operation is considered. The relative importance of radiation effects is agreed to be at least 10% of the total heat transfer when beds operate at temperatures above 800 K.

Experimental data under these conditions are quite sparse so this remains an area of continuing research need.

Models that consider conduction and convection are of essentially two types: the packet or surface-renewal model and the large-particle or gas-convection-dominant model.

The surface-renewal models effectively treat a "packet" of gas and solid particles that moves from the bed to a region near an immersed surface. While near a surface the packet undergoes a temperature change according to a transient-conduction-type analysis over the effective residence time. Following such a heat transfer process in which the packet's temperature change occurs, it is swept away to be replaced by another, and the process continues. The internal surface thus has regular "renewal" of its heat transfer source or sink after some effective residence times.

The gas-convective-dominant models are principally valid for large-particle/short-residence-time cases, where the gas transfers energy between the particles and the solid boundary of interest. Particles will change in temperature very little, if at all, and accurate depictions of the gas flow field both in the interstices between particles and through a bubble are necessary.

Both of these mechanistic approaches to modeling have met with limited success. Any universal model describing the motion of phases and the related heat transfer is, as yet, remote. Much work remains to be done in this area.

3 EXPERIMENTAL PROGRAMS

A decided majority of the experimental work done in the fluidized bed area has been at low temperatures, where radiation effects are negligible. Most work has also been at a relatively small scale, of the "bench top" variety. While much has been learned and much potential remains for such experiments, there is a continuing need for results from larger-scale and higher-temperature experiments. Some experimentation of this sort has been performed in industrial laboratories but only a limited amount of data have been published.

There are always major challenges confronting experimenters. This area possesses more than its share. Among them are

1) The use of a test section with a geometry that is physically meaningful. Thus far most experiments have been conducted in small equipment and usually in a two-dimensional configuration.

2) The use of sensors that are sufficiently rugged to hold up under a dynamic environment yet able to yield precise results.

3) Design of instruments that can provide specific information, i.e., a device that will identify the presence of a bubble or a sensor that will yield independent information concerning radiation heat transfer without convective/conduction coupling.

4) Acquisitions and analysis of continuous data from a large number of sensors; spectral analysis has the potential to provide considerable insight to fluidized bed behavior.

5)   Both long- and short-term description of solid motion is needed. Chen has described some excellent work done in acquiring long-term information of this type.

At the present time, there is no good correlation available for a designer to use in sizing a fluidized bed for a specific heat transfer application. Data that will provide such correlations for heat transfer coefficients as functions of general operating parameters are simply not available at this writing. A comprehensive experimental program to establish such a data base is of absolute highest priority.

4 CONCLUSIONS AND CHALLENGES

Brewster and Chen have presented interesting and helpful papers describing areas where some knowledge exists and somewhere significant research is in progress. The area of gas-solid direct-contact heat transfer is in such a state of need, however, that it represents a fruitful area for large numbers of researchers on many fronts.

In each of the preceding sections I have attempted to relate basic approaches to generally posed problem areas with a common expression that there is still much to be done and a great amount that is yet unknown.

DIRECT-CONTACT EVAPORATION

D. Bharathan

1 INTRODUCTION

Evaporation of a liquid occurs when molecules escape from the main body of the liquid due to thermal agitation. The escaping molecules move with sufficient speed to break through the interfacial surface tension; i.e., they possess kinetic energy exceeding the work function of cohesion at the surface. Since only a small portion of the molecules is at any instant located near enough to the surface and moving in the proper direction to escape, the rate of evaporation is limited.

As the faster molecules emerge, those left behind possess less average energy, thereby lowering the temperature of the liquid. If evaporation takes place in a closed vessel, the escaping molecules accumulate as vapor above the liquid. Many of them return to the liquid; such returns are more frequent as the density and pressure of the vapor increases. The process of escape and return eventually reaches equilibrium when external energy or work transfer ceases. The vapor is then said to be "saturated," its density and pressure no longer increase and the cooling effect ceases.

Evaporation is a major chemical engineering unit operation for separating liquids and solids and, in particular, recovering solute from the solvent (frequently water). The pulp and paper industry is a large user of evaporation equipment. Evaporation is also used extensively in the production of table, industrial, and

203

other salts, in caustic chlorine production, in the phosphate industry, and in food processing. Evaporation is, in principle, the same operation as plain distillation and fractional distillation for a mixture of varied volatile liquids. Vacuum evaporation is frequently used in single or multiple stages with each successive stage operated at an increasing vacuum using the vapor's heat of condensation from the preceding stage. Multiple stage evaporators offer a savings in the operating cost of heat and an increased expenditure for the equipment. Combined high-vacuum and very low-temperature evaporation or drying is used in the final stage of removing water vapor from frozen penicillin, due to the heat-sensitive nature of this material.

Mechanical engineering applications of the evaporation principle are also wide. Cooling towers for rejecting heat from large power plants are perhaps the largest man-made mass-transfer devices found in engineering use. The evaporative cooling principle has been known to man for centuries. Recent energy conservation measures have renewed the interest in direct and indirect evaporative coolers for HVAC applications.

Global temperature extremes are moderated by the evaporative mass flux caused by the incident solar radiation from the equatorial zones toward the polar regions. Atmospheric evaporation and condensation on a major scale affect the day-to-day weather.

The transition from liquid to vapor under nonequilibrium is generally termed evaporation. Sublimation, which refers to the transition from the solid phase directly to vapor, will not be treated in this paper. For large departures from equilibrium, evaporation is often accompanied by discrete vapor bubbles forming within the liquid continuum. This process, termed boiling, is a vast subject in its own right. In this paper we shall confine our discussion to evaporation processes where bubbling does not play a major role.

2 INTERFACIAL PROCESSES

2.1 Limiting Rate of Evaporation

Consider a pure substance liquid and vapor in equilibrium. The pressure exerted by the vapor is equal to the vapor pressure corresponding to the prevailing temperature. No net transport of molecules occurs between the vapor and the liquid. The rate at which the vapor molecules strike the liquid surface is readily calculated from kinetic theory. Because of the high speed of the molecules, only some fraction α remain in the liquid; the remainder rebound back into the vapor space. The fraction α is commonly termed the "accommodation" coefficient.

This reasoning led Hertz (1882), Langmuir (1913), and Knudsen (1915) to the following expression for the maximum possible rate of transport from the surface to the vapor:

$$N_A = 1006 \, \alpha \, (2\pi MRT_s)^{-1/2} \, (p_s - p_g) \text{ g—mol/s cm}^2 \tag{1}$$

where p is in atmospheres and T is in degrees Kelvin.

Assuming it is not influenced by the presence of vapor, the rate of evaporation into absolute vacuum must then proceed at a finite rate.

Schrage (1953) pointed out the importance of this phenomenon as it relates to chemical engineering. He expressed the interfacial resistance $1/k_i$ as

$$\frac{1}{k_i} = \frac{(2\pi MRT_s)^{1/2}}{1006\alpha} \tag{2}$$

where k_i is the mass transfer coefficient for transport across the interface expressed in units of gram moles per second per square centimeter per atmosphere. For water at $20\,^\circ$C with $\alpha = 1$, we note that $k_i = 0.612$ g mol/s cm^2 atm, or 14 700 cm/s. The possible importance of this interfacial resistance clearly depends on the magnitude of other resistances in series. It is generally quite small if the mass-transfer rate is small. Equation (1) becomes important in practice only when the transfer rates are exceptionally high which is uncommon to industrial practice.

The practical application of Eq. (1) requires values for α. There are no useful theories to predict or easy means to experimentally measure the value for α. The surface temperature measurement needed for inferring α using probes leads to errors due to a substantial temperature gradient near the surface. Published values of α for liquids range from 1.0 to 0.02; however, surface temperature measurements leading to these inferences are highly questionable (Sherwood et al., 1975). Maa (1967), using a laminar jet and an ingenious method not requiring a probe to measure the surface temperature, obtained α values of nearly 1.0 for water and a few other simple liquids. It is conceivable that most of the published data are in error, and α is essentially unity for all simple liquids.

2.2 Other Interfacial Phenomena

Interfacial turbulence: Surface ripples and interfacial turbulence have been observed in liquid interfaces in contact with gases as well as in points of contact of two liquids. This form of turbulence is not that induced by bulk fluid motion familiar to most of us. It appears that these phenomena are always associated with simultaneous mass transfer, and the effects are more pronounced when the mass transfer is rapid. They are most common in ternary or multicomponent systems but also noted in some partially miscible binary systems. In some cases the surface activity is strong with mass transfer in one direction but completely absent in the other direction. The most pronounced interfacial turbulence is observed when a chemical reaction occurs simultaneously with mass transfer, as in the extraction of acetic acid from i-butyl alcohol using water containing ammonia (Sherwood and Wei, 1957). A number of cases have been reported where interfacial turbulence increases the mass-transfer rate several fold. These phenomena are by no means well understood. They evidently stem from random variations of interfacial tension, which result from local concentration variations as mass transfer occurs and so depend in part on the rate of interfacial tension change with the solute concentration. The instability that develops causes ripples and sometimes regularly shaped roll cells, which create circulation between the surface and bulk liquid.

This is known as the Marangoni effect (Bikerman, 1948). Levich and Krylov (1969) provide a review of surface-tension-driven phenomena, including the role of Marangoni effect on the mass transfer at the phase boundary.

Surfactants: Many substances in solution tend to concentrate at the liquid surface and change the interfacial tension. Even a monolayer on the surface develops a structure that immobilizes the surface. The presence of such a layer reduces or eliminates surface-tension-driven turbulence and introduces a surface resistance to mass transfer across the interface. The reduction in mass transfer can be large. Evaporation rate from a beaker of water containing a drop of hexa-deconol (cetyl alcohol) at room temperature is reduced by as much as 75% from that of pure water. The role played by surfactants in industrial mass-transfer equipment has been studied extensively (Davies and Rideal, 1963). In gas-liquid contactors with short contact times, the surfactant does not diffuse rapidly to the surface to form an adsorbed barrier. Consequently, agitated systems with rapid surface renewal (such as packed columns) show little effect of added surfactants. However, quiescent evaporative processes, such as in evaporative coolers and residential humidifiers, may suffer a significant reduction in the performance due to the presence of impurities, surfactants, and scale buildup.

3 SIMULTANEOUS HEAT AND MASS TRANSFER

For evaporation of a pure substance into a gas stream, the mechanism of heat transfer can be described in three parts: as the heat transfer from the bulk liquid to the interface at an intermediate temperature, the accompanying molecular crossover mass transfer from the interface to the vapor, and the diffusion of vapor from adjacent to the interface to the bulk stream. Each of these three processes can be quantified using the liquid-side, interfacial, and the gas-side resistances, respectively. In general, the interfacial resistance is small compared with the other two resistances in series. Depending upon the particular evaporator applica-tion, either the liquid-side or the gas-side resistance may dominate. Often only the dominant resistance is used in evaluating the evaporative fluxes in engineering models, ignoring the contribution of the other components. However, considerable improvement in the analytical modeling is possible by using the simultaneous heat and mass-transfer modeling approach.

The original approach to combine the gas-side mass-transfer resistance and the liquid-side heat transfer resistance was proposed by Colburn and Hougen (1934) as a design method for condensing single vapor in the presence of a noncon-densable gas. Their model is illustrated in Fig. 11.1. The heat flux through the wall ϕ is made up of two components, which are the flux ϕ_G for the sensible cool-ing of the gas and the latent heat released at the interface due to the condensing vapor flux G_C. The heat flux from the interface to the coolant is simply expressed as

$$\phi = h^* \left(T_s - T_o \right) \tag{3}$$

Colburn-Hougen 1934.

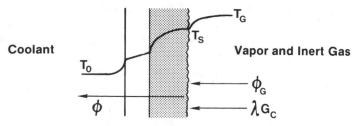

Interface temperature by heat flux balance:

$$h^*(T_S - T_0) = h_{Gn}(T_G - T_S)\frac{H}{1 - e^{-H}} + \lambda K_M \rho_A \ln\left(\frac{p - p_{As}}{p - p_{AG}}\right)$$

Ackerman (1937)

$$\frac{h_G}{h_{Gn}} = \frac{H}{1 - e^{-H}} \quad ; \quad H = \frac{GC_P}{h_{Gn}}$$

Figure 11.1 Model for simultaneous heat and mass transfer for condensation of vapor from a mixture with noncondensable gas.

where T_s and T_o are the interface and coolant temperatures, respectively, and h^* is a combined heat transfer coefficient from the interface to the coolant.

Equating the fluxes on either side of the interface, we get a nonlinear equation in T_s,

$$h^*(T_s - T_o) = h_{Gn}(T_G - T_s)\frac{H}{1 - e^{-H}} + \lambda\, k_M \rho_A \ln\left[\frac{p - p_{AS}}{p - p_{AG}}\right] \quad (4)$$

where h_{Gn} represents the gas-side heat transfer coefficient and λ the latent heat of condensation.

The vapor flux toward the interface G_c is expressed as

$$G_c = k_M \rho_A \ln\left[\frac{p - p_{AS}}{p - p_{AG}}\right] \quad (5)$$

where p, p_{AG}, p_{AS} are the bulk gas pressure, the partial pressure of the vapor in the bulk gas, and the vapor partial pressure at the interface, respectively, and k_M represents the gas-side mass-transfer coefficient.

The Ackerman (1937) correction to account for modified rate of heat transfer due to mass transfer $H/(1 - e^{-H})$ was not included in the original Colburn-Hougen analysis. Here H is GC_{pa}/h_{Gn}.

Standard efficient methods and computer algorithms exist for solving the nonlinear Eq. (4) for the interfacial conditions. (See, for example, Forsythe et al., 1977, Ch. 7.)

For evaporation from a pure liquid the interfacial flux can again be modeled analogous to Eq. (4), noting the reversed directions of heat and mass transfer. One of the advantages of using the interfacial flux equation is that there is no need to ignore any one of the resistance to evaporation a priori. A second advantage here is that the interfacial conditions allow estimation of potential gradients in a direction perpendicular to the main direction of integration, which can be vital information to a designer. One can quickly identify major resistances that may vary along a process path and take appropriate measures to either remedy or account for them during the design. For the increased accuracy with which the process can be modeled using the interfacial flux balance, the price we pay is in terms of added computations, which should be of little concern considering the current availability of powerful and economical computing machines.

For mixtures of liquids, as in distillation applications, provided the vapor-liquid phase equilibrium data are available, equations for concentration gradients analogous to Eq. (4) can be written for the various species and solved for to arrive at interfacial species concentrations.

An approach similar to that of Colburn and Drew (1937) for condensation of vapor mixtures can be adopted for evaporation. For condensation, differential diffusional rates of the species in vapor and liquid may affect the composition of the condensate with the rate of condensation. Similar effects on vapor composition with the rate of evaporation may be expected.

For solute-solvent extraction, concentration gradients within the liquid may result in significant resistance to mass transfer within the liquid. In this case the vapor flux (Eq. (5)) can be modified to consider the added liquid-side resistance to diffusion of solvent through the solution. Reductions in vapor pressure due to solute concentration must also be taken into account in Eq. (5), via a reduction in p_{AS}.

4 EVAPORATION APPLICATIONS

Major applications of the evaporation process occur in the following categories:
- Flash evaporation for vapor production; e.g., OTEC, desalination
- Evaporative heat rejection; e.g., cooling towers, spray ponds
- Evaporative air cooling; e.g., HVAC applications
- Fractional distillation; e.g., petroleum distillation
- Evaporation; e.g., solute extraction

We shall briefly discuss each of the above applications and indicate appropriate research needs.

4.1 Evaporation for Vapor Production

Ocean thermal energy conversion (OTEC) represents a renewable energy technology on the low end of the temperature spectrum. A conventional Rankine cycle is used to generate power between a $25\,^\circ$C source and a $5\,^\circ$C sink temperatures. Because of the low available temperature difference, a seawater flow rate on the

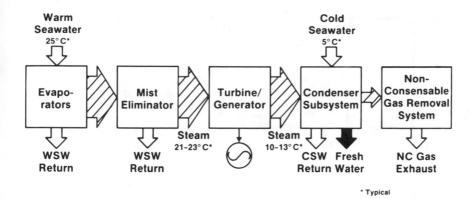

Figure 11.2 Schematic block diagram of an open-cycle ocean thermal energy conversion system.

order of 30,000 gpm through the evaporator is needed per megawatt of power output (Parsons et al., 1984). In a Claude cycle steam evaporated from the warm seawater is used as the working fluid. Figure 11.2 shows a schematic of an open-cycle ocean thermal energy conversion system. In a flash chamber held at approximately 0.3 psia, the warm seawaters cools by 3 °C upon flashing. Spent water is expelled back into the ocean. The flashing process converts about 0.5% of the incoming water mass into steam.

Due to the large resource water flow rates, efficient flashing is of prime importance in the Claude cycle. The overall evaporator efficiency is evaluated in terms of the evaporator effectiveness and pressure losses in the vapor and liquid streams. A variety of flash evaporator geometries were investigated as potential candidates (Bharathan et al., 1984). Due to the high penalty resulting from inefficient evaporation, OTEC perhaps represents one of the critical applications for flash evaporators.

Experimental results for evaporation from free-falling, planar, turbulent water jets of three different initial thicknesses—6.35, 19.05, and 25.4 mm—are shown in Fig. 11.3 as a plot of effectiveness ϵ as a function of the liquid inlet velocity U_o. The effectiveness is defined as the ratio of temperature drop in water stream to the maximum ΔT available, namely, the difference between water inlet temperature and the steam saturation temperature in the flash chamber. Uncertainties in ϵ and U_o are approximately within the size of the symbols. Faired lines through the data points are also included. It should be noted that ϵ is an indicator of the total evaporation from both the jet and the pool below. For the data in this figure, T_i was approximately 28 °C, and the heat transferred was nearly 230 kW. Since in these experiments the heat transferred was held constant, the ratio of the vapor to liquid flow rates and the liquid superheat decreases with increasing thickness at any given U_o and with increasing U_o at any given jet thickness. For all jets, ϵ is observed to decrease initially with increasing U_o, attain a minimum, and then increase gradually at higher velocities. Note that the variation of ϵ with

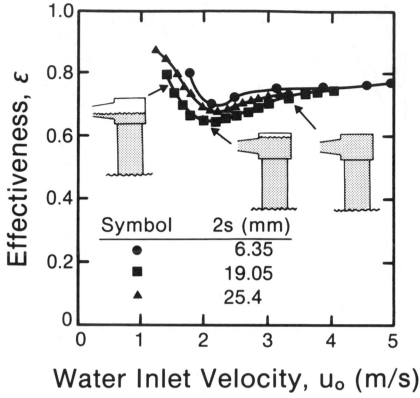

Figure 11.3 Measured variation of evaporation effectiveness ϵ with water inlet velocity U_o for planar turbulent water jets of three thicknesses.

U_o is almost independent of the jet thickness and thus the initial superheat. However, the minimum of ϵ with U_o can be explained based on visual observations of the water level in the upper plenum and the general structure of the jet and is discussed in the following paragraphs.

For all jet thicknesses tested, at low values of U_o near 1.5 m/s, the water level in the upper plenum was nearly 11 cm. (See sketches shown with the data in Fig. 11.3.) Water from the inlet pipe poured on the pool within the upper plenum causing violent mixing in the plenum and resulting in the jet exiting in a spray of droplets right from the plenum. Measured variation of downstream temperature for a 25.4 mm jet, corresponding to this condition at $U_o = 1.41$ m/s, is shown in Fig. 11.4 as case 1. This figure includes a representative uncertainty level in the jet temperature measurement. For this case, a high rate of heat transfer is seen from the jet exit to a dimensionless downstream distance of about 10. At higher downstream distances, based on the slope of the temperature variation with distance, a decrease in the evaporation rate by a factor of more than 18 can be seen.

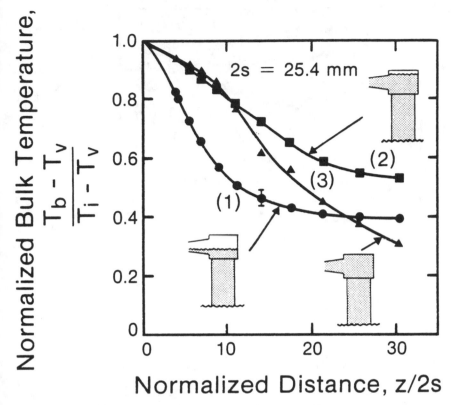

Figure 11.4 Measured variation of normalized liquid jet temperature with downstream distance for a 25.4 mm jet at three water inlet velocities: (1) $U_o = 1.41$; (2) $U_o = 2.1$; (3) $U_o = 3.25$ m/s.

The water level in the upper plenum increases with increasing U_o (see Fig. 11.3). Mixing caused by the incoming water is suppressed by the larger pool water in the upper plenum. The jet exits more and more as a sheet rather than a spray, and the effectiveness decreases. At U_o of nearly 2.2 m/s, the upper plenum is filled completely. The jet exits as a sheet extending approximately 10 cm below the plenum before breaking into droplets. At this point a minimum in ϵ versus U_o is observed for all jet thicknesses tested. A corresponding jet temperature profile for a 25.4 mm jet at $U_o = 2.1$ m/s, shown as case 2 in Fig. 11.4, exhibits nearly three times as low an evaporation rate as does case 1. The associated decrease in the evaporation rate with increasing distance for this case is minimal.

Increasing U_o beyond 2.2 m/s up to 5 m/s (Fig. 11.3) results in an increase in ϵ from nearly 0.65 to 0.75. The jet temperature profile corresponding to a point in this range of U_o is shown in Fig. 11.4 as case 3 for a 25.4 mm jet at $U_o = 3.25$ m/s. Here the initial evaporation rate is similar to case 2, indicating that the jet exits from the plenum as a solid, coherent unbroken sheet, which is confirmed by

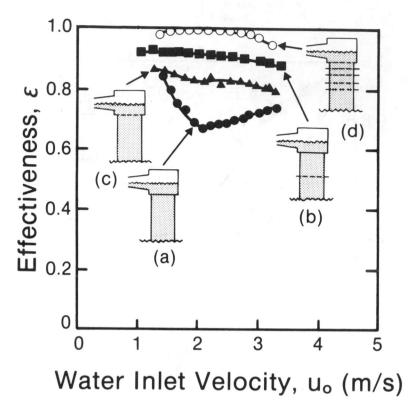

Figure 11.5 Effects of single and multiple screens—variation of effectiveness ϵ with jet inlet velocity U_o for a 25.4 mm jet for (a) no screen, (b) single screen, 40 cm below, (c) single screen at inlet, and (d) one screen at inlet and four others 10 cm apart.

visual observations. However, for this case, at $z/2s > 10$, the evaporation rate increases to a rate nearly equal to that for early evaporation for case 1, indicating a dropwise evaporation. Visual observations of the jet at these distances confirm shattering of the sheet into droplets. The evaporation rate decreases with increasing distance, with the rate at $z/2s$ of 30 nearly three times smaller than a maximum observed at a $z/2s$ of nearly 13. The smaller changes in the evaporation rate observed for this case probably result from decreased residence times for the droplets due to increased jet velocity.

For flash evaporation, screens act as efficient mixers that expose a considerable amount of fresh surfaces. Effects of placing single and multiple screens beneath a 2.54 cm jet are shown in Fig. 11.5. For comparison, evaporator data for the case without screens shown as curve (a) are also repeated (from Fig. 11.3) in this figure.

A single screen (6×25 mm diamond grid, 1 mm thick) placed nominally at midlength (40 cm downstream from the jet exit) yields an effectiveness as high as

0.92, as shown by curve (b). At $U_0 = 2$ m/s, an increase in ϵ of up to 35% can be seen. With increasing U_0, ϵ now decreases more gradually almost to an extent of being insensitive to U_0 and the upper plenum water pool level.

Data for curve (c) correspond to a condition where a screen was placed right at the bottom of the upper plenum. Since this screen (2 mm thick, with 4.8-mm-dia holes at 6.4 mm centers) had a blockage of nearly 50% the width of the opening at the plenum floor was correspondingly increased to yield, at any water flow rate, jet exit velocities nearly the same as the 2.54-mm-wide jet. Note that ϵ for case (c) falls inbetween the data for cases (a) and (b). Thus a screen placed at the liquid exit is not as effective as a screen placed in the liquid free-fall region due to the increased splashing above and below the screen for the latter case. ϵ decreases slightly with increasing U_0 for case (c). A small but finite local minima corresponding to a full upper plenum can be seen in curve (c). Also note that curves (a) and (c) merge at low velocities, because at these low velocities a spray of droplets emerges from the plenum right from the beginning for case (a); similar jet breakup was observed again for case (c).

Curve (d) shows data for the case where a stack of four screens (6 mm $\times$ 25 mm diamond grid, 1 mm thick) was placed in the jet. These screens were located approximately 10 cm apart vertically starting from the jet exit. At all jet velocities, a dense spray of water droplets developed. Within the uncertainty of the data, an effectiveness of unity is observed for U_0 in the range 1.5 to 2.5 m/s. At higher velocities ($U_0 > 3$ m/s), a slight decrease in ϵ due to decreased residence times may be seen.

Since mixing of the liquid jet by screens in the vapor region to generate fresh surfaces proved effective in enhancing evaporation, evaporation from a vertical spout was considered an attractive alternative means for liquid injection. Due to liquid fallback on the incoming liquid, a naturally well-mixed region persists at the liquid entry. Further, a liquid spray forms and distributes itself uniformly over an axisymmetric region around the inlet pipe. With this arrangement, a natural separation with vapor above and the liquid below occurs with no obstructions in the vapor path due to liquid distribution pipes as in the previous configurations.

Since the liquid exits vertically upwards, the spout configuration must be selected suitably for each specific application. Choices of liquid velocity U_0, vapor exit velocity, and the spout height allow proper selection of pipe diameter and the horizontal coverage. For minimal liquid-side pressure losses, the spout height must be kept as low as possible.

For the experiments, the physical size of the test cell and its plumbing layout limited the maximum diameter of the spout to 12.7 cm. A second constraint on the facility is the maximum heat transfer rate of 300 kW. Correspondingly, a water inlet velocity U_0 of 1.5 m/s, and a spout height of 0.5 m were chosen as representative values for a spout in OTEC applications. At the design water flow rate, a temperature drop of 3.8 °C, again typical of OTEC application, could be achieved with the present choices of the spout parameters.

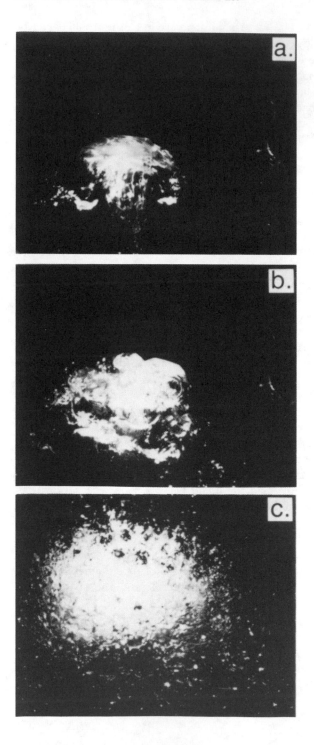

Figure 11.6 Photographs of the spout evaporator (a) without heat transfer, (b) with 100 kW of heat flux, and (c) with 250 kW of heat flux.

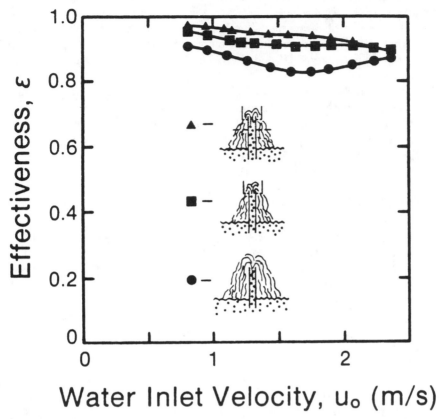

Figure 11.7 Measured variations of effectiveness ϵ with inlet velocity U_o for the spout evaporator for three cases: no screens, a cylindrical enclosures and one screen, and a cylindrical enclosure and two screens.

A series of three photographs of the spout evaporator with and without flashing is shown in Fig. 11.6. Case (a) without evaporation shows that the water jet exists smoothly and distributes itself as an axisymmetric sheet. Upon increasing evaporation (i.e., lowering the condenser inlet water temperature), bubbles begin to emerge from the spout and grow on the falling liquid sheet. Bubbles on the order of 10 cm diameter can be seen at heat fluxes of 100 kW (case (b)). As the evaporation rate is further increased, the jet exit becomes more violent, with the vapor escaping from bursting bubbles. Explosive growth of the vapor shatters the jet into fragments and droplets. Most of the liquid escaping upward falls back on the incoming liquid. The coherent liquid sheet is totally destroyed, becoming a spray of droplets (case (c)).

Plots of the effectiveness ϵ versus the liquid inlet velocity U_o for the spout evaporator with and without screens are shown in Fig. 11.7. For these data, the heat flux from the jet was held constant at nearly 210 kW. For a spout with no screens, the effectiveness ranges from 0.85 to 0.92. Effectiveness is seen to decrease slightly with increasing U_0 up to 1.7 m/s and then increase once again. The initial decrease is a result of decreasing superheat of the inlet water with increasing flow rate. The latter increases result from increased vertical throw or stagnation heights of the jet due to larger U_0 and the resulting extended horizontal coverage of the shattering jet.

For other data shown in this figure, a cylindrical enclosure, 36 cm in diameter and 30.5 cm long, with a screen (2 mm thick, with 4.8-mm dia holes at 6.4-mm centers) at the bottom, was placed around the spout, with the screen 10 cm below the liquid exit level. The purpose of the enclosure was to restrict the horizontal spread of the liquid and to allow further mixing below. The shattering jet collects, remixes, and then exits as droplets through the screen. For this case, the effectiveness is seen to range from 0.9 to 0.95. For U_0 greater than 2 m/s, a considerable amount of liquid spilled over the enclosure and showed a marked decrease in performance.

An additional and similar screen, 40 cm in diameter, placed nominally 10 cm below the bottom screen of the enclosure, yielded slightly increased ϵ up to 0.97 in the range $0.7 < U_0 < 2$ m/s. For U_0 over 2 m/s, due to spilled liquid, the data for the case with and without this second screen do not show any significant difference.

The spout geometry yields effectiveness in the range of 0.9 to 0.97, with a liquid-side pressure loss of about 0.7 m (the spout height plus the exit kinetic energy losses). The corresponding pressure loss for the planar jets in the reported experiments ranges from 0.8 to 1.0 m.

Although effectiveness is an indicator of the evaporator performance, for an effective design many system considerations must be carefully weighed in each application. For example, in design of an open-cycle OTEC plant, considerations that will affect the design include pressure losses on the liquid and vapor paths, simplicity of liquid inlet and exhaust manifolds, evaporator volume, entrainment of droplets, mist elimination needs and losses, immunity to sea states and plant motion for a floating platform, gas desorption, and fabrication costs.

For vapor production the flash evaporation process is primarily controlled by the liquid-side resistance to heat transfer. The liquid-vapor interface cools rapidly to establish steep temperature gradients within the liquid in short times. By using a simplified model and assuming spherical droplets, Bharathan and Penney (1984) showed that bulk mixing of the liquid to expose hotter interiors to the vapor space enhances the overall heat transfer rate and confirmed these conclusions with the series of experiments using a screen placed midway in the flashing jets.

The vertical spout was identified as a promising geometry for the Claude cycle evaporator, due to its inherently easy manifold and low liquid-side pressure loss. Research needs in this area include developing detailed analytical models of evaporation and fluid and thermal coupling between spouts in a multiple field and generating a data base for selecting spout diameter, height, and spacing.

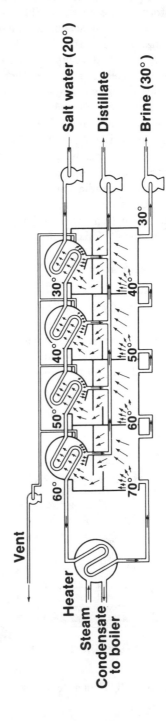

Figure 11.8 Simplified schematic diagram of a multistage flash (MSF) desalination system.

Another application for flash evaporation is in multistage flash (MSF) evaporators for desalination. The primary advantage of the flash evaporator here is the avoidance of scaling problems. A simplified schematic diagram of a MSF desalination system is shown in Fig. 11.8. In each stage the latent heat of condensing steam is recuperated to preheat the incoming feed stream. Advanced heat-exchanger designs together with streamlined vapor flow in commercially available desalination systems are capable of yielding a performance ratio (defined as kg of water produced/kg of steam) as high as 40 compared with a maximum of 12 for conventional MSF processes (Deutsche Verfahrenstechnik, 1985). Typical superheat available per stage for flashing in these processes is merely 0.9 ° C.

Desalination using solar stills also involves producing vapor by evaporation. Typical solar still operation is shown in Fig. 11.9. In solar stills only 50% of the incident solar radiation turns out to be effective due to losses in reflection, loss from basin to cover, and convective losses to air. With the present technology a water production rate of a few liters per day per square meter is possible in favorable climates (Malik et al., 1982).

4.2 Evaporation Heat Rejection

Heat rejection by evaporation is a major application of direct-contact heat exchange. The surrounding atmosphere forms the heat sink to reject heat from power plants using cooling towers, spray ponds, and holding ponds and to reject heat from other sources in case of perspiration cooling and mist cooling.

Cooling tower technology is highly developed. Two-dimensional models of tower heat and mass-transfer processes are available (Majumdar et al., 1983). In

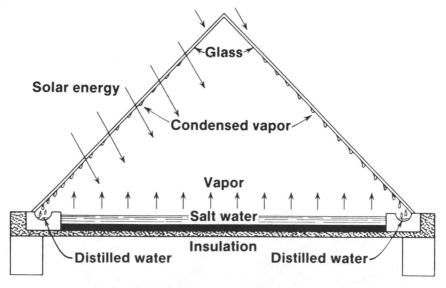

Figure 11.9 Principle of a solar still.

general, for air-water mixture, since the Lewis number is nearly unity, a combined lumped approach for heat and mass transfer is used invoking the humid-air enthalpy as the driving force (the Merkel Approach (1925)). Predicting tower performance within 3% is routinely possible. Ongoing research activities on cooling towers focus on predicting, measuring and improving wet tower thermal performance (Yeager, 1983). Another area of activity involves methods for predicting and measuring the physical distribution of cooling tower emissions in the environment, primarily the visible exhaust plume and associated drift deposition.

Research needs in cooling tower performance predictive modeling include incorporating the simultaneous heat and mass-transfer equation (Eq. (1)) together with independent integration for the gas temperature from the sensible heat flux to predict fog flow rates accurately. Other areas of interest lie in tower fill deterioration, flow maldistribution, and tower icing. Research needs in cooling ponds are in the areas of improved pond thermal predictive capabilities and pond design.

4.3 Evaporative Air Cooling

Cooling air with evaporation of water is cost-effective and energy efficient. Evaporating water directly into the air is accomplished by evaporative coolers, spray-filled or wetted-surface air washers, spray coil units, and other humidifiers.

In a direct evaporative cooler, adiabatic heat exchange between the air and water stream is accomplished. To evaporate water, sensible heat from the air stream is used. The maximum possible reduction in air dry-bulb temperature is the difference between its dry-bulb and wet-bulb temperatures. If the air is cooled to its wet-bulb temperature, the evaporative cooler is said to be 100% effective. In practice, 85% to 90% effectiveness can be achieved. When a direct evaporative cooler cannot provide desired conditions, the recirculating water can be chilled by mechanical refrigeration. This arrangement reduces operating cost by 25% to 40% over the use of refrigeration alone (Watt, 1963).

Indirect evaporative cooling combines heat exchange with a secondary air stream undergoing direct evaporative cooling. Indirect coolers, previously considered not cost-effective, have received increased attention from those interested in conservation. The performance of an indirect unit is expressed similar to that of the direct unit. The performance factor or effectiveness is defined as the depression in dry-bulb temperature of primary air divided by the difference between the dry-bulb temperature of the entering primary air and the wet-bulb temperature of the entering secondary air. Depending on the heat-exchanger design and relative air flows, the effectiveness can be as high as 85%. Normal range is, however, 60% to 70%. An indirect cooler applied as a first-stage upstream of a second direct cooler reduces both dry-bulb and wet-bulb temperatures. The indirect cooler does not increase the absolute humidity of the primary airstream for staged evaporative coolers. Operating cost savings of 60% to 75% are possible over using vapor compression units (Watt, 1963).

Staged systems with indirect and direct coolers with booster refrigeration are attractive for many areas of the United States; i.e., not only areas with a low wet-bulb design temperature but also areas with a high wet-bulb design temperature that are not thought of as suitable for evaporative cooling. An evaporative cooler is a major component in desiccant cooling concepts. The challenge in evaporative cooling lies in achieving high heat and mass-transfer rates in compact, low-pressure drop units. How to distribute liquid at low flow rates and maintain adequately wetted surfaces still remains unsolved. Impurities and scale buildup on the wet side can significantly reduce the evaporative flux and cause higher air pressure losses.

4.4 Fractional Distillation

Distillation remains the most-used method for separating liquid mixtures in chemical engineering. The process of heat and mass transfer in a fractional distillation column is highly complex with simultaneous evaporation of the more volatile components and condensation of the less volatile components in various stages. Steep concentration gradients occur in the liquid and vapor phases. Commonly, the mass-transfer effectiveness of a contacting device such as a bubble-cap tray is defined using a Murphree efficiency based on vapor-liquid equilibrium properties. This approach, while convenient to adopt in analytical calculations, clearly does not provide detailed attention to the complete mass-transfer process occurring in a particular stage.

An excellent summary of research needs in distillation is provided by Fair and Humphrey (1985). Among various recommendations they emphasize the great need for understanding the contacting mechanisms that occur on trays and packings. There does not appear to be enough research in progress to satisfy this need. Understanding the complex two-phase flow behavior occurring on trays and in packings is an obstacle not readily overcome. With the variety of contacting devices used commercially, research focus should perhaps be limited to understanding the mass-transfer process of selected popular geometries, namely, sieve trays and structured packings.

4.5 Solute Extraction

Evaporation is widely used to remove solvent from a solution by vaporization—common in making products such as salt and sugar. The most common solvent is water. Solute is a inorganic or organic solid with relatively low vapor pressure at the temperature of evaporation. This distinguishes solute extraction from distillation. In evaporation the overhead vapor is primarily solvent, contaminated only by small amounts of entrainment. In many cases the dissolved solids exceed their solubility limit in one or more of the evaporating stages and are precipitated as solid materials in the saturated solution.

Many commercial designs of evaporators for solute extraction are available. A major factor of importance is the heating surface. In most cases the liquid flows through tubes with steam condensing outside. An overall coefficient that properly accounts for all resistances is used to evaluate the heat transfer. Boiling-point elevation due to high solute concentration can be significant, 2° to 6°C for salts, 24° to 31°C for acids and alkali. The boiling-point elevation represents a thermal gradient that is largely unavailable for heat transfer.

Considerations must also be given to economy, venting, condenser, scaling, fouling, and entrainment. Increase in viscosity is an important aspect. Wiped-film evaporators are most suitable for highly viscous liquids.

5 CONCLUDING REMARKS

Evaporator applications such as in OTEC, desalination, and heat rejection represent areas for performance improvements with potentially high payoffs.

In OTEC, with only a 20°C temperature difference an improvement of 1°C in the evaporator temperature approach translates into a 10% savings in plant cost. Vapor disengagement with low liquid entrainment and low liquid and vapor pressure losses is critical. Progress in flash evaporator research for OTEC will result in improved designs for various other process applications.

For cooling towers, critical issues are pressure loss in the airstream and the associated fan power. Liquid pumping is also significant. EPRI is conducting cooling tower research. Developing improved packing to enhance air-side heat transfer without increasing pressure losses is important. For improved understanding of evaporation, critical geometries are packed columns and spray columns. In sprays the geometry is very ill-defined. The liquid goes through continuous streams and then breaks up into discrete droplets. The heat transfer during droplet formation can be as high as 30% of the net. The droplets circulate and vibrate, which tends to improve evaporation. However, exact relationships are not available.

A third area of importance is in distilling multicomponent mixtures. Little effort is in progress to improve understanding of the mass-transfer processes occurring in gas-liquid contacting devices. Fundamental studies on hydrodynamic and mass-transfer characteristics of various devices are required for achieving improved, energy efficient columns.

Wide use of efficient evaporative coolers for HVAC can have significant impact on summertime electricity demands.

REFERENCES

Ackermann, G. (1937), "Wärmeübergang und molekulare Stoffübertragung in Gleichen Feld—Grossen Temperatur und Partialdruck-Differenzen," *Forschungsheft*, No. 382, pp. 1-16.

Bharathan, D., Kreith, F., Schlepp, D., and Owens, W. L. (1984), "Heat and Mass Transfer in Open-Cycle OTEC Systems," *Heat Transfer Engineering*, Vol. 5, No. 1-2, pp. 17-30.

Bharathan, D., and Penney, T. (1984), "Flash Evaporation from Turbulent Water Jets," *Journal of Heat Transfer*, Vol. 106, No. 2, pp. 407-416.

Bikerman, J. J. (1948), *Surface Chemistry*, p. 81, Academic Press, New York.

Colburn, A. P., and Drew, T. B. (1937), "The Condensation of Mixed Vapors," *Trans. AIChE*, Vol. 33, pp. 197-215.

Colburn, A. P., and Hougen, O. A. (1934), "Design of Cooler Condensers for Mixtures of Vapors with Noncondensing Gases," *Ind. Eng. Chem.*, Vol. 26, No. 11, pp. 1178-1182.

Davies, J. T. (1963), "Mass-Transfer and Interfacial Phenomena," in *Advances in Chemical Engineering*, T. B. Drew, G. W. Hoopes, Jr. and T. Vermeulen, eds., Vol. 4, pp. 1-50, Academic Press, Orlando, FL.

Deutsche Verfahrenstechnik. (1985), *VTE-MSF Distillation Process*, Graf-Adolf-Strasse 68, D-4000 Düsseldorf *1, West Germany. (Technical Brochure on DVT)*.

Fair, J. R., and Humphrey, J. L. (1984-85), "Distillation: Research Needs," *Separation Science and Technology*, Vol. 19, No. 13-15, pp. 943-961.

Forsythe, G. E., Malcolm, M. A., and Moler, C. B. (1977), *Computer Methods for Mathematical Computations*, Prentice-Hall Inc., Englewood Cliffs, NJ.

Hertz, H. (1882), "Ueber die Verdunstung der Flüssigkeiten, insbesondere des quecksilbers, in luftleeren Raume," *Annalen der Physik und Chemie*, Vol. 17, No. 10, pp. 177-200.

Knudsen, M. (5 Aug 1915), "Die maximale Verdampfungsgeschwindigkeit des quecksilbers," *Annalen der Physik*, Vol. 47, No. 13, pp. 697-708.

Langmuir, I. (1913), "The Vapor Pressure of Metallic Tungsten," *Physical Review*, Vol. 2, pp. 329-342.

Levich, V. G., and Krylov, V. S. (1969), "Surface-Tension-Driven Phenomena," *Annual Review of Fluid Mechanics*, W. R. Sears and M. Van Dyke, eds., Vol. 1, p. 293-316, Annual Reviews, Inc., Palo Alto, CA.

Maa, J. R. (1967), "Evaporation Coefficients of Liquids," *Ind. Eng. Chem. Fundam.*, Vol. 6, No. 4, pp. 504-518.

Majumdar, A. K., Singhal, A. K., and Spalding, D. B. (1983)(Mar), *VERA 2D - A Computer Program for Two-Dimensional Analysis of Flow, Heat and Mass Transfer in Evaporative Cooling Towers*, Vols. 1 & 2, EPRI Report CS-2923, Electric Power Research Institute, Palo Alto, CA.

Malik, M. A. S., Tiwari, G. N., Kumar, A., and Sodha, M. S. (1982), *Solar Distillation*, Pergamon Press, New York.

Merkel, F. (1925), "Verdunstungskuehlung," *VDI Forschungsarbeiten*, No. 275, Berlin.

Parsons, B. P., Bharathan, D., and Althof, J. A. (1984)(Jun), *Open-Cycle OTEC Thermal-Hydraulic Systems Analysis and Parametric Studies*, SERI/TP-252-2330, Solar Energy Research Institute, Golden, CO.

Schrage, R. W. (1953), *A Theoretical Study of Interphase Mass Transfer*, Columbia Univ., New York.

Sherwood, T. K., Pigford, R. L., and Wilke, C. R. (1975), *Mass Transfer*, McGraw-Hill, New York.

Sherwood, T. K., and Wei, J. C. (1957)(June), "Interfacial Phenomena in Liquid Extraction," *Industrial and Engineering Chemistry*, Vol. 49, No. 6, pp. 1030-1034.

Watt, J. R. (1963), *Evaporative Air Conditioning*, The Industrial Press, New York.

Yeager, K. (1983)(Dec), "Coal Combustion Systems Division R&D Status Report," *EPRI Journal*, Vol. 8, No. 10, pp. 45-52.

DIRECT-CONTACT CONDENSATION

Harold R. Jacobs

ABSTRACT

This paper reviews the four basic types of direct-contact condensation schemes, which have been called "drop type," "jet and sheet type," "film type," and "bubble type" and classifies typical equipment that falls under the classifications. Next, it reviews the current state of our ability to analyze the processes and points out the uncertainties related to our knowledge of the basic mechanisms. In doing so, it points out the needed additional research that should be carried out in order to optimize the design of engineering equipment based upon the various condensation processes.

1 INTRODUCTION

Direct-contact condensers have been built and used industrially for well over 80 years. Hausbrand, in his book "Evaporating Condensing and Cooling Apparatus" [1], which appeared in its first German edition in 1900, dealt with the then theoretical aspects of barometric condensers as well as commercial design. Despite this early start, little work of a basic nature had been done prior to the 1960s. In fact, How's 1956 publication on designing barometric condensers [2] was simply an article describing rules of thumb for direct-contact condensers.

In 1972, Fair's article, "Designing Direct Contact Coolers/Condensers," appeared in *Chemical Engineering* [3]. Despite the fact that some experimental data had been published, this article presented techniques primarily based on analogies with mass transfer in a variety of mass-transfer-type equipment including baffle tray columns, spray columns, packed columns, sieve columns, and the like. Most of the work presented was more applicable to cooling gas streams than to condensing vapors.

In 1977, Jacobs and Fannir [4] released the U.S. Department of Energy report "Direct Contact Condensers—A Literature Survey," which reported on the dearth of a theoretical basis for designing direct-contact condensers. In that report the names "drop type," "jet and sheet type," "film type," and "bubble type" were first introduced as a means of classifying direct-type condensers. These classifications will be used in the current review.

Since the mid-1970s, renewed interest in direct-contact condensers has appeared in the United States although, it had never disappeared in other parts of the world. This is particularly true in eastern Europe and the U.S.S.R., where the Heller cycle is important. Oliker [5,6] has reported on the use of the direct-contact condensers as deaereators and their potential for geothermal applications. Goldstick [7] lists 21 companies in western Europe and the United States manufacturing direct-contact devices, of which 13 build condensing apparatus.

In the United States, an accelerated interest in direct-contact condensers was initially driven by desalination schemes and then by alternate energy systems [4]. Today's broad interest includes energy conservation, geothermal energy, solar energy, OTEC systems and even space power plant applications [8–11].

In this paper we primarily review the work since 1980. Work prior to that has been ably reviewed by Sideman and Moalem-Maron [12]. Some reference is given to Section 2.6.8 "Direct Contact Condensers" of *Heat Exchanger Design Handbook* to be released shortly in the Second Supplement [13].

2 DROP-TYPE CONDENSERS

Drop-type condensation refers to condensation on sprays or drops of liquid coolant that are injected into a chamber filled with vapor or a gas vapor mixture (see Fig. 12.1). Early work dealt primarily with the condensation of saturated pure vapors where the liquid and vapor are the same substance. Kutateladze [14] was the first to recommend that the drops be assumed spherical and that the heat transfer be governed by transient conduction within the drops.

In 1973, Ford and Lekic [15] published the results of an experimental study of condensation of steam on single drops of water. Utilizing the equations for transient conduction in a sphere whose surface was suddenly exposed to the saturation temperature of the vapor to predict the instantaneous heat transfer, they found that the growth of the drops was slightly overpredicted. The added resistance due to condensation was neglected. If included, it would of course

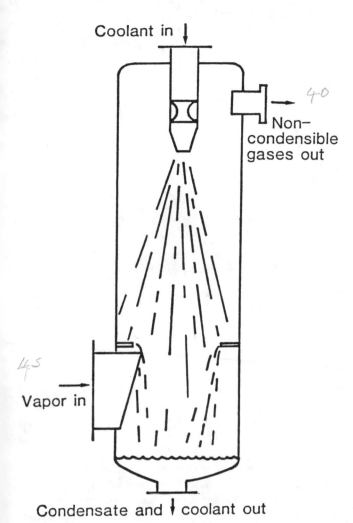

Coolant in

Non–
condensible
gases out

Vapor in

Condensate and coolant out

Figure 12.1 Spray-type condenser.

reduce the heat transfer. Such was the case when Jacobs and Cook [16] developed a theoretical model to account for the added resistance due to drop growth (see Fig. 12.2). Their model recognized that the ratio of final droplet radius to initial radius was given by

$$\frac{R_f}{R_i} = \left(1 + \frac{1}{Ja}\right)^{1/3} \tag{1}$$

where

$$Ja = \frac{\rho h_{fg}}{\rho c_p (T_{SAT} - T_{(t=0)})} \tag{2}$$

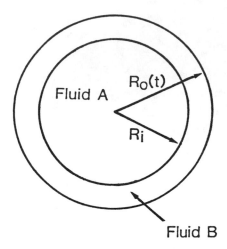

Figure 12.2 Model of variable drop for condensation governed by conduction.

The model assumes that the condensate film added is thin, as would be the case for large values of the Ja defined in Equation (2). Thus they were able to solve the problem of conduction in a sphere subject to the boundary condition

$$q'' = \frac{(T_{SAT} - T_{(Ri)})}{R(t) - Ri} \tag{3}$$

The nondimensional radius at time t was given as

$$\frac{R(t)}{Ri} = \left[1 + \frac{3}{Ja} \int_0^1 \left(\frac{r}{Ri} \right)^2 \left(\frac{T(r,t) - Ti}{T_{SAT} - Ti} \right) \frac{dr}{Ri} \right] \tag{4}$$

Near perfect agreement was found with the experiments of Ford and Lekic [15]. Jacobs and Cook [16] then extended their model to a secondary vapor condensing on an immiscible drop, where the coolant would have been treated such that the vapor condensate would wet the coolant. Typical results were generated for a range of the ratio of thermal conductivities (see Figs. 12.3 and 12.4).

At the same time that Jacobs and Cook [16] were studying the effects of added mass for condensation of pure vapors, Kulik and Rhodes [17] were attempting to model water spray effectiveness in air steam mixtures. Their study showed that for droplets greater than 0.1 mm in diameter, internal resistance was important even when noncondensable gases were present. However, a reliable method of predicting the external resistance with noncondensables present was not adequately dealt with in the opinion of this reviewer. This is also apparently the view of the National Science Foundation which has funded a study by P. S. Ayyaswamy at the University of Pennsylvania [18–21]. This study has been aimed at theoretically modeling both the internal and the external flow around a drop. Except for very large drops, however, the influence of internal circulation is small

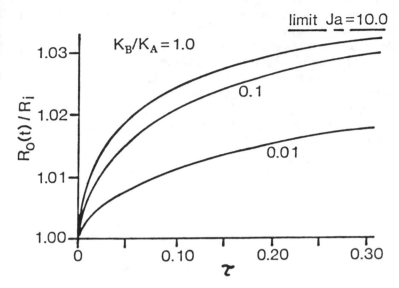

Figure 12.3 Effect of condensate thermal conductivity that of coolant on drop growth.

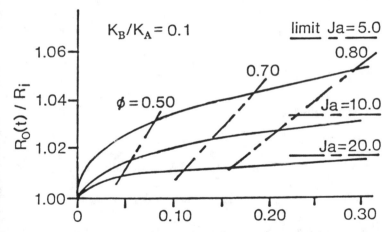

Figure 12.4 Drop growth for different degrees of coolant utilization.

[15–17,22]. Thus it appears that much of the study has been a case of overkill when applied to practical problems where a small diameter would be used to induce a higher heat transfer. Of course, larger-diameter drops may be of importance in cooling high-velocity gas streams. Nevertheless, the methods being developed by Ayyaswamy and coworkers will provide a means of fully understanding the problem of condensation on drops, with two exceptions.

The two areas not dealt with sufficiently are drop spray statistics and the influence of noncondensable gas absorption or loss on drop heat transfer. The prediction of droplet size statistics for a given nozzle is needed to accurately predict heat transfer. More information is needed on the formation of drops in spray nozzles and their size distribution even though such statistics have long been studied [23,24]. In order to design a barometric condenser, for example, the vapor velocity must not exceed the terminal velocity of the smallest drop, yet to fully or nearly fully utilize the cooling capacity of the spray the size of the largest drops must be known. From a practical viewpoint, this means that the diameter of the condenser vessel is dictated by consideration of the smallest drops and the height is dictated by consideration of the largest [13].

The fact that the interface of the droplets with the vapor is not impermeable too noncondensable gases or especially highly soluble ones offers other problems. For example, both CO_2 and H_2S are soluble in water. At the Geysers, direct-contact condensers were used on the first 15 units installed [13]. The condensers were connected to open cooling towers. When the coolant entered the towers, the H_2S was released into the atmosphere. Subsequent use of direct-contact condensers was terminated because the coolant was too low in H_2S concentration to be stripped using existing abatement systems. A better understanding of this mass-transfer problem could have averted a costly refitting, and perhaps an adequate direct-contact design could have been achieved.

In addition to this absorption problem, it is also possible that the coolant drops could have dissolved gas within them. Heating the drops during the condensation process can lead to migration of such gases to the condensate interface and greatly reduce the heat transfer by effectively lowering the fluid conductivity. On the other hand, if gases were to go into solution in the coolant or condensate, the concentration at the interface would be lower than that predicted by assuming it impermeable. In this latter case, the external heat transfer could be higher. These effects need further examination.

3 JET- AND SHEET-TYPE CONDENSERS

Jet- and sheet-type condensers have been widely used commercially. How [2] illustrates many different commercial designs. They can be of the co- or countercurrent variety as shown in Figs. 12.5 and 12.6. Such designs are widely used in the U.S.S.R. [25,26].

An early analysis proposed by S. S. Kutateladze [26] for condensing on jets and sheets, for application to saturated low-pressure vapors, ignored the condensate buildup. It proposed a model based on the Graetz problem with the assumption of a uniform velocity in the coolant stream and the surface of the stream, be it jet or sheet, suddenly exposed to the saturation temperature. More recently Hasson et al. [27,28] extended this to a fan jet. Jacobs and Nadig [29] carried out an integral-type analysis accounting for the condensate film. The solution gave excellent agreement with the Graetz solution for $Ja \rightarrow \infty$, as could be expected. They carried out the analyses for a vapor condensing on its own liquid, but

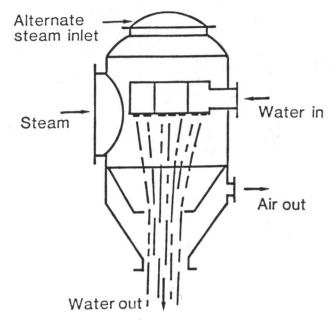

Figure 12.5 Jet condenser.

pointed out that the method of solution was applicable to condensation of vapor on a wettable immiscible coolant.

Taitel and Tamir [30] extended Kutateladze's solution to the problem of a vapor with noncondensable present, but neglected the condensate film. Nadig and Jacobs [31] carried out the more complete analyses and presented data on jet or sheet required lengths for steam-air over a wide range of a Jakob number for non-condensables present in the range of 0.5% to 10%. Jacobs and Nadig [29] give the values for a pure vapor. These works assume the condensate-vapor gas mixture interface is impenetrable to the gas.

Logical and needed extensions of the work, to date, include relaxing the assumption that the interface is impenetrable to the noncondensables. A theory that takes this into account is also missing for surface condensation phenomena.

Experimental data on jet and sheet breakup are also needed. At low flow rates, thin jets and sheets can break up into drops due to surface tension effects. Prior to this, the jets or sheets may become wavy or unstable. The effects on heat transfer are not known and may be important at high pressure where vapor drag could become important also. If the jets and sheets are turbulent, the heat transfer could be increased, yet no studies on turbulent jets or sheets have been documented to this writer's knowledge.

Thus it is clear that considerable work needs to be carried out to optimize the design of jet- and sheet-type condensers depending on their application. How-ever, current work appearing in the literature offers a good basis for design of

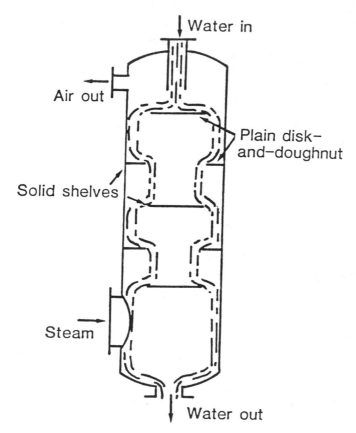

Figure 12.6 Disk- and donut-sheet-type condenser.

pure vapor condensers and a conservative design for those where noncondensables are present if absorption of these gases is not a problem from an environmental or other basis.

4 FILM-TYPE CONDENSERS

Film-type condensers are strongly related to jet and sheet condensers in that the condensation takes place on a thin film of coolant. However, while the jet and sheet are unsupported, the film-type condenser's coolant flows over a solid substrate. Such condensers using a solid substrate to control a thin film are called packed bed condensers. A typical design is shown in Fig. 12.7. Nearly all of the work through 1980 has been described in the review article by Sideman and Moalem-Maron [12].

Recently Bharathan and Althof [35] reported an experimental study of steam-air condensing on water in a packed bed condenser with seven different packing materials. Their work followed the earlier studies of Thomas et al. [36]

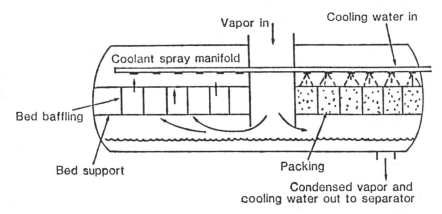

Figure 12.7 Packed bed condenser.

for R-113 vapor condensing on water for packings of 3 cm diameter ceramic spheres and for 2.5 cm Raschig rings. They report a similar correlation of Stanton number as a function of Jakob number, heat capacity ratio, defined as

$$C = \frac{\dot{m}_\ell C p_\ell}{\dot{m}_v C p_v},$$

and $H = L/d_p$, the ratio of packing height to particle diameter. Although Thomas et al. [36] accounted for the percentage of wetting of the packing by the coolant, Bharathan et al. [35] did not. Nonetheless, the correlation was quite similar.

Thomas et al. [36] compared their data with a model based on penetration theory and found it wanting. Thus Jacobs et al. [32–34] attempted to develop a more extensive model based on fluid hydrodynamics as well as heat transfer. The first two papers deal with condensation of a pure vapor on an adiabatic flat plate and sphere, respectively, and are discussed in [13]. They are applicable to the condensation of a vapor both on its own liquid and on an immiscible liquid that is wetted by the condensate.

For a pure vapor condensing on its own liquid or an immiscible wetted liquid for the case of a flat plate, Jacobs and Bogart [32] present a correlation for 99% utilization of the coolant's capacity. It is shown that the required length is a function of $(\nu_f^2/g)^{1/3}$, Ja, Re, Pr_c, Pr_f, as well as the ratio of kinematic viscosities of the two liquid for the case of immiscible fluids. For a vapor condensing on its own liquid, the required length will be a function of only $(\nu_f^2/g)^{1/3}$, Ja, Re, and Pr. The factor C of Thomas did not occur directly. Of course, the models of [32–34] were for single surfaces in an infinite nonflowing vapor.

For condensation on a sphere, the same quantities as for a flat plate are pertinent in defining the degree of coolant utilization. In addition, of course, is the nondimensional sphere radius,

$$\frac{R}{(\nu_f^2/g)^{1/3}}\ .$$

The spherical geometry was chosen by Jacobs et al. [33] because it is the standard with which other packings are compared and to the existence of some experimental data [37,38]. Unfortunately, the data are not well defined and are subject to inaccurate reporting of the instrumentation locations and lack of care in removing noncondensables. Nonetheless, much of the data are in general agreement with the theory. Claims of a large interfacial resistance [38] are extremely unlikely and are probably representative of the amount of noncondensables present.

Because packed bed condensers are of interest for use in the presence of non-condensables, Jacobs and Nadig [34] studied condensation on a film flowing over a vertical plate in the presence of noncondensables. A study of their effect on plate length for complete coolant utilization is given in [13]. Nadig is currently extending his model for condensation with noncondensables to an adiabatic sphere.

Other related theoretical studies account for heat transfer at the wall. Murty and Sastri [40,41] studied condensation of a pure vapor on its own liquid. Rao and Sarma [42] studied the same geometry and boundary conditions, but dealt with a pure vapor condensing on an immiscible wettable liquid. This same problem was also treated by Nadig [43], who in addition studied the same problem in the presence of a noncondensable gas. Nadig then extended the problem to condensation on a thin film flowing over a tube.

In all of the above theoretical studies, the coolant film was assumed laminar. Extension to wavy films, turbulence, and vapor flow are logical. Further, relaxation of the boundary condition that the interface is impenetrable to noncondensable gases is desirable.

Experimental data are also needed so that more reliable correlations can be obtained. Of particular interest is the condition where the immiscible substrate is not wetted by condensate. It is quite likely that for nonwetting conditions, a much higher heat transfer rate can be achieved. This problem is analogous to condensation of vapor mixtures on ordinary condensers when the fluids are nonwetting. Lenses of the more volatile fluid can form on the surface. For binary or mixture systems, impurities may or may not alter the interfacial tension allowing for either lens formation or near complete sheets.

5 BUBBLE-TYPE CONDENSING

By "bubble-type" condensing we refer to the injection of vapor as jets or bubbles into a continuous stream or pool of coolant (see Fig. 12.8 for example). A wide range of experiments on this phenomenon have been carried out in the past by Sideman and coworkers [12]. These studies include works for bubble trains as well as single bubbles. Works dealing with condensation of jets are less plentiful [12].

In 1978, Jacobs et al. [44] extended the earlier modeling of single bubble collapse, (e.g., Chao et al. [45] and then Isenberg et al. [46,47], to account for the wetting of the inside of a condensing bubble by the condensate. Prior analyses

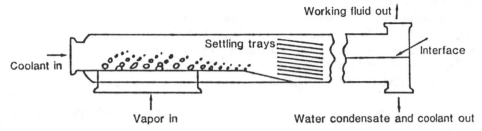

Figure 12.8 A possible design for a bubble-type condenser for use with immiscible fluids.

had ignored this resistance to heat transfer. More recently Jacobs and Major [48] developed a model to account for noncondensables. For small bubbles, they found that the collapse is governed by diffusion of the noncondensables away from the interface. For bubbles larger than 4.5 mm in diameter, fluctuation of the ellipsoidal bubbles produces mixing of the vapor-gas, and bubble collapse is described by the uniformly mixed model of Isenberg and Sideman [47]. In the work of Isenberg et al. [46,47], a correction factor was used to correct for the bubble hydrodynamics, while Jacobs and Major assumed the liquid moved at potential flow velocities. Letan [49,50] has indicated that a slip velocity should occur between the condensate and coolant for immiscible liquids. However, Grace [22] indicates that this is a function of the Eötvos number and the presence of impurities. Impurities tend to immobilize the surface. Without significant impurities and for large drops, interfacial slip is probably small. In examining the various models, it is clear that a variety of effects can explain the data and that each investigator has shown nearly equally rational arguments for their case and equal agreement with existing data. Surely, further experiments are needed that are more precise.

In applying single bubble data to heat-exchanger systems, i.e., bubble trains, Sideman and Moalem-Maron [12] review the various models. In a recent M.S. thesis at the University of Utah, Golafshani [51], using a model proposed by Moalem-Maron et al. [52], developed a computer program to predict the collapse of a series of uniformly spaced bubble trains. Golafshani used the model of Jacobs and Major [48] for single drops instead of that of Sideman et al. [46,47]. The results showed little difference with that of [52], which used the model of [46]. Thus it appears that the potential heat-exchanger systems for multiple bubbles are relatively insensitive to small changes in the single bubble theory.

Despite the relative acceptability of the theories, Johnson et al. [53] and Sudhoff [54] have found the experiments are plagued by a socalled persistent bubble. Johnson carried out experiments in a very deep heat exchanger. He reports that visual observation indicates that the persistent bubbles finally disappear. He concludes that this is due to absorption of the noncondensable gases into the liquid. (The initial bubble collapse appears to be well predicted by the theory of Jacobs and Major [48]). So far, none of the other investigators have offered any hypothesis for this phenomenon although all of the experimental investigations have reported it.

If Johnson's [53] arguments concerning the persistent bubble are valid, it may be that a heat exchanger could be designed of a depth that condensation can be accomplished and gases easily separated. This requires investigation of essentially the same problem as for other direct condensation schemes mentioned. What is the *real* process of condensation with noncondensable but slightly soluble gases present?

Other work of importance deals with condensation of jets. This is mentioned in [12] and [13]. No new work has been reported since 1982, to this writer's knowledge; although, some studies are currently being conducted [55].

REFERENCES

1.   Hausbrand, E. *Evaporating Condensing, and Cooling Apparatus.* Fifth English Edition, Van Nostrand, New York (1933).
2.   How, H. How to Design Barometric Condensers. *Chemical Engineering*, pp. 174-182 (Feb. 1956).
3.   Fair, J. R. Designing Direct Contact Coolers/Condensers. *Chemical Engineering*, pp. 91-100 (June 1972).
4.   Jacobs, H. R., H. Fannir. Direct Contact Condensers—A Literature Survey. *Report DGE/1523-3* to the U.S.E.R.D.A., Division of Geothermal Energy, University of Utah (Feb. 1977).
5.   Oliker, I., V. A. Permyakov. Thermal Deareation of Water in Thermal Power Plants, Energiya, Leningrad, U.S.S.R. (1971).
6.   Oliker, I. Application of Direct Contact Heat Exchangers in Geothermal Systems. *ASME Paper No. 77-HT-3* (1977).
7.   Goldstick, R. J., KVB, Inc. Survey of Flue Gas Condensation Heat Recovery Systems. *GRI 80/0152*, The Gas Research Institute, Chicago, IL (1981).
8.   Vallario, R. W., D. E. DeBellis. State of Technology of Direct Contact Heat Exchanging, *U.S.D.O.E. Report No. PNL-5008*, Pacific Northwest Laboratories, Battelle Memorial Institute, Richland, WA (May 1984).
9.   Fisher, E. M., J. D. Wright. Direct Contact Condensers for Solar Pond Production. *U.S.D.O.E. Report SERI/TR-252-2164*, Solar Energy Research Institute, Golden, CO (May 1984).
10.   Bharathan, D., J. A. Althof, B. K. Parsons. Direct Contact Condensers for Open-Cycle Ocean Thermal Energy Conversion. *U.S.D.O.E. Report No. SERI/RR-252-2472*, Solar Energy Research Institute, Golden, CO (April 1985).
11.   Bharathan, D., J. Althof. An Experimental Study of Steam Condensation on Water in Countercurrent Flow in the Presence of Inert Gases. *ASME Paper 84-WA/Sol-25.*
12.   Sideman, S., D. Moalem-Maron. Direct Contact Condensation. *Advances in Heat Transfer*, Vol. 15, Academic Press, Inc., New York, NY, pp. 227-281 (1982).
13.   Jacobs, H. R. Direct Contact Condensers. Section 2.6.8., Supplement No. 2 of *Heat Exchanger Design Handbook*, Hemisphere Press (1985).
14.   Kutateladze, S. S., V. H. Borishanskii. *A Concise Encyclopedia of Heat Transfer*, Academic Press (1966).
15.   Ford, J. D., A. Lekic. Rate of Growth of Drops During Condensation. *Internat. Journal of Heat and Mass Transfer* 16:61-66 (1973).
16.   Jacobs, H. R., D. S. Cook. Direct Contact Condensation on a Non-Circulating Drop. *Proceedings of the 6th Internat. Heat Transfer Conf.*, Heat Transfer 1978, 3:389-393, Toronto, Canada (Aug. 1978).

17. Kulik, E., E. Rhodes. Heat Transfer Rates to Moving Droplets in Air/Steam Mixtures. *Proceedings of the 6th Internat. Heat Transfer Conf.*, Heat Transfer 1978, 1:469-474, Toronto, Canada (Aug. 1978).

18. Chung, J. N., P. S. Ayyaswamy. Material Removal Associated with Condensation on a Droplet in Motion. *Internat. Journal of Multiphase Flow* 7:329-342 (1981).

19. Sundararajan, T., P. S. Ayyaswamy. Hydrodynamics and Heat Transfer Associated with Condensation on a Moving Drop: Solutions for Intermediate Reynolds Numbers. *Journal of Fluid Mechanics*, 149:33-58 (1984).

20. Sundararajan, T., P. S. Ayyaswamy. Numerical Evaluation of Heat and Mass Transfer to a Moving Liquid Drop Experiencing Condensation. To appear in *Numerical Heat Transfer* in 1985.

21. Sundararajan, T., P. S. Ayyaswamy. Heat and Mass Transfer Associated with Condensation on a Moving Drop: Solutions for Intermediate Reynolds Numbers by a Boundary Layer Formulation. To appear in the *ASME Journal of Heat Transfer* in 1985.

22. Grace, J. R. Hydrodynamics of Liquid Drops in Immiscible Liquids. *Handbook of Fluids in Motion*, Chapter 38, Ed. N. P. Cheremisinoff and R. Gupta, Ann Arbor Science, The Butterworth Group, Ann Arbor, MI (1983).

23. Brown, G. Heat Transmission During Condensation of Steam on a Spray of Water Drops. Institution of Mechanical Engineers, General Discussion on Heat Transfer, pp. 49-51 (1951).

24. Isachenko, V. P., V. I. Kushnyrev. Condensation in Dispersed Liquid Sprays. *Fifth Internat. Heat Transfer Conf.*, Vol. III, pp. 217-220 (1974).

25. Oliker, I. On Calculation of Heat and Mass Transfer in Jet Type Direct Contact Heaters. *ASME Paper No. 76-HT-21, St. Louis, MO (Aug. 1976).*

26. Kutateladze, S. S. *Heat Transfer in Condensing and Boiling*. Chapter 7, Moscow, U.S.S.R. (1952).

27. Hasson, D., D. Luss, R. Peck. Theoretical Analyses of Vapor Condensation on Laminar Jets. *Internat. Journal of Heat and Mass Transfer* 7:969-981 (1964).

28. Hasson, D., D. Luss, V. Navon. An Experimental Study of Steam Condensing on a Laminar Water Sheet. *Internat. Journal of Heat and Mass Transfer* 7:983-1001 (1964).

29. Jacobs, H. R., R. Nadig. Condensation on Coolant Jets and Sheets. *ASME Paper No. 84-HT-29*, Niagara Falls, NY, (Aug. 1984).

30. Taitel, Y., A. Tamir. Condensation in the Presence of a Noncondensable Gas in Direct Contact. *Internat. Journal of Heat and Mass Transfer* 12:1157-1169 (1969).

31. Nadig, R., H. R. Jacobs. Condensation on Coolant Jets and Sheets in the Presence of Non-Condensable Gases. *ASME Paper 84-HT-28*, Niagara Falls, NY (Aug. 1984).

32. Jacobs, H. R., J. A. Bogart. Condensation on Immiscible Falling Films. *ASME Paper No. 80-HT-110*, Orlando, FL (July 1980).

33. Jacobs, H. R., J. A. Bogart, R. W. Pensel. Condensation on a Thin Film Flowing Over an Adiabatic Sphere. *Proceedings of the 7th Internat. Heat Transfer Conf.*, Heat Transfer 1982, 5:89-94, Munich, Germany (1982).

34. Jacobs, H. R., R. Nadig. Condensation on an Immiscible Falling Film in the Presence of a Non-Condensible Gas. *Heat Exchangers for Two-Phase Applications*, HTD-ASME 27:99-106 (July 1983).

35. Bharathan, D., J. Althof. An Experimental Study of Steam Condensation on Water in Countercurrent Flow in Presence of Inert Gases. *ASME Paper 84-WA/Sol-25*, New Orleans, LA (Dec. 1984).

36. Thomas, K. D., H. R. Jacobs, R. F. Boehm. Direct Contact Condensation of Immiscible Fluids in Packed Beds. *Condensation Heat Transfer, ASME*, pp. 103-110 (Aug. 1979).

37. Tamir, A. and Rachmilev. Direct Contact Condensation of an Immiscible Vapor on a Thin Film of Water. *Internat. Journal of Heat and Mass Transfer* 17:1241-1251 (1974).

38. Finklestein, Y., A. Tamir. Interfacial Heat Transfer Coefficients of Various Vapors in Direct Contact Condensation. *The Chemical Engineering Journal* 12:199-209 (1976).

39. Nadig, R. Private communication with R. Nadig, 1985.

40. Murty, N. S., V. M. K. Sastri. Condensation on a Falling Laminar Liquid Film. *Proceedings of the 5th Internat. Heat Transfer Conf.*, Heat Transfer 1974, 3:231-235 (Sept. 1974).

41. Murty, N. S., V. M. K. Sastri. Condensation on a Falling Laminar Liquid Sheet. *Canadian Journal of Chemical Engineering* 54:633-635 (1976).

42. Rao, V. D., P. K. Sarma. Condensation Heat Transfer on Laminar Liquid Film. *ASME Journal of Heat Transfer* 106:518-523 (Aug. 1984).

43. Nadig, R. Design Studies for Direct Contact Condensers With and Without Noncondensable Gas. Ph.D. Dissertation, University of Utah, Salt Lake City, UT (Dec. 1984).

44. Jacobs, H. R., H. Fannir, G. C. Beggs. Collapse of a Bubble of Vapor in an Immiscible Liquid. *Proceedings of the 6th Internat. Heat Transfer*, Heat Transfer 1978, 3:383-388, Toronto, Canada (Aug. 1978).

45. Florschuetz, L. W., B. T. Chao. On the Mechanics of Vapor Bubble Collapse. *ASME Journal of Heat Transfer* 87:209-220 (1965).

46. Isenberg, J., D. Moalem-Maron, S. Sideman. Direct Contact Heat Transfer with Change of Phase: Bubble Collapse with Translatory Motion in Single and Two-Component Systems. *Proceedings of the 4th Internat. Heat Transfer Conf.*, Vol. 5, Paper B2.5 (1970).

47. Isenberg, J., S. Sideman. Direct Contact Heat Transfer with Change of Phase: Bubble Condensation in Immiscible Liquids. *Internat. Journal of Heat and Mass Transfer* 13:997-1011 (1970).

48. Jacobs, H. R., B. H. Major. The Effect of Noncondensable Gases on Bubble Condensation in an Immiscible Fluid. *ASME Journal of Heat Transfer* 104:487-492.

49. Lerner, Y., H. Kalman, R. Letan, "Condensation of an Accelerating-Decelerating Bubble: Experimental and Phenomenlogical Studies," *Basic Aspects of Two Phase Flow and Heat Transfer, ASME Symposium Volume G00250* (1984).

50. Letan, R. Dynamics of Condensing Bubbles: Effect of Injection Frequency. *ASME/AIChE National Heat Transfer Conf.* (Aug. 1985).

51. Golafshani, M. Bubble Type Direct Contact Condensers. M.S. Thesis, University of Utah (1983).

52. Moalem-Maron, D., S. Sideman, et al. Condensation of Bubble Trains: An Approximate Solution. *Progress in Heat and Mass Transfer* 6:155-177 (1972).

53. Johnson, K. M., H. R. Jacobs, R. F. Boehm. Collapse Height for Condensing Vapor Bubbles in an Immiscible Liquid. *Proceedings of the Joint ASME/JSME Heat Transfer Conf.* 2:155-163, Honolulu, Hawaii (March 1983).

54. Sudhoff, B. *Direkter Warmubergang bei der Kondensation in Blapfensaulen.* Ph.D. Dissertation, Universitat Dortmund, F. R., Germany (May 1982).

55. G. Faeth. Private communication with G. Faeth, Dept. of Aeronautical Engineering, University of Michigan, Ann Arbor, MI (July 1985).

DISCUSSION OF DIRECT CONTACT CONDENSATION AND EVAPORATION

A. F. Mills

1 INTRODUCTION

This chapter is a summary of the discussion that followed the presentations on direct-contact evaporation and condensation. Some detailed comments on these two topics are presented, which we hope will add to their value to the reader. Next is a discussion of important research and development needs. Finally, some brief conclusions are presented.

2 SUMMARY OF THE DISCUSSION

2.1 Evaporation

Maclaine-Cross noted that in equipment such as evaporative coolers and cooling towers, the gas-side resistance to heat and mass-transfer controls, and in reducing this resistance, there is usually a trade-off with gas phase pressure drop. Parallel plate packings have a low pressure drop, but it is difficult to obtain a uniform liquid distribution on these packings.

Jacobs noted that the thrust of the paper was toward obtaining improvements in the performance of OTEC evaporators, and of distillation columns, but in his view there are many fundamental problems that need attention. As examples he gave the problem of designing a boiler that contacts two immiscible fluids such as isobutane and brine, and the consequences to the environment of a large spill of propane on the ocean.

2.2 Direct-Contact Condensation: What We Know and What We Don't

Mills noted that the cited analyses of the effect of noncondensables on the performance of jet and film condensers all assumed a quiescent vapor, but that in reality there was always a significant vapor flow. Jacobs replied that the vapor velocity was always very low in the particular applications of interest to him.

Bharathan asked for a comment on the possibility of fog formation in direct-contact condensers. Jacobs replied that when a super-saturated condition exists there are usually ample nucleation sites for fog to nucleate, practical equipment being dirty. If the number of sites are assumed, analysis of the condensation process is straightforward.

Welty asked for a comment on when the droplet configuration was preferable to films or jets. Jacobs replied that it depended on the particular application. One difficulty with droplet condensers is the large nozzle pressure drop required to get a fine spray: if coolant pressure drop is an important constraint, jet-type condensers are preferable. Thus the economics of the coolant pumping plays an important role in choosing the type of direct-contact condenser.

Pesaran asked which condenser type was more sensitive to the effects of noncondensables, droplet or falling films. Jacobs replied that most people do not concern themselves with this question when they design a direct-contact condenser, but they should. Again it depends on the particular application, and the complete system.

3 FURTHER COMMENTS ON THE CONTRIBUTED PAPERS

3.1 Direct-Contact Evaporation

Contrary to the statement made in the Introduction, there is no significant relation between surface tension and volatility, e.g., water is more volatile than n-butyl alcohol, although its surface tension is three times greater.

Referring to the subsection, Limiting Rate of Evaporation, a key statement is missing from the first paragraph, as can be seen by examining the source for this material (Sherwood, Pigford, and Wilke, 1975). In order to complete the reasoning leading to their Eq. (1), one must note that evaporation flux equals the incident flux at equilibrium, and then postulate that for nonequilibrium the evaporation flux is unchanged.

The interfacial resistance to evaporation is more usefully expressed in terms of a temperature driving force and an interfacial heat transfer coefficient, (Silver, 1946). When the effect of bulk motion in the vapor phase is accounted for the result is (Schrage 1953, Mills and Seban, 1967)

$$h_i = \frac{1}{k_i} = \frac{\alpha}{1 - 0.5\alpha} \frac{h_{fg}^2}{\left(2\pi R T_s\right)^{1/2} T_s v_g} \quad W/m^2 K \qquad (1)$$

where h_{fg} is the latent heat of vaporization, and v_g is the vapor specific volume. For an accommodation coefficient $\alpha = 1$, Eq. (1) gives an interfacial resistance equal to 1/2 of the value given Barathan in his Eq. (2). Also contrary to his statement, the associated resistance is not dependent on transfer rate, as is any linear resistance, and can be important at low as well as high mass-transfer rates. The important dependence is on pressure level: in Eq. (1) this feature enters through the vapor specific volume. For clean water there are no reliable experimental data that indicate a value of α less than unity (Mills and Seban, 1967). For an OTEC evaporator at $25\,^\circ$C, $h_i = 1.1 \times 10^6$ W/m^{2}K, which is very large, and hence the interfacial resistance can be ignored in the evaporator design (Wassel and Ghiaasiaan, 1985). At $5\,^\circ$C, $h_i = 0.36 \times 10^6$ W/m^{2}K, and in an OTEC condenser the interfacial resistance may play a small role, particularly if the feed to the evaporator is deaerated.

Figure 1 and Eq. (4) in the paper do not pertain to a *direct-contact* process since there is an exchanger wall and a cooling or heating fluid. In direct-contact evaporation (or condensation) the liquid-side heat and mass-transfer coefficients are for transfer between the bulk liquid and interface. For falling films sample correlations are given by Wassel and Mills (1982). There is a paucity of experimental data for such coefficients, and no reliable theories owing to the complex nature of the effects of surface waves, and in the case of turbulent liquid flows, owing to the effect of surface tension on damping of turbulence near the interface (Won and Mills, 1982). Modeling of simultaneous heat and mass transfer as recommended by Bharathan has already been used for direct-contact evaporation and condensation: applications include OTEC falling film exchangers (Wassel and Mills, 1982) and scrubbing of radioactive nuclides in a boiling water nuclear reactor pressure suppression pool (Wassel et al., 1985).

In the section, Evaporation for Vapor Production, the author refers to an analysis by Bharathan and Penney (1984) to support their experimental results, which showed how screens could be used to improve the performance of flashing jet evaporators. The analysis cited incorporates the interfacial resistance, and the consequences of the results are explained in terms of the effect of interfacial resistance on interfacial temperature. However, as noted above, the interfacial resistance has a negligible impact on evaporator performance under OTEC conditions. Wassel and Ghiaasiaan (1985) ignored the interfacial resistance in their model and obtained results identical to those of Bharathan and Penney, as expected.

3.2 Direct-Contact Condensation: What We Know and What We Don't

In describing the current status of direct-contact condensation on droplets the author shows how Jacobs and Cook (1978) modified the simple conduction in a sphere model of Ford and Lekic (1973) to account for the effect of the moving boundary associated with droplet growth. The correction was small but did indeed improve agreement with experimental data. However, of more concern is surely the effect of shape oscillations on heat transfer inside the droplet. Ford and Lekic obtained data for water droplets of $\approx$ 1.5 mm diameter, and if oscillations were present they were apparently not observed in high-speed photographs, and clearly there was no apparent effect on heat transfer. On the other hand Hijikata et al. (1984) obtained data for methanol and refrigerant R 113 droplets of $\approx$ 1 mm diameter and did observe marked shape oscillations; they obtained heat transfer rates more than 10 times larger than indicated by conduction analyses, and more than four times larger than the circulation analysis of Kronig and Brink (1949). The question of under what conditions droplets oscillate and circulate clearly needs to be answered before direct-contact droplet condensers can be reliably designed.

In connection with both droplet and bubble condensers the author indicates a need for further examination and understanding of the effects of slightly soluble gases. This problem has already been examined for OTEC exchangers by Wassel and Mills (1982), for pressure suppression pools by Wassel et al. (1985), and for droplets by Chung and Ayyaswamy (1981). The analysis is straightforward and is generally not subject to any uncertainties over and above those associated with the noncondensable gas problem, since the transfer processes involved are identical.

In the case of jet and film condensers the theoretical analyses cited by the author are subject to the severe limitation of laminar liquid flow with wave free surfaces and a quiescent vapor. The author suggests that the results of Jacobs and Nadig (1983) are relevant to packed bed film condensers, but in the analysis the vapor is quiescent and in a packed bed there is forced vapor flow. It is well known that the vapor flow configuration and parameters play a critical role in determining the effect of noncondensables. The analyses performed to date do not supply more useful information than could be obtained by simple scaling of the governing equations.

Recent work relevant to condensation on turbulent jets includes that of Mills et al. (1982), Bharathan et al. (1982), Kim (1983), and Sam and Patel (1984). It is clear from this work that there are no satisfactory models for liquid phase transport in a turbulent jet available at this time.

4 RESEARCH AND DEVELOPMENT NEEDS

It is the view of this writer that research and development needs for direct-contact evaporators and condensers fall into two broad categories experimental work to characterize key hydrodynamic phenomena, and development of computer codes for component and system design. Each of these categories will now be discussed.

4.1 Hydrodynamic Phenomena

Direct-contact processes very often have complex hydrodynamic phenomena playing a key role. For example, for droplets and bubbles, there is the question of under what conditions they oscillate or have internal circulation, and the resultant effect on transfer rates: the effect of surfactant level on these phenomena is presently very poorly understood. As a second example, there is the role played by interfacial waves (both gravity and capillary) on liquid-side transport in turbulent falling films and jets. A third example is the question of under what conditions jets break up and the nature of the resulting dispersed phase. In most such situations it is more important to reliably know the gross hydrodynamic features than to have accurate models or correlations of the associated heat, mass-, and momentum-transfer rates. Given the hydrodynamics, transfer rates can be estimated sufficiently accurately for most design purposes.

4.2 Computer Codes

Direct-contact processes involve at least two phases and often more than one chemical species. Thus analysis is inherently complex (but not necessarily difficult) owing to the number of conservation equations to be solved. Given the present stage of development in the use of computers by engineers, it is suggested that the *minimum* level of approach should involve simultaneous solution of one-dimensional forms of the governing conservation equations. Of course such an approach was pioneered by Colburn and Hougen (1934) for surface condensers, before computers were available. Lee (1971) was perhaps the first to use a computer for this purpose.

With this approach the couplings between heat-, mass-, and momentum-transfer processes can be properly accounted for and the complex transport and thermodynamic properties properly evaluated. Concerns about partially soluble gases disappear since species conservation equations for each such gas species are easily added (Wassel and Mills, 1982). In some exchangers aerosol transport and deposition is of concern, and additional equations for aerosol transport can be then added (Wassel et al., 1985). In some situations two-dimensional analyses are warranted and feasible, e.g. for cooling towers (Mujumdar et al., 1983).

In one-dimensional codes all transfer processes are, of necessity, described by correlation formulas. Thus the purpose of more fundamental research should be to improve the accuracy of key correlations. Of course, availability of such codes allows sensitivity studies to be made that can identify the key transfer processes. When the hydrodynamics are well established it is usually possible to supply correlations of sufficient accuracy. Testing of scale model or prototype equipment then can be used to make adjustments in the correlations to obtain an improved comparison between prediction and test data.

5 CONCLUSIONS

The most important research and development needs identified are as follows:
1.   Experimental research to characterize key hydrodynamic phenomena.
2.   Development of computer codes for design purposes, which, as a minimum, solve one-dimensional forms of the governing mass, momentum, energy, and species conservation equations.
3.   More attention should be given to direct-contact evaporation and condensation phenomena involved in systems other than those commonly found in the process and chemical industries.

REFERENCES

Bharathan, D., Olson, D. A., Green, H. J., and Johnson, D. H., *Measured Performance of Direct Contact Jet Condensers*, SERI/TP-252-1437, Solar Energy Research Institute, Golden, Colorado, January, 1982.

Bharathan, D., and Penney, T., "Flash Evaporation from Turbulent Water Jets," *Journal of Heat Transfer*, Vol. 106, 1984, pp. 407-416.

Chung, J. N., and Ayyaswamy, P. S., "Material Removal Associated with Condensation on a Droplet in Motion," *Internat. Journal of Multiphase Flow*, Vol. 7, 1981, pp. 329-342.

Colburn, A. P., and Hougen, O. A., "Design of Cooler Condensers for Mixtures of Vapors with Noncondensing Gases," *Ind. Eng. Chem.*, Vol. 26, 1934, pp. 1178-1182.

Ford, J. D., and Lekic, A., "Rate of Growth of Drops During Condensation," *Internat. Journal of Heat and Mass Transfer*, Vol. 16, 1973,, pp. 61-66.

Hijikata, K., Mori, Y., and Kawaguchi, "Direct Contact Condensation of Vapor to Falling Cooled Droplets," *Internat. Journal of Heat and Mass Transfer*, Vol. 27, 1984, pp. 1631-1640.

Jacobs, H. R., and Cook, D. S., *Direct Contact Condensation on a Non-Circulating Drop* Proceedings of the 6th Internat. Heat Transfer Conf., Toronto, Canada, Aug. 1978, Vol. 3, pp. 389-393.

Jacobs, H. R., and Nadig, R., *Condensation on an Immiscible Falling Film in the Presence of a Non-Condensable Gas*, Heat Exchangers for Two-Phase Applications, HTD-ASME Vol. 27, July, 1983, pp. 99-106.

Kim, S., *An Investigation of Heat and Mass Transport in Turbulent Liquid Jets*, Ph.D. Dissertation, 1983. School of Engineering and Applied Science, University of California, Los Angeles.

Kronig, R., and Brink, J. C., "On the Theory of Extraction from Falling Droplets" *Appl. Scient. Res*, 1949, Vol. A2, pp. 142-155.

Lee, S. J., "Effect of Noncondensable Gas on Condenser Performance," MS Thesis, 1971. School of Engineering and Applied Science, University of California, Los Angeles.

Majumdar, A. K., Singhal, A. K., and Spalding, D. B., *VERA 2D-A Computer Program for Two-Dimensional Analysis of Flow, Heat and Mass Transfer in Evaporative Cooling Towers*, Vols. 1 and 2. EPRI Report CS-2923, March, 1983. Electric Power Research Institute, Palo Alto, California.

Mills, A. F., and Seban, R. A., "The Condensation Coefficient of Water," *Internat. Journal of Heat and Mass Transfer*, Vol. 10, 1967, pp. 1815-1828.

Sam, R. G., and Patel, B. R., "An Experimental Investigation of OC-OTEC Direct-Contact Condensation and Evaporation Processes," *Journal of Solar Energy Engineering*, Vol. 106, 1984, pp. 120-127.

Schrage, R. W., *A Theoretical Study of Interphase Mass Transfer 1953*. Columbia University Press, New York.

Sherwood, T. K., Pigford, R. L., and Wilke, C. R., *Mass Transfer 1975*. McGraw-Hill, New York, pp. 182-184.

Silver, R. S., "Heat Transfer Coefficients in Surface Condensers," *Engineering*, London, V. 161, 1946, p. 505.

Wassel, A. T., and Mills, A. F., *Turbulent Falling Film Evaporators and Condensers for Open Cycle Ocean Thermal Energy Conversion*, Advancement in Heat Exchangers, 1982, Hemisphere Press.

Wassel, A. T., and Ghiaasiaan, S. M., "Falling Jet Flash Evaporators for Open Cycle Ocean Thermal Energy Conversion," *Int. Comm. Heat Mass Transfer*, Vol. 12, 1985, pp. 113-125.

Wassel, A. T., Mills, A. F., Bugby, D. C., and Oehlberg, R. N., "Analysis of Radionuclide Retention in Water Pools," *Nuclear Engineering and Design*, Vol. 19, 1985, pp. 764-781.

Won, Y. S., and Mills, A. F., "Correlation of the Effects of Viscosity and Surface Tension on Gas Absorption into Freely Falling Turbulent Liquid Films," *Internat. Journal of Heat and Mass Transfer*, Vol. 25, 1982, pp. 223-229.

RESEARCH NEEDS IN DIRECT-CONTACT HEAT EXCHANGE

R. F. Boehm and Frank Kreith

1 INTRODUCTION

In the preceding chapters of this book, direct-contact heat transfer phenomena have been reviewed, and available theoretical and empirical information has been summarized. In this chapter, the current research needs are summarized under several general categories.

There are, simultaneously, marked distinctions and amazing similarity in the many facets that make up the field of direct-contact heat exchange. While on the one hand there are differences between situations where solids, liquids, and vapors are individually interacting in complex flow patterns with a surrounding fluid, these flow patterns also exhibit some of the same general characteristics. More-over, a key ingredient of almost all of the heat transfer processes described is a fundamental dependence on the fluid mechanics of swarms of particles in a bulk fluid. Sometimes the effects of the fluid mechanics interact with change-of-phase phenomena, but there is still a dependence upon the fluid motion to drive the direct-contact heat transfer process.

The *most critical need* identified in this workshop is the *development of simplified design techniques for direct-contact heat transfer devices.* If direct-contact heat transfer processes are to be given design considerations in a manner similar to closed heat exchangers, pertinent design tools must be developed.

2 AREAS WHERE RESEARCH IS NEEDED

2.1 Analogy Between Heat and Mass Transfer

Many of the early insights about direct-contact heat transfer processes have been gained from the application of analogies between heat transfer and mass transfer phenomena. The mass-transfer field was developed primarily to meet the needs of the chemical process industry, and a great deal of data exists for direct-contact mass transfer.

There are many similarities between heat transfer and mass transfer. Both are driven by gradients, are highly dependent on the effective contact areas, and are heavily influenced by the fluid mechanics present. But there are also a number of distinctions between heat and mass transfer processes. Usually the mass transfer occurs in a nearly isothermal environment, whereas by definition heat transfer must be nonisothermal. A variety of nonisothermal conditions can exist in a heat transfer process, and in addition there are a number of situations that involve simultaneous heat and mass transfer.

The application of mass transfer data that exist for a variety of direct-contact devices and processes is straightforward in some heat transfer situations, but not in others. Hence, the use and applicability of the mass transfer/heat transfer analogy must be more completely delineated. Particularly important is a careful mapping of limitations of this valuable tool. The heat transfer/mass transfer analogy may play an important role in a fundamental development of generalized design methods, but before this approach is used its generality needs to be evaluated.

Basic work is needed on the effects of simultaneously imposed temperature and concentration gradients. Much might be learned from single-bubble experiments about this situation

2.2 Influence of Internal Configurations in Heat Transfer Devices

Internal configurations such as packings, trays, and screens have an important role in the performance of columns and other types of direct contactors. The influence of these components on the flow pattern and the heat transfer performance needs to be understood. Available data need to be analyzed, and areas in which additional experiments are needed should be identified.

Valuable contributions to basic understanding can come from a general characterization of the mass transfer/heat transfer pressure-drop performance of commercial packings. While some of this information is available from manufacturers, standardization of data correlation is needed for designers. The field could benefit from an independent evaluation and data tabulation. Of particular importance is a better understanding of the performance of new structured and high conductivity packings. In particular, heat transfer versus pressure drop design tradeoffs between spray columns and towers with internals of various packing density and structure are needed for engineer design and optimization.

2.3 Influence of Additives, Contaminants, and Inert Gases

The existing direct-contact heat transfer design information is generally given without considering the possible influence of other substances besides those that make up the two primary streams. But there are few industrial applications that are "clean," and studies are therefore required to define the design implications of contaminants on the performance of heat transfer systems. For example, the effects of trace amounts of gases in liquid systems and their influence on heat transfer need too be defined. This includes understanding the effect of inert or noncondensable gases in change-of-phase direct-contact processes. A body of knowledge that applies to the related situation in closed heat exchangers leads to two conclusions: noncondensables have a negative effect on performance of condensers and evaporators, and a complete understanding of these effects is very difficult to attain. Similar situations are found in direct-contact heat transfer. Inerts can exist either in the continuous phase or the in the dispersed phase. Little information is available to define what effects these two situations have. Attempts to analyze these effects have assumed the interface is not penetrated by the inerts. Modifications to theories need to be made to include the effects of inerts penetrating the interface. Certain types of heat exchanger configurations perform better than others in the presence inerts, and the tradeoffs between exchanger configuration must be analyzed.

Effects of surfactant additives, particularly in change-of-phase systems, need to be explored. While some understanding of the role of surface tension in direct-contact systems exists, these effects need to be examined in detail. If the designer had information on the quantitative effects of surfactants with various combinations of fluids, the performance of direct-contact applications could be improved.

2.4 Fluid Mechanical Considerations

All direct-contact processes rely upon some aspects of fluid mechanics to accomplish a desired operation. In this sense, a basic understanding of the fluid flow patterns in a given application will always be of value. There are some complicated fluid mechanical aspects that might yield fruitful results on a broad front if general insights could be gained.

The interactions between bubbles in columnar flow are not well understood. This includes the droplet breakup and agglomeration that can occur in a variety of direct-contact processes. While some progress has been made in recent years to develop physical models for these situations, much room remains for improvement.

A fundamental problem that combines several aspects of fluid flow is a description of the droplets' area in terms of the variables influencing their geometric configuration. An understanding of this relationship could greatly facilitate the description of the performance of a variety of direct-contact devices. If variations in droplet shape as a function of droplet size and external fluid environment could be accurately predicted, this could lead to a better understanding of the overall heat exchanger performance.

The overall size of a device (e.g., a spray tower's height and diameter) has a profound influence on the manner in which it should be optimally operated and the resulting overall heat transfer effectiveness that the device can attain. Droplet coalescence and velocities that vary radially in spray columns, for example, are phenomena that demonstrate the effect of complex fluid mechanics. Many of these phenomena are not easily predicted in a quantitative manner. Of particular importance is an understanding of those fluid mechanics phenomena that require increases in the overall size of the heat transfer device to achieve a given heat duty.

Most work of a fundamental nature on the fluid mechanics of direct-contact heat exchangers has been for laminar flow. Hence, studies to determine when turbulent effects are important, as well as work to analyze the result of these turbulent effects, are necessary. Related to this is a need to be able to predict the effects of interfacial waves generated by gravity and capillary effects.

A large number of special fluid mechanics effects are not well understood. Examples are agitation, pulsation, and acoustics wave interaction. An understanding of which of these can improve the performance of heat-transfer devices could be quite valuable.

Fluid mechanics can have a very large impact on the performance of a nozzle during droplet formation. There have been a large number of nozzle injection and jet breakup studies performed for a wide variety of applications, including some that are applicable to direct-contact systems. This portion of the overall direct-contact heat exchanger process is extremely important because the heat transfer rates are relatively large in a region of the injection process. Hence, appropriate design of the fluid injection device is critical to maximizing performance. Definition of appropriate design practice based on examining previous research as well as performing new work could pay off in major performance improvement. Also, the movement of gas bubbles through the bed and the effects of these bubbles on the overall heat transfer performance should be studied.

Description of partial motion (both short- and long-term) in solids-fluidized-bed devices is only now beginning to show progress. This process is complicated because of the erratic motion and the interaction with immersed tubes. As with liquid and vapor systems, an understanding of this motion is fundamental to the development of meaningful models.

2.5 High-Temperature Effects

In high-temperature gas-solids direct-contact heat transfer systems radiation is an important mechanism of energy exchange. The theory describing this mode of exchange is generally well developed, but the methods required to calculate the performance information for typical systems can be very complicated. Hence, the frontiers are somewhat different in this area than for the other topics discussed in this chapter.

Valuable contributions to design techniques can be made by developing methods that incorporate the complex radiation information in a simplified form. Although there have been some attempts to do this, a great deal of room for improvements exists. Critical issues to be addressed include simplified rules that allow the designer to determine when radiation effects will be important compared convection and/or phase change phenomena, and short-cut methods for calculating the contribution of the radiant flux in specific physical situations.

Previous work in radiative transport could be used to optimize direct-contact heat transfer designs by affecting the radiant heat transport. This can be done by controlling the parameters that can be varied, including the optical properties of the solids and the enclosing walls, the particle size, and the temperature distribution within the solids' stream. In many applications, an enhancement of lateral and/or transverse mixing of particles may increase the particular effect sought.

2.6 Experimental Techniques

Many of the research needs listed above require experimental measurements. Most of the direct-contact heat transfer processes currently conceived are characterized by complex interactions between two heat-exchanging streams. Hence, the development of experimental techniques to visualize fundamental aspects of these phenomena is important.

Experimental studies can focus on the macroscopic behavior of the device. Determination of the temperatures of the various input and output streams is an example of this. Another possibly more difficult problem is the determination of average holdup in a fluid-fluid heat exchanger under a given set of operating conditions.

Microscopic details could be quite valuable in trying to understand the overall performance of a device, but inferring these details may range from difficult to impossible. Consider the complexities of determining the individual particle motion and thermal environment for solids in a fluidized bed. Here a given particle interacts with a variety of other particles, all of which are moving at different velocities and are at different temperatures. Superficially both the velocities and the temperatures appear to be totally random. Creative, experimental solutions and flow visualization could be of great benefit to this field.

Measurement of heat transfer rates to bubbles in seemingly randomly moving swarms can be equally difficult. The situation is often complicated by the presence of trace amounts of other materials (e.g., noncondensable gases) that could have a profound effect on the overall performance.

The determination of the temperatures of the continuous and dispersed phases separately at a given point within an exchanger has been an extremely perplexing problem. A Lagrangian frame of reference for measurements has normally been considered too difficult. Use of an Eulerian basis has resulted in problems of associating real meaning to fluctuating temperature traces.

High-temperature systems, usually with solids flow, offer a particularly complicated environment in which to determine microscopic aspects of performance. In order to understand thermal performance, it may be desirable to measure separately the following items at a given point: the particle temperature, the continuous phase temperature, and the radiation flux in a number of directions. Good instruments are needed to achieve these measurements.

2.7 Numerical Modeling

In the years to come, numerical modeling will play a key role in tailoring direct-contact processes for applications in heat transfer. The physics of the processes may be treated as the bricks and numerical modeling as the mortar, which together build design solutions.

There have been many attempts at developing numerical models to be direct-contact processes. Some have been quite successful but several have failed, and it is not always clear why. In nearly all cases, however, the documentation of these models has been insufficient to allow easy application by others. Assumptions made during code development are often not stated clearly enough to enable a user to determine whether the situation represented in one case can be applied to others.

Code development will continue to fall into three categories. The most detailed models will represent the microscopic aspects of the problem, either on a local basis or globally. This approach will be formulated in a two-dimensional (axisymmetric) or fully three-dimensional representation of the physical processes that control the performance of a given direct-contact device. These codes will be used to develop simplifications that can be used in the next level of numerical modeling.

The work-horse of the various codes will continue to be the one-dimensional models. In these models, the influences of the physical processes in two of the dimensions are combined and represented as an averaged effect locally, varying only with the longitudinal dimension. Some types of devices will not be well described by this approach. Examples of configurations difficult to represent one-dimensionally are truly crossflow configurations. Usually, though, appropriate incorporation of the physical processes locally will enable the engineer to examine the effects of key variables on the overall performance of the device at relatively small code development and computation cost.

Finally, the simplest codes will continue to use the results of other studies in an empirical formulation to predict the overall performance of the device in a direct manner. The heat transfer/mass-transfer analogy is one example of this level. Output from more complicated codes can usually be correlated and used in these simpler forms. Work is needed to develop the codes necessary to reduce the application and design of direct-contact devices to common engineering practice. This process is needed for many of the applications described in earlier chapters.

As indicated in Chapter 3, the fundamentals of modeling are generally well understood. The major need is to incorporate the physics of the processes into the various codes. Also, the codes must be completely documented to allow the wide application of the code in evaluating performance of devices and in developing design information.

The high-temperature applications, while complicated by the presence of radiative transport, should follow approximately the same type of development. Here the fluid mechanics, convective transport, radiative transport, and possibly change of phase physics will need to be incorporated. The use of simplifying codes to evaluate the output of more complete models may enable approximate but accurate incorporation of radiative transport into empirically based design codes.

3 CONCLUSION

The overriding need to increase the use and application of direct-contact heat transfer processes in engineering is the development of reliable and simplified design techniques. The currently available design tools are empirically derived and lack generality. Design methods that are based on a solid understanding of the physical processes and whose range of application is clearly delineated are needed. Analogies are useful, but until their limitations are known quantitatively their application can lead to erroneous conclusions in some cases.

There are a lot of data on direct-contact processes in the possession of proprietary research organizations such as FRI, EPRI, and HTRI. Many of these data are outdated or of unknown reliability. It would be of great value if these proprietary data could in some way be made available to scrutiny and analysis by engineering scientists so that a solid basis for further work can be established rapidly.

After the research outlined above has been completed, one can expect that generalized approaches similar to the effectiveness-NTU or the modified log mean temperature difference techniques for closed heat exchangers will become available for a wide range of direct-contact heat transfer processes. When that occurs, industry and the consumer will reap the benefits of products being produced more efficiently and at lower cost. In the meantime, the use of direct-contact processes will continue to develop as appropriate applications are specified by resourceful engineers.

EXAMPLE CALCULATIONS FOR MASS TRANSFER EFFECTS

J. J. Perona

Consider a direct contact boiler in which isobutane is vaporized by contact with hot brine. Estimate the concentration of isobutane dissolved in the effluent brine.

Let us choose boiler conditions similar to those of the 500 kw geothermal pilot plant at East Mesa:

> Brine flow rate 97,200 lb/hr
> Inlet brine temperature 340 ° F
> Boiler pressure 467 psia
> Isobutane flow rate 99,400 lb/hr

The rate of dissolution of isobutane into the brine is given by a material balance over the boiler:

$$R = B(X_1 - X_0) \tag{1}$$

where R = rate dissolved, lb/hr

B = brine flow rate, lb/hr

X_1 = mass fraction of isobutane in brine phase leaving the boiler

X_0 = mass fraction of isobutane in brine entering the boiler

Of course, X_0 is zero. Observation of experimental direct contact boilers indicates that good mixing takes place. Let us assume that the boiler is perfectly mixed; that is, the temperature and compositions of the bulk brine and isobutane phases are uniform throughout the boiler, and therefore are the same as the exit streams.

With the assumption of perfect mixing, the rate of transfer of isobutane to the brine phase may be written in terms of the mass transfer coefficient as follows:

$$R = (k_z a) \, V \, (X^* - X_1) \tag{2}$$

where $k_z a$ = the volumetric mass transfer coefficient
for the brine phase, lbs/(hr)(cu. ft of boiler volume)

X^* = equilibrium concentration of isobutane in brine, mass fraction

V = boiler volume, cu. ft.

The isobutane phase is essentially pure isobutane, and there is no resistance to mass transfer in the isobutane phase.

Equations (1) and (2) may be combined by eliminating R, and solving for X_1

$$X_1 = \frac{(k_z a) \, (V)(X^*)}{(k_z a) \, (V) + B} \tag{3}$$

Equation (3) provides the answer for the example, but it requires a value for a mass transfer coefficient in a direct contact boiler. No experimental correlations are available for this; therefore, we will make an estimate from experimental heat transfer coefficients. Thus, from the Colburn (or j-factor) analogy between heat and mass transfer

$$k_z a = \frac{1}{C_p} \cdot \left[\frac{N_{Pr}}{N_{Sc}} \right]^{2/3} U_v \tag{4}$$

where U_v = volumetric heat transfer coefficient, Btu/hr ft^3 °F.

A well-accepted general correlation for predicting values of U_v is not yet available. Values for U_v for the isobutane-water boiler as East Mesa are reported by Lawrence Berkeley Laboratory to be in the 30,000 to 40,000 range.

The saturation temperature is 250°F and the bulk brine temperature is slightly higher, at about 260°F. Under these conditions the following property values are found:

$N_{Pr} = 1.4$

$N_{Sc} = 45$

$X^* = 360$ ppm

Using a value for U_v of 35,000 Btu/hr ft^3 °F, we obtain

$$k_z a = 3560 \text{ lb/hr ft}^3$$

and $V = 15 \text{ ft}^3$

Finally, $X_1 = 130$ ppm

AIR/MOLTEN SALT DIRECT-CONTACT HEAT TRANSFER ANALYSIS

Mark S. Bohn

NOMENCLATURE

a	interfacial surface area per unit volume, m^{-1} (ft^{-1})
A_a	finned-tube heat exchanger surface area, m^2 (ft^2)
AC	levelized annual cost, \$/yr
C	specific heat, J/kg °C (Btu/lbm °F)
C_f	parameter in H_d correlation
C_f	plant capacity factor
C_p	specific heat at constant pressure, J/kg °C (Btu/lbm °F); 2.394 x 10^{-4}
CI	capital cost, \$

Originally prepared under Task No. 4250-00, WPA No. 431-83, for the U.S. Department of Energy Contract No. DE AC02 83CH10093, SERI/TR-252-2015, UC Category: 59c, DE 84000080, November, 1983.

D	mass diffusivity, m^2/h (ft^2/h); 10.76
d_p	packing size, cm (in.)
d_t	column diameter, m (ft)
f_1	viscosity function
f_2	density function
f_3	surface tension function
G	gas-flow rate per unit bed area, $kg/h\ m^2$ ($lbm/h\ ft^2$); 0.2044
H	column height or finned-tube exchanger height, m (ft)
H_d	height of a transfer unit, m (ft)
h	heat transfer coefficient, W/m^2 °C ($Btu/h\ ft^2$ °F); 0.1761
k	thermal conductivity, W/m °C ($Btu/h\ ft$ °F); 0.5777
k_g	gas side mass-transfer coefficient, kg mol/h m^2 atm (lb mol/h ft^2 atm); 0.2044
L	liquid flow rate per unit bed area, $kg/h\ m^2$ ($lbm/h\ ft^2$); 0.2044
L	length of finned-tube heat exchanger in flow direction, m (ft)
$\dot{m}$	mass flow rate, kg/h (lbm/h)
m,n	exponents in H_d correlation
OM	operating cost, \$/yr
P	total pressure, atm
P_{BM}	logarithmic mean partial pressure of component B, atm
Pr	Prandtl number
Q	heat transfer or heat duty, W (Btu/h); 3.412
R	universal gas constant, m^3 atm/°C kg mol (ft^3 atm/°R lb mol); 8.918
Sc	Schmidt number
T	absolute temperature, °C (°R)
Ua	overall volumetric heat transfer coefficient, W/m^3 °C ($Btu/h\ ft^3$ °F); 0.05368
V	flow velocity, m/h (ft/h)
V_p	volume of packing bed, m^3 (ft^3)
W	finned-tube heat exchanger width, m (ft)

Greek

α	heat transfer area per volume, m^{-1} (ft^{-1})
Δp	heat exchanger pressure drop, N/m^2 (psi); 1.451×10^{-4}
ΔT_m	log mean temperature difference, °C (°F)
μ	absolute viscosity, N h/m^2 (lbm/ft h); 8.690×10^6
ν	kinematic viscosity, m^2/s (ft^2/s); 10.76
ρ	density, kg/m^3 (lbm/ft^3); 0.0623
σ	surface tension, dyne/cm
ϕ,ψ	parameters in H_d correlation

Subscripts

a	air
g	gas
i	inlet
o	outlet
s	salt
ℓ	liquid

NOTE ON SYSTEM OF UNITS

Since a majority of the chemical engineering literature, especially product literature, continues to use the English system of units, we have not attempted to con-
vert such information into the SI system of units. Any data generated in this
study, however, is presented in the SI system. To facilitate conversion between
the two systems, the factor for converting SI units to English units follows the
more important quantities in the preceding list. For example, $Ua = 1000$ W/m^3
$^\circ$C $= 53.68$ Btu/h ft^3 $^\circ$F for a volumetric heat transfer coefficient.

1 INTRODUCTION

Direct-contact heat transfer is the transfer of heat across the phase boundary of
two immiscible fluids, either two liquids or a liquid and a gas. Conventional heat
exchange technology involves heat transfer across a solid boundary such as the
wall of a steel tube in a shell-and-tube exchanger or across a plate in plate heat
exchangers. Where the two fluids do not react and can be separated after the heat
exchange has been effected, direct-contact heat exchange (DCHX) has several
advantages over conventional heat exchangers: (1) without the intervening wall, a
lower thermal resistance is present, and there is not heat exchange surface to foul;
(2) intimate mixing of the two fluid streams can produce very high rates of heat
transfer; and (3) the heat exchanger design can be simpler, require fewer materials
of construction, and allow more flexibility in choice of materials.

Figure 1.1 depicts a conventional finned-tube heat exchanger, and Fig. 1.2
shows a DCHX with a packed column. In the finned-tube heat exchanger one
fluid is pumped through the tubes, and the other fluid is pumped over the outside
of the tubes. The entire tube bundle is enclosed in the shell, which contains the
latter fluid. Sufficient heat transfer area is provided so, given the temperature
differential between the fluids and the heat transfer coefficients, the required heat
duty can be met. If one or both of the fluids is a gas, it is common to provide fins
on the gas side to increase the heat transfer surface because of the poor heat
transfer characteristics of gases. The DCHX is a column substantially filled with a
packing material. The packing material consists of rings or saddles (Fig. 1.3) that
are generally 2-3 in. in size for large columns and are dumped in the column in a
random arrangement. As shown in Fig. 1.2, one fluid enters the top of the vessel
and flows countercurrent up through the vessel. It is also possible to have a
crossflow configuration.

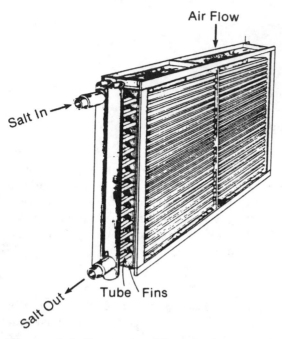

Figure 1.1 Conventional finned-tube heat exchanger.

When the DCHX uses a gas and a liquid, the liquid flows downward by gravity and the gas flows upward in the countercurrent configuration. By properly distributing the liquid at the top of the packing, the liquid forms many small rivulets that flow over the packing. These rivulets give a large surface area between the two phases and increase the time during which the liquid stream is exposed to the gas, greatly increasing the rate of heat transfer per unit volume of heat exchanger. We can eliminate the packing by simply spraying the liquid downward and having the gas flow upward in an empty column (i.e., a spray column). Although we did not study the spray column, results of the economic analysis (Section 4) indicate that it could offer some advantages over the packed bed at very high temperatures and, therefore, it warrants further investigation.

Applications in which DCHX is especially attractive include those in which it is necessary to transfer heat between a gas and a liquid because large heat transfer rates can be achieved without the added expense of finned tubes. In solar thermal technology, two examples include high-temperature process air and the Brayton cycle (shown in Fig. 1.4). In both examples, solar energy provides a heat source at a central receiver in which molten salt cools the receiver and transfers the solar energy to a storage device. Molten salt is the logical heat transfer fluid at high temperatures because it exhibits very low vapor pressure, has high sensible heat storage, has excellent heat transfer characteristics, and is relatively benign in relation to the receiver containment materials (at temperatures below 600 °C for state-of-the-art nitrate salts).

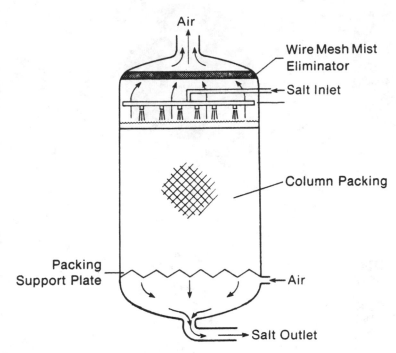

Figure 1.2 Direct-contact heat exchanger.

There are two configurations for storing the energy in the molten salt to accommodate the diurnal variation in energy supply or variations in the load. In the first configuration, the molten salt is stored in an insulated vessel providing storage by the sensible heat in the salt. In the second, the salt is used to heat air, which in turn heats a rock bed to provide sensible heat storage. The first method will be necessary for very high temperature applications in which the only material that can tolerate the temperature cycling in the storage is the salt itself. The second storage method is attractive for lower temperature applications ($<600\,^\circ$ C) because it provides economical, long-term storage. In either storage concept it is necessary to transfer heat from the molten salt to air.

Using packed columns is very common in the chemical process industry for mass-transfer operations. One example is the removal of carbon dioxide from a gas stream by contacting the gas stream with monoethanolamine (an organic liquid) or with a hot carbonate solution. The gas is blown up through the bottom of the column, and the monoethanolamine or carbonate solution enters through the top of the column and is distributed over the packing. As a result of this common application and other similar applications, numerous data and design correlations are available for mass-transfer applications, but very little of this information is available for heat transfer applications.

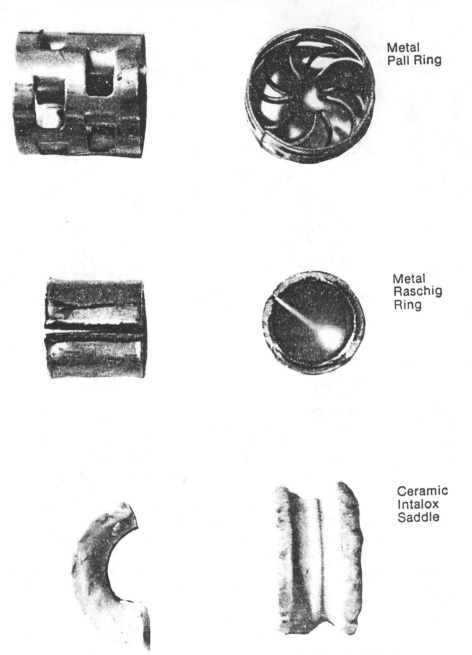

Metal
Pall Ring

Metal
Raschig
Ring

Ceramic
Intalox
Saddle

Figure 1.3 Three types of packing for a direct-contact heat exchanger.

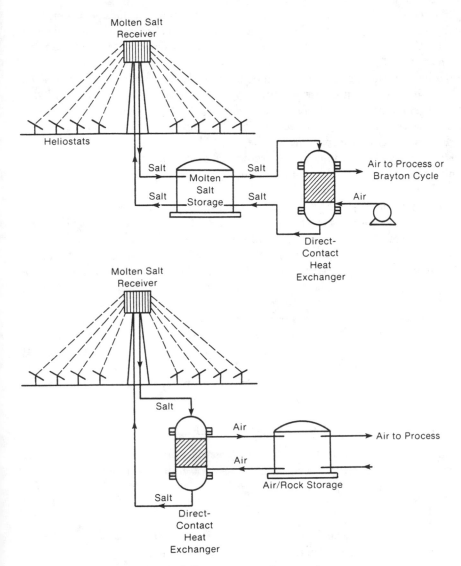

Figure 1.4 Applications of direct-contact heat exchangers.

Because of this lack of heat transfer data or design correlations, we cannot accurately assess the economic potential of direct-contact heat exchange. Such an assessment requires one to determine the rate of heat transfer per unit volume in the DCHX. This determines the required size and cost of the column to deliver the required amount of heat to the air. It also helps one in determining the costs associated with operating the equipment, primarily the cost of blowing the air through the column.

It is possible to use mass-transfer data by invoking the mass-transfer/heat transfer analogy (Fair, 1972). However, there are several reasons to suspect this approach. Some mechanisms of heat transfer have no analogy to mass transfer. Fair's mass-transfer data and correlations are generally for experiments on water/carbon dioxide systems or water/sodium hydroxide systems. The wetting of the packing by the molten salt probably differs from that of water, and this affects how much interfacial surface area is created by the flow down the packing. Heat may be transferred by conduction in the packing, thereby transferring heat from the dry parts of the packing to the air. This fin effect has no analogy in mass transfer. At high temperatures, radiation heat transfer may be significant, and this mechanism also has no mass-transfer analogy. One may conclude that the calculations of heat transfer based on the mass-transfer analogy as given in Fair (1972) may underestimate the heat transfer coefficients.

The objective of the present work is threefold: (1) to experimentally determine the heat transfer coefficients in direct-contact heat exchange between molten salt and air, (2) to calculate these heat transfer coefficients based on the mass-transfer analogy and compare them with the experimental data, and (3) to analyze the economics of this system by using the experimental data and comparing DCHX with conventional finned-tube heat exchangers. In general, we want to determine if, and in what applications, DCHX is a cost-effective technology.

In the following sections, we describe calculations of the heat transfer coefficient based on the mass-transfer analogy; describe the experimental apparatus, methods, and results; compare the results with the calculated values; and, finally, describe an economic analysis that compares the cost effectiveness of DCHX and finned-tube heat exchangers in several applications.

2 THE HEAT-TRANSFER/MASS-TRANSFER ANALOGY

The dimensionless heat transfer coefficient (Stanton number) may be related to the dimensionless mass-transfer coefficient (Sherwood number) (Kreith, 1976) by

$$\frac{h}{GC_p} = \frac{k_g RTP_{BM}}{VP} , \tag{2.1}$$

which is based on the Reynolds analogy and holds only if the Prandtl number (Pr) and the Schmidt number (Sc) both equal unity as shown here:

$$Pr = \frac{\mu C_p}{k} = 1 = \frac{\mu}{\rho D} = Sc . \tag{2.2}$$

In the case where Eq. 2.2 does not hold, heat transfer can often be related to mass transfer by

$$\frac{h}{GC_p} Pr^{2/3} = \frac{k_g RTP_{BM}}{VP} Sc^{2/3} \tag{2.3}$$

from which we can calculate the heat transfer coefficient

$$h = \left(\frac{Sc}{Pr}\right)^{2/3} \frac{GC_p}{\left(\dfrac{VP}{k_g RTP_{BM}}\right)} \tag{2.4}$$

For a packed column, transfer coefficients are commonly presented as volumetric coefficients by multiplying the surface coefficients by the interfacial surface area per unit volume a

$$ha = \left(\frac{Sc}{Pr}\right)^{2/3} \frac{GC_p}{\left(\dfrac{VP}{k_g aRTP_{BM}}\right)} . \tag{2.5}$$

The denominator in parentheses in Eq. 2.5 has dimensions of length and is called the height of a transfer unit H_d. Correlations of mass transfer are often expressed in terms of H_d. Fair (1972) gives such a correlation expression for H_d for packed columns

$$H_{g,d} = \frac{\psi Sc_g^{1/2} d_t^n}{(Lf_1 f_2 f_3)m} , \tag{2.6}$$

and

$$H_{\ell,d} = \phi C_f (Sc_\ell)^{1/2} \tag{2.7}$$

for the gas side and liquid side, respectively. Note that Eqs. 2.6 and 2.7 are dimensional. Dimensions of $H_{g,d}$ are in ft, L is in lbm/h ft^2, and d_t is in ft. We can then calculate the gas side and liquid side heat transfer coefficients from

$$h_g a = \left(\frac{Sc_g}{Pr_g}\right)^{2/3} \frac{C_p G}{H_{g,d}} \tag{2.8}$$

$$h_\ell a = \left(\frac{Sc_\ell}{Pr_\ell}\right)^{2/3} \frac{C_\ell L}{H_{\ell,d}} . \tag{2.9}$$

The overall volumetric heat transfer coefficient is then calculated from

$$Ua = \left[\frac{1}{h_g a} + \frac{1}{h_\ell a}\right]^{-1} . \tag{2.10}$$

The value of the parameters m, n, Ψ, ϕ, and C_f in Eqs. 2.8 and 2.9 may be found in Table 2.1. The parameters f_1, f_2, and f_3 are functions of the liquid viscosity, density, and surface tension, respectively, and are defined as

$$f_1 = \mu_\ell^{0.16} \tag{2.11}$$

$$f_2 = \rho_\ell^{-1.25} \tag{2.12}$$

$$f_3 = (\sigma_\ell/72.8)^{-0.8} . \tag{2.13}$$

Dimensions are: μ_ℓ ($\sim$cp), ρ_ℓ ($\sim$g/cm^3), σ_ℓ ($\sim$dyne/cm).

Table 2.1 Parameters for packed-column heat and mass transfer

	Raschig Rings		Berl Saddles	
	1-in.	2-in.	1-in.	2-in.
m	0.6	0.6	0.5	0.5
n	1.24	1.24	1.11	1.11
Ψ				
40% flood[b]	110	210	60	95
60% flood	105	210	60	95
80% flood	80	--[c]	--	--
ϕ				
L = 2450	0.045	0.059	0.032	--
L = 4900	0.048	0.065	0.040	--
L = 24,500	0.048	0.090	0.068	--
L = 49,000	0.082	0.110	0.090	--
C_f				
<50% flood	1.00	1.00	1.00	1.00
60% flood	0.90	0.90	0.90	0.90
80% flood	0.60	0.60	0.60	0.60

[a]Fair 1972.
[b]Flooding is defined as a column pressure drop of 1.5 in.
(water column) per foot of bed height.
[c]Data not available or extrapolated.

Experimental conditions can then be used to calculate the heat transfer coefficient based on the heat transfer/mass-transfer analogy. The air and salt inlet and outlet temperatures determine all property values. The air and salt flow rate and column diameter determine G and L for Eqs. 2.8 and 2.9, respectively. For the experimental apparatus it will be necessary to use Table 2.1 data for the 1-in. Raschig ring packing (because there is not data for 0.5-in. rings) even though the experiment used 0.5-in. Raschig rings. Extrapolation from the 1-in. to 0.5-in. rings from the 1-in. and 2-in. ring data would be nothing more than guesswork. However, note that Ψ nearly halved in changing from 1-in. to 0.5-in. rings. This, in turn, increases the gas side heat transfer coefficient. Thus, using Ψ for the 1-in. rings should give conservative values of heat transfer coefficients. We also need to estimate the percentage of flooding to determine ϕ and C_f. Constant values of ψ = 110, ϕ = 0.048, and C_f = 1 were used. We determined the property values as discussed in section 7. Figure 2.1 gives results for air and salt flows and temperatures typical of the experimental apparatus.

The film coefficient on the gas side is much smaller than that on the liquid side, and, therefore, the overall coefficient Ua very nearly equals $h_g a$. From Eqs. 2.6 and 2.8 we see that

$$Ua \sim \dot{m}_a \dot{m}_s{}^{0.6} . \tag{2.14}$$

It will be useful to compare the exponents in Eq. 2.14 with the experimental

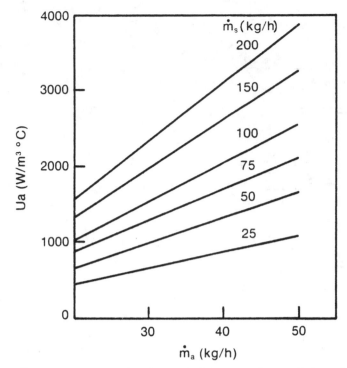

Figure 2.1 Overall volumetric heat transfer coefficients based on mass-transfer data.

results. In addition to comparing the relative magnitudes of the calculated and experimental results, comparing the trends is important to determine whether the transfer phenomena are actually equivalent.

The range of experimental parameters (air flow and salt flow) is restricted (see Section 3.4). Within this range we should expect heat transfer coefficients Ua from about 1000 to 3250 W/m^3 °C.

3 EXPERIMENTAL MEASUREMENTS OF VOLUMETRIC HEAT TRANSFER COEFFICIENTS

3.1 Purpose

As previously mentioned, there are uncertainties associated with calculating the heat transfer coefficients from mass-transfer data. In addition, there are mechanisms of heat transfer for which no mass-transfer data exist. Therefore, we have developed an experimental program to determine actual volumetric heat transfer coefficients in direct-contact heat exchange between air and molten salt. Using these data to determine the economic value of DCHX should give us more confidence in the results than if only the calculated heat transfer values were used.

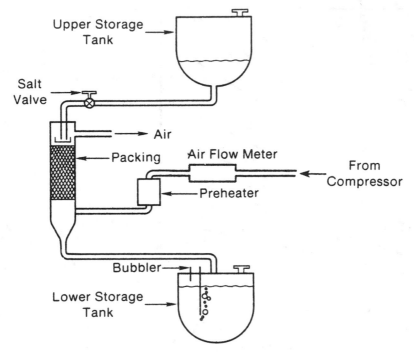

Figure 3.1 Flow diagram of DCHX test loop.

3.2 Description of the Apparatus

A flow diagram of the experimental apparatus is given in Fig. 3.1, a detailed diagram of the packed column is shown in Fig. 3.2, and a photograph of the apparatus is shown in Fig. 3.3. The test loop is a batch operation with regulated air pressure on the upper tank providing regulated salt flow through the salt valve. The upper tank is filled with molten salt and pressurized to approximately 50 kPa. In this way, the salt flow is affected minimally by loss of salt head in the upper tank. The salt flows through the salt valve into the top of the column and into a salt distributor (a can with three holes in its bottom) that distributes the salt uniformly over the top of the bed. Salt flows from the salt inlet pipe directly into the distributor where it flows out the three holes. The distributor is slightly smaller than the column inside diameter, allowing air to flow in the resulting annulus.

The packing bed is supported by a gas-injection support plate that allows the salt to flow downward while providing a uniform air distribution at the bottom of the packing. After the air passes up through the bed, it flows around the annular gap between the salt distributor and the column inside diameter. The air then flows through a wire-mesh mist eliminator that removes any small salt droplets present before the air flows out of the column. Salt flowing out of the bottom of the bed is collected at the bottom of the column and flows to the lower salt tank.

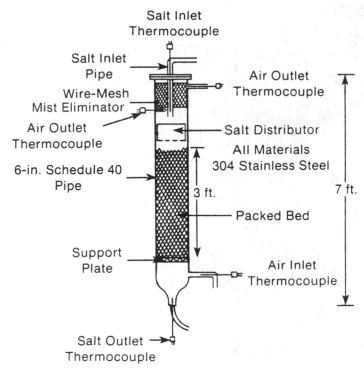

Figure 3.2 Details of the DCHX packed volume.

The column size used in this experiment is typical of pilot-scale studies. We determined the column design (size, distributors, bed height, packing size) with the assistance of Norton Chemical Company, Rolling Meadows, Ill. The entire test loop, with the exception of the salt valve, is constructed of 304 stainless steel. The salt valve is 316 stainless steel.

We tested two types of commercially available packing: stainless Raschig rings, and stainless Pall rings (see Fig. 1.2). Data on the Raschig rings were useful because of the large amount of mass-transfer data available for them. Supplemental data on the Pall rings were taken because the Norton Chemical Company determined that Pall rings would be the most effective packing for heat transfer duty. For proper liquid distribution, a general guideline is to use a packing size about 1/10 of the column diameter. The Raschig rings we tested were 0.5 in., and the Pall rings were 0.6 in. (both the height and diameter of the packing).

Air supplied at the bottom of the column is preheated by a 9-kW electric preheater powered by a silicon-controlled rectifier (SCR) power supply. A proportional-integral process controller supplies the control signal to the SCR power supply based on the desired air temperature and the measured air temperature at the preheater outlet. A two-cylinder, 10-hp compressor supplies air to the preheater.

Figure 3.3 Photograph of the DCHX test apparatus.

Two heat transfer salts were tested in the apparatus. Both were supplied by the Park Chemical Company, Detroit, Mich. The first salt tested is called Parth-erm 430, which has a melting point of 222°C (430°F). The nominal molar composition of the salt is 43% potassium nitrate and 57% sodium nitrate. Although the melting point is listed as 222°C by the manufacturer, most of the salt melts in the 240°-305° range, according to differential scanning calorimetry tests. Since this high melting point caused considerable problems in operation (see Section 3.5), a salt with a lower melting point was used for most of the tests. This salt was Partherm 290 with a molar composition of 40% sodium nitrite, 52% potassium nitrate, and 8% sodium nitrate. The melting point of Partherm 290 is listed as 143°C (290°F) by the manufacturer. One would assume that considerable melting does not occur until approximately 200°C for this salt.

3.3 Instrumentation

Primary instrumentation is shown in Figs. 3.1 and 3.2. Air flow rate is measured by an inline mass flow transducer manufactured by Datametrics, Inc.

A bubbler system determines the salt flow rate by continuously monitoring the level of salt in the lower salt tank. This system consists of a tube that passes through the top of the tank to within a few centimeters of the tank bottom and a similar tube short enough so its end is always above the salt surface. By measuring the pressure required to force a bubble of air out the bottom of the long tube, one can determine the salt depth because it determines the pressure head at the end of the long tube. If the lower tank is pressurized, this pressure is sensed by the short tube and subtracted from the pressure at the long tube. The major advantage of this type of system is that molten salt does not come into contact with any parts of the flow measuring system. The output of the bubbler is proportional to the salt depth, and differentiating this output gives the salt flow rate.

The lower tank was calibrated by filling it with water in 5-l increments and recording the bubbler voltage output. In this way we could directly measure the liquid flow rate because mass flow rate determined by this type of measurement is independent of liquid density. When attempting to measure the salt depth, one must be sure that the salt is completely molten and free of entrained air bubbles.

All thermocouples were Chromel-Alumel (type K). A probe inserted into the vertical portion of the pipe from the upper tank measured the salt inlet temperature (see Fig. 3.2). The probe should be an accurate measure of salt inlet temperature since it is totally immersed in salt just before it flows into the salt distributor. This temperature was typically within $2°C$ of the upper-tank salt temperature.

A probe inserted in the pipe leading out of the bottom of the column measured the salt outlet temperature. We inserted the probe just to where the cone at the bottom of the column begins to expand. The probe was exposed to rivulets dripping from the packing support plate and is the best compromise for measuring salt outlet temperature. Constraints on this measurement include the trace heating on the column wall, which could affect the temperature of salt flowing along the wall, and air entering the column at a lower temperature than the salt leaving the column, which could reduce the outlet salt temperature reading if the probe were inserted further into the column. We observed the responses of this probe to sudden changes in the salt flow, air flow, and air inlet temperature and determined that the probe gives a good indication of salt outlet temperature.

A probe inserted into the horizontal portion of the air inlet pipe measured the air inlet temperature, sensing the temperature of the air about 20 cm from the column. A probe inserted through the column just below the mist eliminator measured the air outlet temperature. Secondary measurements included upper-tank salt and surface temperatures, lower-tank surface temperature, column surface temperature, bed temperature, and the pressure differential between the preheater outlet pipe and the column air outlet pipe.

Operation of the primary thermocouples was checked by placing the probes in condensing steam at local atmospheric pressure (622 mm Hg) corresponding to a saturation temperature of $94.3°C$. The five primary probes read $94.4°$, $94.7°$, $95.1°$, $94.2°$, and $95.1°$ for the salt in, salt out, air in, air out (column), and air out (outlet pipe) temperatures, respectively.

Readings from the outlet air probe indicated a close approach (~10 °C) to the salt inlet temperature. This implies that either the heat exchanger is very effective or that some of the salt flowing in the top of the bed is carried up beyond the distributor and contacts the probe causing it to read too high. Any salt trapped in the mist eliminator could also drip onto this probe. To resolve this we inserted a second air outlet probe through the air outlet pipe to just inside the mist eliminator (see Fig. 3.2). This probe read about 7 °C lower than the probe near the mist eliminator. Pulling the probe out about 2 cm so it was not inside the mist eliminator increased the difference to about 14 °C. With the probe pulled out to about 20 cm from the mist eliminator the discrepancy was about 50 °C. Therefore, a large temperature gradient exists in the air outlet port area, and it is difficult to measure the air outlet temperature. For other reasons (see Section 3.6) we used the temperature measured by the probe inserted in the column just below the mist eliminator as the actual air outlet temperature.

Data were recorded by a Hewlett-Packard Model 85 computer that gave printed, displayed, and plotted information. A Leeds and Northrup strip chart also recorded surface temperatures, air flow rate, and bubbler output.

3.4 Column Sizing

Flooding constraints determine the column design in terms of allowable liquid and gas flow. Flooding occurs when a large quantity of the liquid is entrained with the gas and carried upward with it. This situation is caused by increasing the liquid flow rate at a fixed gas flow rate. Decreasing the gas flow allows more liquid flow before flooding. Flooding produces excessive pressure drop and must be avoided in commercial applications. The generalized pressure drop correlation (Norton Chemical, 1977), seen in Fig. 3.4, gives this relationship in general terms. This correlation gives lines of constant pressure drop across the column bed as a function of flow rates and properties of the gas and liquid. When this correlation is made specific to the experimental design with air and molten salt and for 0.5-in. Raschig rings, the pressure drops shown in Fig. 3.5 results. Also shown in Fig. 3.5 are the limits of salt flow and air flow for the apparatus and the points where actual data were taken.

Table 3.1 gives nominal design values for the experimental apparatus. We chose the maximum operating temperature of 350 °C because common nitrate salts do not cause excessive corrosion with the stainless steel alloys at this temperature. We felt that from what is known about materials compatibility, adequate operating time could be expected from the apparatus by limiting operation to 350 °C.

It is clearly necessary to maintain all surfaces with which the salt comes in contact at a temperature above the freezing point of the salt, including the exterior surface of the column. In practice, the column surface would probably be heated only for start-up; and when operating conditions were reached, the heat tracing would be turned off. Then, the insulation applied to the outside surface of the heat exchanger would control the heat losses. In this experimental apparatus

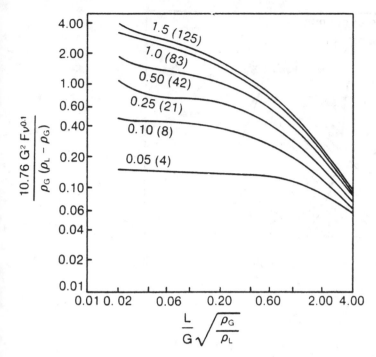

Figure 3.4 Generalized pressure drop correlation.

we left the heat tracing on during heat transfer experiments to control the tendency of cold spots to form frozen salt, which blocks the column. In addition, the heat tracing was set to maintain the column surface near the operating temperature to minimize start-up transients and to act as a guard heater, minimizing losses from the salt to ambient.

Heat tracing required to initially bring the loop up to operating temperature and to maintain this temperature was a high-temperature tracing supplied by Nelson Electric (type A-846K-016-07). The tracing has a stainless steel shell (0.25-in. outside diameter) with nichrome wire inside that is protected from the shell with a refractory insulation. This heat tracing was secured to all exterior surfaces of the test loop (including both tanks, piping, and the column) with baling wire supplied with the heat tracing. Approximately 50 m of the heat tracing was required to provide adequate heating. We then insulated the test loop with a Johns-Manville Cerawool blanket to a thickness of approximately 15 cm.

3.5 Operational Problems

Most operational difficulties were caused by localized cold spots in the transfer piping. It was difficult to apply the heat tracing, because it was rather stiff and could not be easily formed to fit all contours uniformly especially near valve bodies, on the transfer pipes, and on the bottom conical portion of the column.

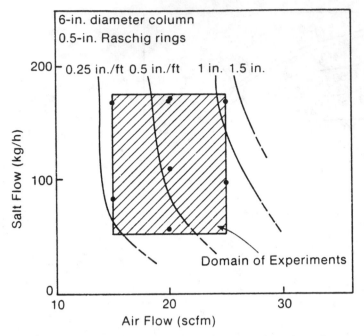

Figure 3.5 Pressure drops and limits of salt and air flow.

We alleviated some of these problems by replacing the original salt with one having a lower melting point (143 ° C versus 221 ° C for the first salt).

Only one materials-related failure occurred. The heat tracing overheated on the connecting tube about 30 cm from the upper tank, corroding the tube, and the entire contents of the upper tank leaked out. On examination we found the tubing had a dark brown discoloration about 3 cm long on either side of the hole. We subsequently replaced all the tubing with some that had a larger diameter and a heavier wall (0.75-in. schedule 40 pipe with a 0.113-in. wall versus 0.5-in., 0.035-in.-wall tubing). Using a larger diameter tubing allowed better application of the

Table 3.1 Nominal column operating conditions.

Salt inlet temperature	$350^{\circ}C$
Air inlet temperature	$200^{\circ}C$
Salt flow rate	$\dot{m}_s = 80$ kg/h $L = 4390$ kg/h m^2
Air flow rate	$\dot{m}_a = 40$ kg/h $G = 2190$ kg/h m^2
Heat duty	$Q = 2$ kW

Table 3.2 Analysis of upper tank residue

Residue	Salt after ~3600 h	As-received Salt
% Fe 0.70	--[a]	--
% Mg 1.50	--	0.005
% Ca 0.50	0.003	0.003
% Ti 0.03	--	--
% Si 1.0	--	--
% Al 0.2	--	--
ppm Mn 100	--	--
B 20	--	--
Ba 30	--	--
Cr 70	5	--
Cu 15	--	--
Mo 15	--	--
Ni 150	--	--
Pb 20	--	--
Sr 30	--	--
V 20	--	--

[a]None detected.

heat tracing because it was easier to form the tracing to the contours of the larger tubing. In addition, the salt tends to freeze more easily in small-diameter tubing. Most operational difficulties were eliminated after installation of the larger-diameter tubing.

A brownish residue collected in the bottom of the upper salt tank. Analysis of this residue is given in Table 3.2 along with an analysis of the as-received salt. It appears that at some point in the loop, the salt is reacting with the containment materials; that even at 350°C, long life could be a problem; or that temperatures in parts of the test loop are substantially above 350°C. Operational problems resulted, as this viscous residue tended to plug the salt valve, making it difficult to maintain a contant salt flow. Foreign particulate matter also became trapped in the valve orifice. These particles were very hard and brittle (similar to small pebbles) and were either related to the viscous residue or in the salt as delivered.

The apparatus shown in Fig. 3.1 is somewhat simplified from the original design, which had two features that caused operational problems and were ultimately abandoned. The salt valve originally was electrically actuated and could be operated remotely from the control room. This provided a way to control the salt flow as the salt head in the upper tank was reduced. Unfortunately, the valve did not operate as smoothly as required, and, if salt froze in the valve, it was difficult to diagnose the lack of salt flow because of the remote location of the actuator. Finally, the weight of the entire valve/actuator assembly caused one of the tubing welds to break. We replaced the valve with a manual bellows valve, and the pneumatic system described previously provided good salt flow control.

The second feature we abandoned was a pneumatic system for transferring the salt in the lower salt tank to the upper salt tank. The system involved a pipe from the lower tank to the upper tank and associated valves for isolating the lower tank. Applying air pressure to the lower tank forced the salt into the upper tank. Problems with this system were primarily related to salt freezing in the return line; plus, the extra valves provided more locations where we could not apply the heat tracing. We removed the additional valves when we replaced the small-diameter tubing, and we manually transferred the salt to the upper tank thereafter.

3.6 Heat Transfer Measurements and Procedures

Using the inlet and outlet salt and air temperatures and the salt and air flow rates, we can determine the rate of heat transfer from

$$Q_s = \dot{m}_s C_s (T_{si} - T_{so}) , \tag{3.1}$$

$$Q_a = \dot{m}_a C_{pa} (T_{ao} - T_{ai}) . \tag{3.2}$$

Equation 3.1 gives the rate of heat transfer from the salt, and Eq. 3.2 gives the rate of heat transfer to the air. We determined the specific heat for the air and the salt using the method described in section 7. A comparison of Q_s and Q_a gives a quantitative measure of the quality of the heat transfer data since in the absence of heat losses and measurement errors we would have $Q_s = Q_a$. Therefore, we will refer to the absolute value of the quantity $100(1 - Q_s/Q_a)\%$ as the heat balance for the experiment.

We can then calculate the volumetric heat transfer coefficient from

$$Ua = \frac{Q}{V_p \, \Delta T_m} , \tag{3.3}$$

where V_p is the volume of the packing bed [15 cm (inside diameter) x 0.914 m = 0.0167 m^3] and ΔT_m is the log-mean temperature difference, defined as

$$\Delta T_m = \frac{\left(T_{si} - T_{ao}\right) - \left(T_{so} - T_{ai}\right)}{\ln\left(\dfrac{T_{si} - T_{ao}}{T_{so} - T_{ai}}\right)} . \tag{3.4}$$

The value of Q in Eq. 3.3 can be either Q_s or Q_a, and the error in Ua is therefore equal to the heat balance for the experiment.

From Eq. 3.4 it is clear that for a close approach ($T_{ao} = T_{si}$), which is typical of these experiments, large errors in ΔT_m and therefore in Ua can result. Table 3.3 demonstrates this from the baseline of actual measured data for one run; the value of T_{ao} was perturbed to show the effect on Q_a and Ua. This shows that for values of T_{ao} lower than the measured value (typical of the probe in the air outlet line), the heat balance is poor. For values of T_{ao} larger than the measured value at the bottom of the mist eliminator, the heat balance is good, but Ua

Table 3.3 Measured data showing effect on Q_a and Ua

	T_{ai}	T_{ao}	T_{si}	T_{so}	$\dot{m}_s$	$\dot{m}_a$	Q_s	Q_a	Ua
		($^\circ$ C)			(kg/h)		(W)	(W)	W/m^3 $^\circ$ C
aseline	193.7	334.8	341.9	309.9	167.8	50.9	2305	2065	3536
	"	294.0	"	"	"	"	"	1468	1791
	"	300.0	"	"	"	"	"	1556	1895
	"	340.0	"	"	"	"	"	2141	4966
	"	341.8	"	"	"	"	"	2167	8389

increases very rapidly as T_{ao} approaches T_{si}. For a 5.2 $^\circ$ C increase in T_{ao}, Ua increases by 40%.

Although the thermocouple should not generate errors greater than ± 1 $^\circ$ C (see Section 3.3), placing the air outlet probe where it is influenced by salt draining from the mist eliminator, heat tracing on the column walls, etc., could cause large errors. The solution is to totally separate the two phases to eliminate the influence of salt in the measured air temperature and at the same time to place the probe close enough to the top of the bed to get a true air outlet temperature. Based on the two air outlet temperature probes (one inserted through the column wall just below the mist eliminator and the other inserted through the knockout pot just inside the mist eliminator), it appears that one probe may read slightly high because of entrained salt and that the other may read lower by a few degrees. The best temperature measurement to use, therefore, is the one just below the mist eliminator.

Experimentally, it is possible to vary the salt flow rate, the air flow rate, and the salt and air inlet temperatures. The last two variables are of secondary importance (as long as the air inlet temperature is above the salt freezing point), so we did not vary them in any systematic way. The salt flow rate was varied from 50 to 200 kg/h, and the air flow was varied from 30 to 50 kg/h (see Fig. 3.5). We could attain higher salt flow rates, but this would result in run times too short to establish steady conditions—a crucial requirement for good data; i.e., small values of the heat balance. Figure 3.5 shows that air flow rates much larger than 50 kg/h ($\sim$25 scfm) produces column flooding.

The system was not temperature-cycled but was left at operating temperature for about six months continuously with the exception of downtime for repairs, as described previously. To minimize the time required to reach steady state and to minimize losses, we set the heat tracing so the bed temperature was fairly close to the upper tank salt temperature. Pressure was applied to the upper tank from the regulated air supply, and the salt valve was opened. To achieve a constant salt flow rate, as indicated by the bubbler output trace, generally required 20 minutes. (As long as nothing lodged in the valve, this flow rate was steady until the upper tank was empty.) We then set the air flow to the desired value (it was

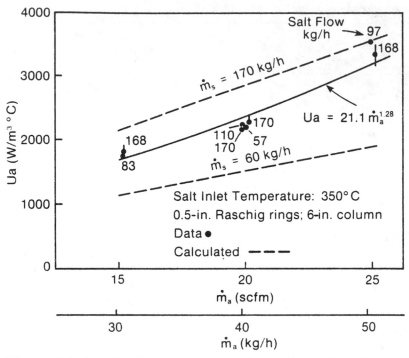

Figure 3.6 Overall volumetric heat transfer coefficients based on experimental data.

helpful to heat the preheater to about $200°C$ before turning on the air), and, when steady state was achieved, we could adjust the air flow to a new setting.

Examination of the data indicated that the best heat balances were achieved when the salt flow was the most uniform and when no adjustments of the salt valve or tank pressure were necessary. A typical run of two hours provided data on one salt flow rate and five air flow rates.

3.7 Results and Discussion

Experimental data for Raschig rings are presented in Fig. 3.6 in the form of volumetric heat transfer coefficient versus air flow rate with salt flow as a parameter. The data are also shown in Table 3.4 with the heat transfer coefficient calculated from the mass-transfer analogy, Eqs. 2.6 through 2.9.

The heat transfer coefficients do not appear to depend on salt flow rate, as all the data for $m_a \simeq 40$ kg/h and for $57 \leq \dot{m}_s \leq 170$ kg/h vary by only a few percentage points. The variation with air flow is relatively strong—a best fit produces

$$Ua = 21.1 \ \dot{m}_a^{1.28} \tag{3.5}$$

Table 3.4 Heat-transfer data and comparison with mass-transfer calculations

Salt Inlet Temperature (°C)	Air Flow (kg/h)	Salt Flow (kg/h)	Measured Ua (W/m³ °C)	Heat Balance (±%)	Calculated Ua (W/m³ °C)
342	30.8	168	1820	3	2171
348	30.9	83	1771	2	1429
341	40.7	171	2252	6	2896
353	40.5	57	2203	3	1478
349	40.3	170	2164	4	2854
348	40.2	110	2228	1	2189
348	50.5	96	3520	5	2535
342	50.9	168	3351	5	3574

where $\dot{m}_a$ is in kg/h and Ua is in W/m³ °C. This is clearly at variance with Eq. 2.14.

As shown in the last column of Table 3.4, heat transfer coefficients calculated from mass-transfer data underestimate measured heat transfer coefficients except at large salt flows. Because the experimental data do not correlate with Eq. 2.14, it is doubtful that using the mass-transfer/heat transfer analogy will work. Apparently, the heat transfer mechanism does differ significantly from the mass-transfer mechanism, as discussed in Section 1.

Data for the Pall rings are compared with the Raschig ring data correlation, Eq. 3.5 in Fig. 3.7. Data were for a salt flow rate of only ~130 kg/h; we did not test the dependence of heat transfer on salt flow rate. The three data points fall fairly close to the Raschig ring curve, although the point for the highest air flow is somewhat below (~20%) the curve.

Overall heat transfer coefficients calculated from mass-transfer data are shown in Fig. 3.6 for two salt flow rates, 170 and 60 kg/h. These results further demonstrate the lack of sensitivity to salt flow rate for the heat transfer data compared to the mass-transfer data. There are several possible explanations as to why the heat transfer data do not depend on salt flow rate while the mass-transfer data do, as explained in Section 1.0. These explanations are the (1) different wetting characteristics of the packing by the salt, (2) heat conduction through the metal wall of the packing, and (3) radiation heat transfer.

If the salt totally wets the packing, any increase in salt flow beyond some minimum will not provide more heat transfer surface area per volume of packing, causing the volumetric heat transfer coefficient to be insensitive to salt flow. Since $Ua \simeq h_g a$, the volumetric coefficient Ua may only be affected by changing the gas-side film coefficient h_g or the surface area per unit volume a. This is consistent with the similarity between the Pall ring data and the Raschig ring data

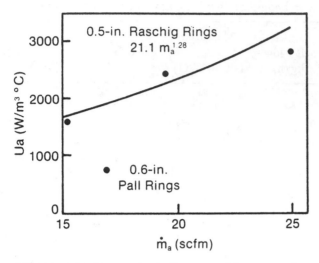

Figure 3.7 Comparison of measured heat transfer coefficients for Pall rings and Raschig rings.

(Fig. 3.7). From Fig. 1.3 one can see that the two packing types should provide similar surface areas per unit volume since the only difference is that the Pall rings have the spokes punched in from the periphery of the ring.

Peters and Timmerhaus (1980) give surface areas per unit volume for several types and sizes of packings. For the 0.5-in. metal Raschig ring (with 0.06-in. wall), approximately 118 ft^2 of surface are provided per ft^3 of packing volume. For the Pall ring, the value is 104 ft^2/ft^3. The two types of packing provide similar heat transfer areas and, therefore, if fully wetted by the salt, should exhibit about the same heat transfer performance.

On the other hand, if conduction heat transfer in the packing is important, then it would not be important how much of the surface of the packing is wet by the salt because heat could then be transferred from the dry areas of the packing to the air.

A better understanding of the heat transfer mechanism is required. By performing tests at various temperatures with liquids having various wetting properties and with packings of various thermal conductivities and surface areas, it should be possible to separate the different heat transfer mechanisms.

Overall system pressure drop is plotted in Fig. 3.8. Recall that this is a measure of the differential pressure from the column air inlet pipe to the column air outlet pipe. Therefore, it includes not only pressure drop across the bed (Fig. 3.5), but expansion and contraction losses at the column inlet and column outlet and loss across the air distributor, salt distributor, and the mist eliminator. We took additional pressure drop data with zero salt flow to determine the contribution of all these column components. These data allow only a qualitative assessment of the bed pressure drop because it is only about 30% or 40% of the

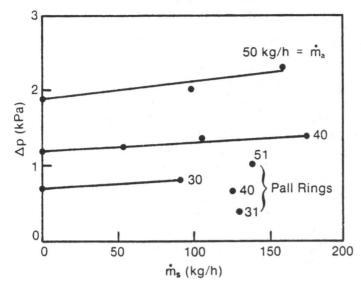

Figure 3.8 Overall system pressure drop.

measured system pressure drop. We could not find a satisfactory method for measuring bed pressure drop because of the difficulties associated with isolating the high-temperature salt from a pressure-sensing port or isolation diaphragm.

The data in Fig. 3.8 clearly show the benefits associated with using Pall rings. At a given air flow the overall system pressure drop for the Pall rings is about half that of the Raschig rings. From Fig. 3.4 we see that constant bed pressure drop, $G \sim \sqrt{F}$, in the region of the map where the Δp lines are level. Since the packing factor F for Pall rings is about 0.17 that of Raschig rings, we could operate with approximately $\sqrt{6} = 2.4$ times as much air flow with Pall rings compared to Raschig rings in the experimental column. The ratio for large (2-in.) packing is close to 2, so we could operate with about 41% more air in a large column with Pall rings than with Raschig rings. From Eq. 3.5 the Pall rings could provide about $1.41^{1.28} = 1.55$ more heat transfer per unit volume than Raschig rings in a large column. Extrapolation of Eq. 3.5 to such large air flows should be tested experimentally to determine if flooding is approached. This would most likely cause the volumetric heat transfer to fall below that predicted by Eq. 3.5.

4 ECONOMIC ANALYSIS

4.1 Purpose

With experimental and calculated values for the heat transfer coefficient, we can now determine the economic value of DCHX. Rather than basing economic calculations solely on the mass-transfer data, using actual heat transfer data should give us more confidence in the results. As expained in Section 4.2, the experimental data cannot be applied directly to a commercial-size DCHX, and, therefore, even these results require some caution in interpretation.

4.2 Method of Analysis

The economics of DCHX and finned-tube exchangers are compared by considering all capital and operating costs associated with the heat exchanger. Using a methodology described by Bohn (1983), one can calculate the annual levelized cost. This is the constant annual cost (in fixed dollars) that, if paid over the lifetime of the heat exchanger, would have a present value equal to the present value of the actual costs incurred over the lifetime of the heat exchanger. In computing one single cost this method can easily consider: escalation rates, depreciation, discount rates, lifetime, and tax rates, among other parameters. Bohn (1983) describes the method and also gives the values for these parameters used in the analysis.

We will assume that only one capital cost in incurred and that the only operating cost is that associated with the power required to pump the air through either heat exchanger. Maintenance costs will be taken as a constant annual cost equal to 3% of the capital cost. With these assumptions the annual cost may be computed from

$$AC = 0.2299 \times CI + 1.886 \times OM , \qquad (4.1)$$

where CI is the capital cost expressed in 1981 dollars and OM is the annual cost of pumping the air, also expressed in 1981 dollars. We will generally give results in the form of AC/Q, which is the annual cost per unit heat transferred in $/GJ$. Note that this is quite close to $/10^6$ Btu transferred.

Considering the pumping cost first, for the cost of electricity given in Bohn (1983) ($12.89/GJ = $0.0464/kWh), we see that

$$OM = 1.68 \times 10^{-4} \ C_f \ \frac{\Delta p}{\rho} \ \dot{m}_a(\$/yr) , \qquad (4.2)$$

which assumes an isentropic efficiency of 0.70 for the compressor, an electrical motor efficiency of 0.96, and a plant capacity factor C_f of 0.8. We can calculate the pressure drop Δp through the heat exchanger once we know the air flow rate and the characteristics of the heat exchanger (friction factor versus Reynolds number plot for the finned-tube exchanger and the generalized pressure drop correlation curve for the packed column).

To determine capital cost, we must know the size of the heat exchanger and the materials of construction. The first is determined by the required heat duty, overall heat exchange coefficients, and log-mean temperature differences. Materials of construction are determined by operating temperature and working fluids (see Section 4.3). For consistency, we used the data given by Peters and Timmerhaus (1980) for all the capital costs.

For the finned-tube heat exchanger, the size is best expressed as heat exchange area including fins. Peters and Timmerhaus (1980, p. 669) give a graph of the cost of carbon steel, finned-tube, and floating-head heat exchangers operating at 10 atm. This curve has been generalized to a correlation curve

$$CI = 2051 \, A_a^{0.6622} \quad (\$) \tag{4.3}$$

that includes a factor of 2.3 for installation cost, a factor of 1.8 for the use of stainless steel in the entire heat exchanger (see Peters and Timmerhaus, 1980, p. 677), and a factor of 1.12 to escalate the 1979 cost to 1981 dollars. For atmospheric pressure operation the capital cost is reduced by a factor of 0.92 (Peters and Timmerhaus, 1980, p 673). For materials other than stainless steel the cost is adjusted according to the table given in Peters and Timmerhaus (1980, p. 677), which lists relative cost factors for entire heat exchangers of several different materials of construction. For carbon steel the factor is 0.56; and for Incoloy, it is 1.67. The factor between carbon steel and stainless steel (0.56) is consistent with cost data in Dubberly et al. (1981), which gives 0.60. It is also consistent with a rule-of-thumb (0.5) used by Mercury Fin Tube Products to scale the cost of a stainless steel, finned-tube heat exchanger to a carbon steel unit.

Peters and Timmerhaus (1980, p. 772) also give installed costs for packed towers (excluding cost of packing) as a function of the height and diameter of the column. For a stainless steel column from 1 m to 5 m in diameter, the correlation curve is

$$CI = 10{,}762 \, Hd_t^{1.29} \quad (\$) \, , \tag{4.4a}$$

and for carbon steel it is

$$CI = 2620 \, Hd_t^{1.34} \quad (\$) \, , \tag{4.4b}$$

which includes a factor of 1.03 to account for the cost of installing insulation and the same 1.12 factor to escalate costs to 1981 dollars. For operation at pressures other than atmospheric the cost is escalated by the same factor used to rate shell-and-tube heat exchangers (Peters and Timmerhaus, 1980, p. 673) applied to the fraction of column cost attributable to the shell, ~55%. Note that we did not increase the cost of the finned-tube heat exchanger to include insulation costs because we assumed that such a unit would ordinarily be insulated, and the insulation would be part of the 2.3 factor for installation costs. Since packed columns are not ordinarily insulated, we included the factor of 1.03 for the packed-column capital cost calculation. For the high-temperature DCHX a more complex column design required costing of individual components related to the insulation (see Section 4.3).

For the finned-tube heat exchanger we used the characteristics of the heat exchange core, denoted CF-8.8-1.0J by Kays and London (1964), which consists of 1-in. tubes (outside diameter) on 1.96-in. spacing using spiral-wound fins with 8.8 fins/in. and a 0.012-in. fin thickness. Data on the core includes the Colburn j factor (the heat transfer coefficient on the air side) and the friction factor as a function of the air-side Reynolds number. The salt-side Reynolds number was chosen as a constant (10,000) because the overall heat transfer coefficient is not a strong function of the salt-side Reynolds number as long as the salt flow is turbulent. Also, salt-side pumping work is negligible, so we need to consider only the salt-side pressure drop from the standpoint of tube stress at elevated temperatures.

The layout of the core is shown in Fig. 4.1. Salt flows through the tubes, and air flows across the finned tube banks (crossflow arrangement). Given the flow rates (determined from the heat duty and terminal temperatures), we can determine the heat exchanger effectiveness and the required number of transfer units (NTU) from equations for crossflow exchangers. Beginning with the lowest air-side Reynolds number for which the heat exchanger core data are given, we can determine the value of W (H is arbitrarily set equal to W), calculate the air-side heat transfer coefficient and the fin efficiencies, and then calculate the required total heat transfer surface required, which determines the core dimension in the air flow direction L, which also determines the core pressure drop. With the surface area and pressure drop we can calculate the cost AC. This procedure is repeated for increasing air-side Reynolds numbers until a minimum heat transfer area (900 m^2) for which cost data exist in Peters and Timmerhaus (1980). This corresponds to a shell diameter of 2.77 m (9.09 ft) for 4.87-m (16-ft) long, 2.54-cm (1-in.) (outside diameter) tubes. Multiple heat exchangers are specified when heat duties require more than this maximum heat transfer area.

Optimization of the DCHX is somewhat different because of flooding constraints. Outlet air temperature cannot be specified a priori. Beginning with a column diameter of 1 m (the smallest diameter for which cost data are available), we calculated the volume of packing. (We used the shortest practical column height, equal to the diameter, because this always minimized annual cost. Column diameters larger than the column height produce problems with uniform salt and air distribution in the column.) For an assumed value of the overall heat transfer coefficient, we determined the log-mean temperature difference, which gives us the air outlet temperature. This determines the air flow rate, and from the generalized pressure drop correlation we can then determine the pressure drop. We rejected diameters that produce operating conditions off the generalized pressure drop map (Fig. 3.4). We used a maximum column diameter of 5 m since this is the largest for which Peters and Timmerhaus (1980) give cost data. Knowing the column size and pressure drop, we could then calculate the annual cost.

We repeated the procedure for increasing column diameters giving annual cost as a function of approach temperature (air outlet temperature). The overall volumetric heat transfer is taken as a parameter; for most calculations we used the values Ua = 2000, 3000, 4000 W/m^3 °C. We can then use the experimental values of Ua or the calculated values of Ua along with the results of this economic calculation (which also goves sensitivity to Ua) to appraise the economic viability of DCHX relative to finned-tube heat exchangers.

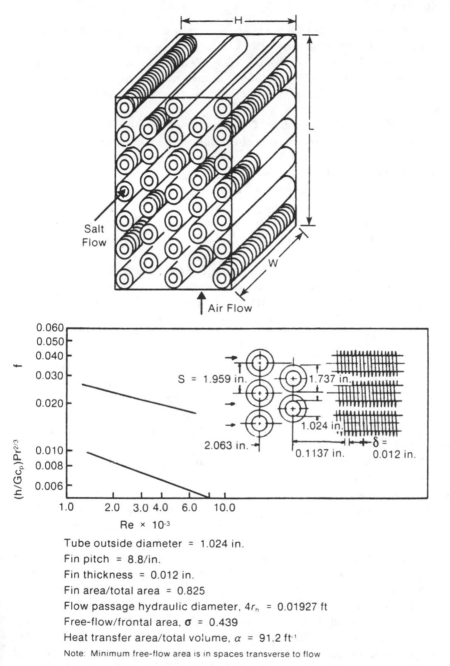

Tube outside diameter = 1.024 in.
Fin pitch = 8.8/in.
Fin thickness = 0.012 in.
Fin area/total area = 0.825
Flow passage hydraulic diameter, $4r_h$ = 0.01927 ft
Free-flow/frontal area, σ = 0.439
Heat transfer area/total volume, α = 91.2 ft^{-1}
Note: Minimum free-flow area is in spaces transverse to flow

Figure 4.1 Layout of the finned-tube heat exchanger core.

4.3 Materials

From exposure tests of up to 4500 hours at temperatures up to about 600 ° C, Tor-torelli and DeVan (1982) demonstrate that stainless steel alloy 316 oxidizes at the rate of 5 mil/yr or less, typically 2 mil/yr. A 90-mil tube wall thickness, there-fore, should be adequate for a 30-year life. Fin material will also have to be stain-less steel because of manufacturing constraints and materials compatibility con-siderations.

For the packed column at 600 ° C or lower, we specify stainless steel Pall rings for the packing and a stainless steel column with external insulation. Inter-nal insulation with a liner to contain the salt and support the packing would allow us to use a carbon steel column, greatly reducing the cost. This insulation method has been tested by Martin Marietta (1979) but is not commercially available at this time; therefore, it involves some technical risk. An alternative design that may prove economical at intermediate temperatures is an externally insulated, carbon-steel column with an Inconel liner. This configuration was not examined in this study.

At temperatures above 600 ° C, common nitrate heat transfer salts decom-pose. SERI, Oak Ridge National Laboratory, and other laboratories are presently researching candidate heat transfer salts and compatible containment materials at temperatures above 600 ° C. Based on state-of-the-art knowledge it appears that Incoloy 800 alloy is a reasonable choice for finned-tube heat-exchanger materials for temperatures $\leq$800 ° C.

The DCHX for temperatures $\leq$800 ° C consists of an internally insulated, carbon steel column similar in design to the storage tank described in Martin Marietta (1979). A high-purity (99% alumina) packing is required to resist attack by the salt. To seal the insulating firebrick from the salt, an Inconel liner is neces-sary. A layer of fiberglass insulation will cover the outside of the column, and aluminum lagging will weatherproof the assembly.

High-purity alumina packing is not commonly available. We determined the cost of this packing from the cost ratio (3:1) between 99% and 57% alumina catalyst support from Norton Chemical Company. To account for unknown manufacturing difficulties that could arise when fabricating saddles from the 99% alumina, a cost factor of 4 (suggested by Norton Chemical Company) was applied to the 1983 second quarter price ($19.80/ ft^3, 100 ft^3 order) quoted by Norton Chemical Company, giving an equivalent 1981 cost of $70.49/ft^3. The packing is the major cost item for the DCHX, which suggests that a spray column could be a viable, high-temperature alternative.

The Inconel liner cost was taken from Martin Marietta (1979) who priced an Incoloy liner with a waffled design to accommodate thermal cycling. Since the DCHX does not operate in a cycling mode, the liner can be a simpler design that allows for thermal expansion. Allowing a 2 cost factor between Incoloy and Inconel and a 0.5 factor for a simpler liner design, the cost of the installed liner is $283/m^2 internal column area.

The firebricks (Krilite 30) need to be thick enough to allow the carbon-steel column to operate below 316°C for an internal temperature of 760°C and to keep thermal losses to less than 1% of the transferred heat. The cost of the firebrick from Martin Marietta (1979) is \$809/m³.

The cost of the carbon-steel shell is calculated in Eq. 4.4b. For the fiberglass outer insulation (sufficient thickness to keep losses to 1% of the transferred heat when the ambient temperature is 20°C and the carbon-steel column is 316°C) Martin Marietta (1979) used \$265/m³ of insulation and \$25/m² of aluminum lagging.

Temperatures above 800°C will most likely require ceramic finned-tube heat exchangers. Although the cost of these ceramic heat exchanger tubes is not prohibitive (most likely less than the cost of higher alloy tubes), fabricating the tubes into a heat exchanger is a relatively unknown art. Therefore, cost projections are difficult. Construction of the packed column, however, would not change drastically for temperatures above 800°C and could be costed with the same level of confidence as the 760° application.

Table 4.1 summarizes the assumed construction of both types of heat exchangers used for the calculation as a function of salt inlet temperature. Assumed air inlet temperature is also given for the three salt temperatures that were run. Table 4.2 summarizes the component costs for the 760°C DCHX with internal insulation. These are installed costs for large storage vessels. Costs for smaller units, such as a 1 m x 1 m DCHX, would be greater for field erection, but if the units could be fabricated in a shop, the costs given in Table 4.2 would probably be conservative.

Table 4.1 Materials of Construction

Temperatures at Salt Inlet		Direct-Contact Heat Exchanger	Finned-Tube Heat Exchanger
Salt In	Air In		
360°C	200°C	Stainless steel Pall rings in an externally insulated, carbon steel column	Carbon steel tubes and fins
560°C	250°C	Stainless steel Pall rings in an externally insulated, stainless steel column	Stainless steel tubes and fins
760°C	550°C	99% alumina saddles in an internally insulated, carbon steel column with Inconel liner	Incoloy 800 tubes and fins

Table 4.2 Installed Materials Cost, 760 ° C DCHX

Component	Material	Cost 1981$
Packing	99% alumina saddles	$2489/m^3
Liner	Inconel	$283/m^2
Firebrick	Krilite 30	$809/m^3
Column	Carbon steel	(Eq. 4.4b)
Insulation, outer	Glass fiber	$265/m^3
Lagging	Aluminum	$25/m^2

4.4 Effect of Packing Size and Type

We need to relate the experimentally measured, heat transfer coefficients (Section 3.0) to those expected at full-scale. The experimentally measured, heat transfer coefficients were for a pilot-scale experiment with 0.5-in. stainless Raschig rings, 0.6-in. stainless Pall rings, and a 350 ° C salt inlet temperature. The economic calculations assumed 2-in. stainless pall rings or 2-in. ceramic Intalox saddles and salt inlet temperatures of 360 ° C, 560 ° C, or 760 ° C.

The pressure drop characteristics of all these packings are well understood. Since the viscosity of molten salt (1.5-3.9 cp) does not differ greatly from that of water (1 cp), using the generalized pressure drop correlating Fig. 3.4 with appropriate property values for salt should be adequate for pressure drop calculations for any packing.

Heat-transfer performance, however, is not well characterized. Even if one wishes to use mass-transfer data, results will be restricted to 1-in. or 2-in. Raschig rings or Berl saddles. As mentioned in Section 2.0, we had to extrapolate down to 0.5-in. Raschig rings to calculate expected heat transfer coefficients for the experiment from mass-transfer data. For the 2-in. Pall rings or 2-in. Intalox saddles used in the economic calculations, no information is available either in the form of mass-transfer or heat transfer data. Therefore, we must comment on the applicability of 0.5-in. Raschig ring heat transfer data (Section 3.0) to the 2-in. Pall ring or 2-in. Intalox saddle data used in the economic calculation and 0.6-in. Pall ring data.

For a given packing type, increasing the size reduces the pressure drop at fixed G and L because the column flow area is less restricted. For fixed G and L, however, the mass-transfer coefficient is reduced for a larger packing, presumably because the larger packing cannot provide as much interfacial surface area between the gas and the liquid. However, the higher flow capacity of the larger packing offsets this, and the maximum mass-transfer coefficient (which occurs near column loading) is only weakly dependent on packing size. This is illustrated in

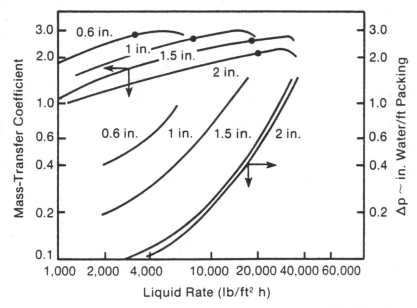

Figure 4.2 Mass-transfer coefficient for various sizes of Rashig rings.

Fig. 4.2 where mass-transfer and pressure-drop data from Norton Chemical (1977) on metal Raschig rings 0.6 in., 1 in., 1.5 in., and 2 in. in size were superimposed on one plot. The maximum mass-transfer coefficient for the 2-in. rings is about 75% of the maximum for the 0.6-in. rings. The four points are the mass-transfer coefficients for DCHX operation at a pressure drop of 0.5-in. of water per foot. With this comparison, the largest rings again have a mass-transfer coefficient about 75% of that of the 0.6-in. rings (mass-transfer data for 0.5-in. Raschig rings was not available). It appears that heat transfer coefficients for 2-in. Raschig rings should be about 75% of that measured in the present column at a given pressure drop.

Comparing the measured heat transfer coefficients with calculated values based on mass transfer (see Fig. 3.6), we found that the mechanisms of heat transfer and mass transfer appear to differ and that the mass-transfer/heat transfer analogy does not apply. We have some experimental evidence (Fig. 3.7) suggesting that heat transfer does not depend on packing type, at least when changing from Raschig rings to Pall rings. Fig. 4.3 depicts the effect of packing type on mass transfer. However, since the analogy with heat transfer is suspect, we must assume that the 2-in. Pall ring (as used in the economic calculations) transfers the same amount of heat as the 0.5-in. Raschig ring at a given air mass velocity G. Pressure-drop performance for the 2-in. Pall rings is accounted for in the calculation procedure, so the reduced pressure drop caused by the Pall rings is taken into account.

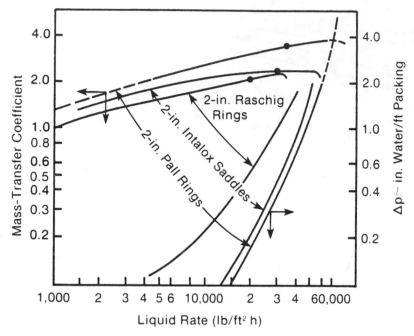

Figure 4.3 Mass-transfer coefficient for various types of packing.

Based on the experimental data presented in Section 3.0, an appropriate range of heat transfer coefficient Ua for the 0.5-in. Raschig rings and therefore the 2-in. Pall rings is 1800-3500 W /m^3 ° C. From the previous discussion it is reasonable to perform the economic calculations based on a range of Ua from 2000-4000 W /m^3 ° C. We can then assess the sensitivity of the economics on Ua until a full range of data at full scale is made available.

4.5 Results

Figures 4.4 through 4.9 present the results of the economic analysis. Each figure shows the cost of transferring 1 GJ of energy as a function of air-outlet temperature. For each temperature range two graphs give results for 1 atm and 5 atm operating pressure. The 1 atm case represents process-heat applications, and the 5 atm case represents a Brayton-cycle application. The Brayton cycle probably will not apply to the two lower temperatures, but higher pressure operation is generally more economical because of increases in air density, and, therefore, the low-temperature, high-pressure case may have application in process heat.

As the air-outlet temperature approaches the salt-inlet temperature, the finned-tube heat exchanger needs a larger surface area and the DCHX requires more packing. This is because the log-mean temperature difference (Eq. 3.4) is reduced, and this increases the packing volume for a fixed, heat transfer coefficient

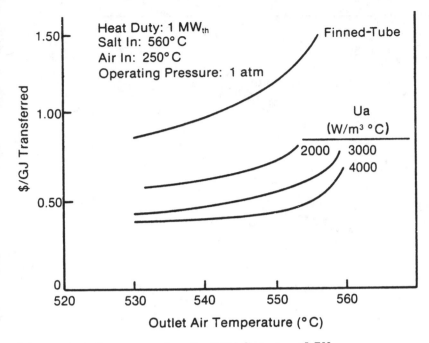

Figure 4.4 Cost comparison for 360° C, 1 atm, 1MW$_{th}$.

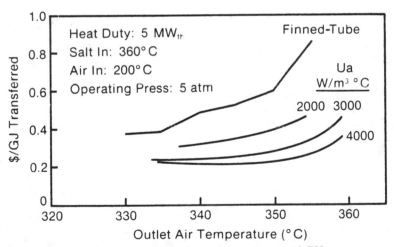

Figure 4.5 Cost comparison for 360° C, 5 atm, 5MW$_{th}$.

and heat duty (Eq. 3.3). We can compensate for this by increasing the air flow, which increases the heat transfer coefficient, but this drives up operating costs. Generally, DCHX provides closer temperature approaches than the finned-tube

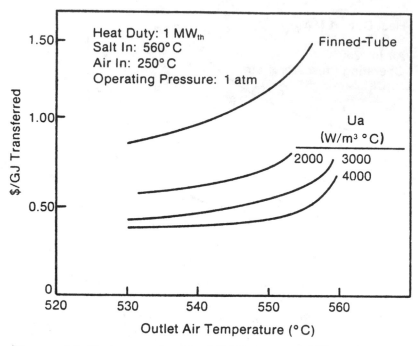

Figure 4.6 Cost comparison for 560°C, 1 atm, 1MW$_{th}$.

heat exchanger before the costs increase rapidly. [The curves for the DCHX for a given Ua increase for low outlet air temperature because the volumetric heat transfer coefficient has been artificially fixed, and the only way to reduce outlet temperature is to increase air flow, which drives up the cost. For $Ua = 2000 W/m^3$ °C this effect is generally not seen for the temperature approaches presented (less than 30°C).]

Calculations at the higher operating pressure (5 atm) were for 5 MW$_{th}$ or 2 MW$_{th}$ heat duty, while those for 1 atm pressure were for 1 MW$_{th}$. These generally resulted in maximum-sized packed columns (5 × 5m) at the close approaches. Larger columns can be built, but since the cost data were restricted to 5-m-diameter columns, we restricted the calculations to this diameter for consistency. Also, the finned-tube heat exchanger tended to reach maximum size (900 m^2 before the DCHX. This is seen in the curves for the finned-tube heat exchanger, which show the change of slope (Fig. 4.5, for example). These slope changes occur because multiple heat exchangers are used to meet the heat duty. Therefore, one major conclusion we can make is that DCHX provides substantially more heat transfer capacity for a given size of equipment.

The cost advantage of DCHX relative to finned-tube heat exchangers is not a strong function of approach temperature (difference between salt inlet and air outlet temperatures) except for small temperature approaches where the cost of the finned-tube heat exchanger increases more rapidly. Table 4.3 gives the cost

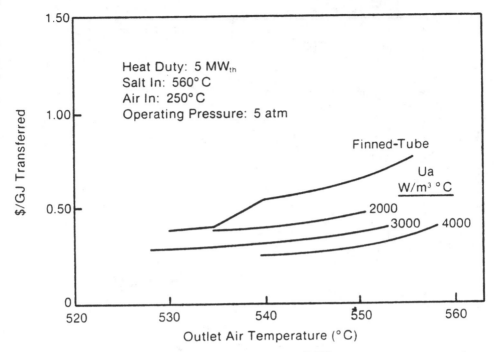

Figure 4.7 Cost comparison for 560°C, 5 atm, 5MW$_{th}$

ratios from all six graphs at a 10°C approach. We used the DCHX curve for $Ua = 3000$ W/m^3 °C in each case.

There appears to be no great decrease in this ratio in going from 360°C to 560°C. This is because the cost of construction materials increased substantially for both types of heat exchangers. Using an internally insulated, carbon steel column with stainless steel Pall rings for 560°C operation significantly improves the cost ratio. (Since this represents a larger technical risk than the externally insulated, stainless steel column, it is not a fair comparison to make with commercially available, stainless steel, finned-tube heat exchangers.)

The large reduction in relative cost in going from 560°C to 760°C occurs because the DCHX does not need high alloy steels, while the finned-tube exchanger does require such materials. Even though the high-purity alumina packing is more costly than stainless steel packing, using a carbon steel column provides a very large cost advantage over the Incoloy finned-tube heat exchanger.

5 CONCLUSIONS AND FUTURE RESEARCH

We measured volumetric heat transfer coefficients in the range of 1800-3500 W/m^3 °C in a 6-in.-diameter column with a 3-ft bed of 0.5-in. metal Raschig rings and 0.6-in. metal Pall rings. The heat transfer coefficient depends on air flow rate but not on salt flow rate. Heat-transfer coefficients based on mass-

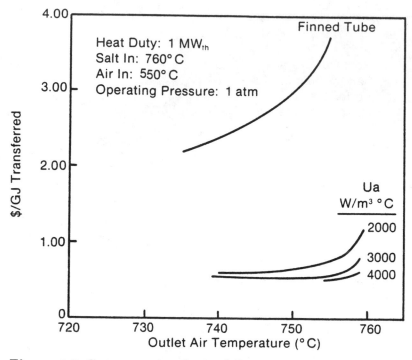

Figure 4.8 Cost comparison for 760°C, 1 atm, 1MW$_{th}$.

transfer data show dependence on both air flow and salt flow. Thus, the mechanisms controlling heat transfer appear to differ from those controlling mass transfer.

The measured heat transfer coefficients are large enough so one can say with confidence that direct-contact heat exchangers are more cost-effective than conventional finned-tube heat exchangers. At low- to mid-temperatures (360°-560°C), the cost (capital and operating) ratio should be about one-half, while at high temperatures (600°-800°C), where high alloy steels are required in the finned-tube heat exchanger, the cost ratio is about one-fifth. The cost advantage occurs because of the high rates of heat transfer and the ability to use materials other than high alloy steels to contain the salt in the DCHX.

Future research should be directed toward experimentally determining the effect of packing size and type, since in lieu of such data one must project this effect based on mass-transfer data. Since we showed that the heat- and mass-transfer mechanisms are different, this analogy is not a satisfactory approach. At high temperatures, radiation heat transfer will become important. Thus, high-temperature testing, perhaps with internally insulated columns, will be necessary.

To ultimately produce heat transfer correlations valid over a wide range of operating conditions that will aid designers, it is necessary to understand the mechanisms of heat transfer. A study on separating the effects of radiation, fin-effect, packing wetting, etc., will be helpful.

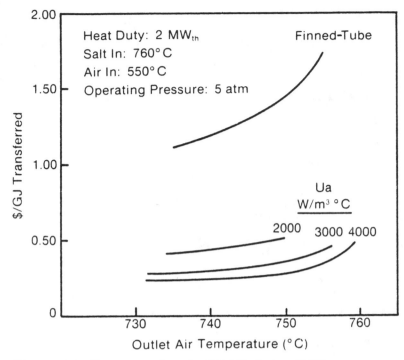

Figure 4.9 Cost comparison for 760° C, 5 atm, 2MW$_{th}$.

Any high-temperature (>600° C) experiments on heat transfer must be delayed until materials research has identified compatible heat transfer salts and containment materials.

Table 4.3 Cost ratios of transferring heat via DCHX relative to finned-tube heat exchanger.

Temperature (°C)	Operating Pressure	
	(1 atm)	(5 atm)
360	0.44	0.46
560	0.46	0.57
760	0.18	0.26

6 PROPERTY VALUES

Values

We took the constant values for specific heat density and thermal conductivity of the salt from data provided by Park Chemical Company (1983). Constant values of the diffusion coefficient and viscosity and surface tension as a function of temperature were taken from NBS (1981). The values or functions are

$$C = 1553 \;\; J/kg \; K$$

$$\rho = 1820 \;\; kg/m^3$$

$$k = 0.573 \;\; W/m \; K$$

$$D = 2.91 \times 10^{-9} \;\; m^2/s$$

$$\mu = 90.811 - 0.3517T + (4.665 \times 10^{-4})T^2 - (2.086 \times 10^{-7})T^3 \;\; (cp)$$

$$\sigma = 155.678 - 0.0627T - (2.315 \times 10^{-7})MT^2$$

$$\quad + (5.9877 \times 10^{-7})M^2T \;\; (dyne/cm)$$

$$M = mol \; \% \; KNO_3$$

The temperature used in the equations for viscosity and surface tension is the average of the salt inlet and outlet temperatures.

We evaluated air properties at the average of the air inlet and outlet temperatures by the following equations derived from tabular data in Kreith (1976).

$$\rho = 350.8 \; P/T \;\; (atm, K, kg/m^3)$$

$$\mu = (3.5158 \times 10^{-6}) + (4.8240 \times 10^{-8})T - (9.2908 \times 10^{-12})T^2 \;\; (kg/ms)$$

$$k = (2.719 \times 10^{-3}) + (7.8017 \times 10^{-5})T - (1.1598 \times 10^{-8})T^2 \;\; (W/m \; K)$$

$$Cp = 997.9 + 0.143T + (1.10 \times 10^{-4})T^2 - (6.776 \times 10^{-8})T^3 \;\; (J/kg \; K).$$

We derived the diffusion coefficient following the procedure in Sherwood, Pigford, and Wilke (1975).

$$D = \frac{(3.555 \times 10^{-5})T^{1.5}}{0.00285\left(\frac{T}{78.6}\right)^2 - 0.06063\left(\frac{T}{78.6}\right) + 1.0739}$$

In the above equations for air properties, we used the average of the inlet and outlet temperatures. All temperatures are in kelvin.

REFERENCES

Bohn, M. S., 1983 (Nov.), *Air/Molten Salt Direct Contact Heat-Transfer Experiment and Economic Analysis*, SERI/Tr-252-2015, Golden, CO: Solar Energy Research Institute.

Dubberly, L. J., J. E. Gormely, J. A. Kochmann, W. R. Lang, and A. W. McKenzie, 1981(Dec.), *Cost and Performance of Thermal Storage Concepts in Solar Thermal Systems*, SERI/TR-XP-0-9001-1-B, Golden, CO: Solar Energy Research Institute.

Fair, J. R., 1972(June), "Designing Direct Contact Coolers/Condensers", *Chemical Engineering*, pp. 91-100.

Kays, W. M. and A. L. London, 1964, *Compact Heat Exchangers*, New York: McGraw Hill.

Kreith, F., 1976, *Principles of Heat Transfer*, New York: Harper and Row.

Martin Marietta Aerospace, 1979(Dec.), *Internally Insulated Thermal Storage System Development Program*, MCR-79-1369, Denver, CO: Martin Marietta Aerospace.

National Bureau of Standards, 1981, *Physical Properties Data Compilations Relevant to Energy Storage, Vol. II: Molten Salts Data on Single and Multi-Component Salt Systems*, NSRDS-NBS 61, Washington, DC: NBS.

Norton Chemical Company, 1977, "Design Information for Packed Towers", Bulletin DC-11, Akron, OH: Norton Chemical Company.

Park Chemical Company, 1983, Technical Bulletin J-9, Detroit, MI: Park Chemical Co.

Peters, M. and K. Timmerhaus, 1980, *Plant Design and Economics for Chemical Engineers*, New York: McGraw Hill.

Sherwood, T. K., R. L. Pigford, and C. R. Wilke, 1975, *Mass Transfer*, New York: McGraw Hill.

Tortorelli, P. F. and J. H. DeVan, 1982(Dec.), *Thermal Convection Loop Study of the Corrosion of Fe-Ni-Cr Alloys by Molten* $NaNO_3-KNO_3$, ORNL/TM-8298, Oak Ridge, TN: Oak Ridge National Laboratory.

DESIGN OF DIRECT-CONTACT PREHEATER/BOILERS FOR SOLAR POND POWER PLANTS

John D. Wright

1 INTRODUCTION

Salt-gradient solar ponds may provide the simplest and least expensive method of converting solar to thermal energy. The pond combines the functions of both collection and storage. Because the salt gradient suppresses thermally induced convection, temperatures of 75°-100°C may be achieved in the storage layer. Thermal energy from ponds may be used to generate electricity, because the low collection cost offsets the inherently poor efficiency of the low-temperature power cycle. One promising method of power production is the organic Rankine cycle. Because of the low efficiency of the conversion cycle, the shell-and-tube heat exchangers for heat addition and rejection are the major capital expense. Replacing these heat exchangers with direct-contact heat exchangers can result in major cost reductions.

Originally prepared under Task No. 9123.00, WPA eo. 21-999-99, for the U.S. Department of Energy Contract No. EG-77-C-0104042. SERI/TR-252-1401, UUC Categories, 62c, 63e, May, 1982.

This report critically reviews the methods available for sizing direct-contact heat exchangers used to couple an organic (pentane) Rankine cycle to a solar pond. Conceptual heat exchanger designs are developed, and areas requiring further research are identified. Section 2.0 describes the overall operation of the pond, power cycle, and heat exchanger. Section 3.0 describes methods for determining the cross-sectional areas of the liquid/liquid and vaporization zones, while Sec. 4.0 discusses heat transfer in each zone. Section 5.0 describes complete heat exchanger designs, and Sec. 6.0 describes the research required for confident sizing of such devices. For a more detailed discussion of the overall system design, choice of working fluid, and system economics, see the earlier report, *An Organic Rankine Cycle Coupled to a Solar Pond by Direct-Contact Het ExchangeSelection of a Working Fluid* (Wright, 1981).

2 SYSTEM DESCRIPTION

2.1 Salt-Gradient Solar Ponds

In a salt-gradient solar pond, salt is dissolved in high concentrations at the bottom, decreasing to low concentrations near the surface. Solar radiation enters the pond, and most of the energy which is not absorbed on its way down is absorbed on the dark bottom, warming the storage layer. The salt concentration gradient establishes a density gradient. The warmer bottom waters typically exhibit a specific gravity of 1.2, while the cooler, nearly salt-free surface waters have a specific gravity of approximately 1.0. Pure water becomes less dense when warmed. In the absence of the salt gradient, warm water from the bottom would continually rise to the surface and lose its heat. However, the density gradient prevents thermally induced convection. In the absence of convection, heat loss to the surface is by the much slower process of conduction, and high temperatures (60°-100°C) may be achieved at the bottom of the pond.

An actual pond has three layers (Fig. 2.1). The virtually salt-free top layer is vertically mixed by wind and evaporation and should be kept as thin as possible. The next layer, approximately one meter thick, contains the salt concentration gradient. It is essentially salt-free at the top and saturated at the bottom. The lowest layer is saturated with salt and provides thermal storage. Since the salt concentration is similar throughout the storage region, convection can occur within the layer, and the temperature throughout the region is constant.

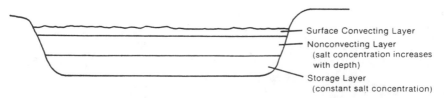

Figure 2.1 Salt-gradient solar pond.

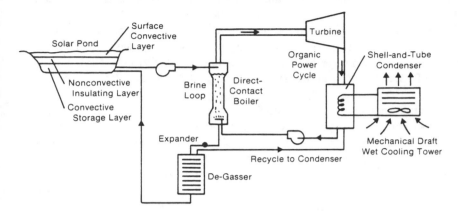

Figure 2.2 Organic Rankine cycle coupled to a solar pond.

Temperatures of up to 100°C have been achieved in the storage layer of solar ponds. However, as in all collectors, the heat losses are proportional to the operating temperature, and, therefore, the collection efficiency decreases with increasing temperature (Jayadev, 1980).

2.2 Organic Rankine Cycles

The thermal energy contained in the storage layer may be converted to electricity using a generator driven by an organic Rankine cycle engine. In an organic Rankine cycle, the organic working fluid vaporizes as it absorbs heat from the hot pond fluid. The organic vapors are passed through a turbine and condensed. Five to ten percent of the energy absorbed is converted to electricity in the turbine, and the remainder is rejected to the condenser. The liquid working fluid is then pumped back to boiler pressure and the cycle repeated (Fig. 2.2). Organic fluids are used instead of steam because of their much higher vapor densities at the low temperatures prevailing in the cycle.

The efficiency of a Rankine cycle engine increases with increasing inlet temperature. The theoretical Carnot-cycle efficiency for a power cycle operating between 100° and 30°C is approximately 19%. When the limitations of real working fluids, equipment, internal power consumption, and operating conditions are considered, a maximum cycle efficiency of 10% is reasonable.

2.3 Cost Considerations

The relatively low cycle efficiency of an organic Rankine cycle engine coupled to a solar pond (conventional fossil fuel plants have thermal efficiencies of 30%-40%) leads directly to three observations. When a power cycle of 10% thermal efficiency is coupled to a pond with a collection efficiency on the order of 12%, the overall efficiency of converting sunlight to electricity is 1.2%. This suggests that the pond surface areas required will be very large, and, therefore, the pond cost per unit surface area must be very low. The capital costs of solar ponds are not well established and are strongly dependent on location.

The most optimistic cost projection is $5/m² by Ormat Turbines (Israel) for use of preexisting bodies of water and a local salt supply, Jayadev (1980) estimates $15/m² plus salt expense for artificaial ponds lined with Hypalon. Capital costs on the order of $5/m² will be necessary for ponds to be practical for power generation.

The major capital cost in a fossil fuel power plant is the turbine. In the less efficient low-temperature cycles, much larger amounts of heat must be transferred to produce a similar quantity of electrical energy. Consequently, the heat exchangers which supply heat to and reject heat from the power plant become the major cost items. Furthermore, because the size and capital cost of both the pond and generating plant are inversely proportional to the net cycle efficiency, efficiency will be of paramount importance in plant design.

2.4 Direct-Contact Heat Exchangers

Direct-contact boilers and condensers have the potential to significantly reduce the capital cost of the power conversion cycle. For example, a shell-and-tube boiler in a 5-MW$_e$ plant would contribute between $2 and $2.5 million out of the total $7 million direct capital cost of the plant (Wright, 1981). A direct-contact boiler, to perform the same function, should cost between $100,000 and $400,000. The savings could be considerably greater if it were necessary to overdesign the shell-and-tube heat exchanger to compensate for scaling caused by the high concentration of salt in the pond storage layer, or if spare shell-and-tube exchangers were needed for periods of time when the main exchangers were out of service for cleaning. Because extremely large surface areas are achieved in direct-contact equipment, proper design of the boiler can result in small temperature differentials between the water and organic fluids and in increased plant efficiency.

One potential design for a direct-contact heat exchanger (DCHX), combining the functions of preheater and boiler, is shown in Fig. 2.3. Liquid pentane drops are injected into the column at the bottom, while hot brine enters at the top. The pentane drops rise through the brine continuous phase and absorb heat. When the vapor pressure of the pentane reaches the column pressure, the drops begin to vaporize. By the time the pentane reaches the top of the active volume, it has completely vaporized. The pentane vapor disengages from the brine and is piped out to the turbine. The column is baffled or filled with packing to minimize large-scale mixing of the aqueous phase.

This report is concerned with the design and sizing of the direct-contact heat exchanger. It is necessary to predict the height required for heat transfer and the cross-sectional area required to allow the two fluids to pass through. The height of the preheater and boiler are calculated separately and added together to give the total height. The volumetric heat-transfer coefficient in the preheater, which is set by the interfacial area and the mechanism of heat transfer from the continuous phase to the drops, is calculated from the flow rates and the physical

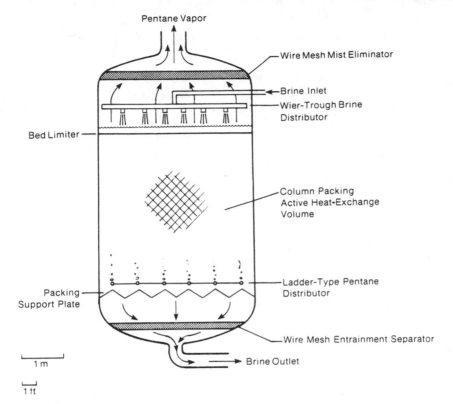

Figure 2.3 Direct-contact preheater/boiler.

properties. The coefficient, with units of $(W/^{\circ}Cm^2)(m^2/m^3)$, is the product of the heat-transfer coefficient at the surface of a single drop and the surface area per unit volume. From a knowledge of the preheater duty Q_p, the log mean temperature difference ΔT (driving force), and the volumetric heat-transfer coefficient U_v, the volume V required in the preheater can be calculated:

$$Q_p = U_v V \Delta T \ .$$

The cross-sectional area of the liquid/liquid preheater is determined by the flow rates and physical properties of the two phases. For a given ratio between the flow rates in the continuous and dispersed phases, there is a minimum cross-sectional area through which the two phases will flow stably. Dividing the volume by the cross-sectional area, we determine the required height.

The height and cross-sectional area of the boiling zone are determined by similar methods. Models to predict the volume and cross-sectional area of the preheater exist, but they do not agree well. Models for the boiling zone are for the most part inadequate.

Throughout this report, the various correlations are applied at the heat exchanger entrance and exit. The flow rates and physical properties of the fluids in the 50-MW$_t$ heat exchanger are shown in Table 2.1. Figure 2.4 shows the

Table 2.1. Flow rates and physical properties in a 50-MW$_t$ Preheater/Boiler

Property	Continuous Phase Brine (NaCl-H$_2$O)	Dispersed Phase Pentane (C$_5$H$_{12}$)	State
Mass flow rate	2120 kg/s	115	
Volumetric flow rate	1.84 m^3/s	0.193	Liquid
		15.8	Vapor
Temperature	77°C	72	Top
	70°C	27	Bottom
Density	1150 kg/m^3	596	Liquid
		7.25	Vapor
Heat of vaporization		324.7 kJ/kg	
Specific Heat	3.55 kJ/kg	2.4	Liquid
Viscosity	1.01 cp	0.185	Liquid
		0.0075	Vapor
Surface tension	73 dyne/cm		Water-Air
Interfacial tension	51 dyne/cm		Water-Liquid Pentane

temperature versus percent energy exchange achieved in an optimized counter-flow solar pond heat exchanger (Wright, 1981). The important points are the small temperature drop in the brine, the relatively small enthalpy change occurring in the preheat section, and the low driving force for heat exchange available in the boiling section.

3 CALCULATION OF COLUMN CROSS-SECTIONAL AREA

A single drop or bubble of pentane will rise through the brine continuous phase with a terminal velocity set by a balance between buoyancy and drag. If many drops are rising simultaneously through the brine, they interfere with each other, and their upward velocity is decreased. As the flow of the dispersed phase increases, the drops crowd closer together and are further slowed. When a critical flow rate is passed, the downward velocity of the continuous phase is greater than the upward velocity of the drops, and drops are swept out the bottom with the brine. *This critical velocity is the flooding velocity and represents the maximum throughput which can be achieved in a given system.* The flooding velocity of a phase is a function of the physical properties of the two phases, the drop size, and the flow rate of the other phase.

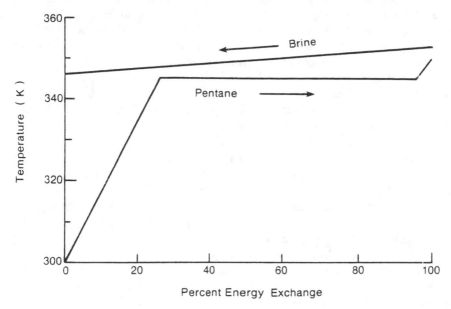

Figure 2.4 Temperature vs. percent energy exchange in boiler.

Column cross-sectional area is determined from prediction of the flooding velocity. From a knowledge of physical properties and the ratio of the continuous and dispersed phase flow rates, the maximum possible velocity of each phase through a given cross-sectional area is calculated. From this, the cross-sectional area of the column is determined.

Interest in direct-contact heat exchange for desalination has generated a large number of publications on the hydrodynamics of liquid/liquid spray columns, and a much smaller body of literature on gases dispersed in liquids and on systems in which the dispersed phase is vaporizing. The DOE Geothermal Program (Jacobs and Boehm, 1980) has tested several preheater/boilers, but the primary emphasis has been on heat exchange and on obtaining operational experience. A small body of literature exists on liquid/liquid flow in packed columns, while an extensive literature describes gas/liquid flows in packed columns.

3.1 Liquid/Liquid Spray Columns

Spray columns are simply empty towers in which the heavier continuous phase (brine) flows downward, while drops of the lighter dispersed phase (pentane) rise upward. Spray columns with length/diameter (L/D) ratios of less than 10 are usually subject to severe backmixing and rarely provide the equivalent of more than one or two theoretical stages. In industrial practice, units with straight sides and low L/D ratios are used, but most experimental data have been gathered in laboratory-scale columns of the Elgin design with high L/D ratios (Jacobs and Boehm, 1980).

The earliest attempts at defining the flooding velocity in spray columns as a function of the physical properties of the fluids and the drop diameter were by Minard and Johnson (1952) and by Sakiadis and Johnson (1954). In each case, the Bernoulli equation was written for each phase, frictional losses were described in terms of flow around submerged objects, and the equations were solved simultaneously to determine the form of the flooding relationship. The constants in the equations were then determined from experimental data. The Minard-Johnson correlation was fitted to a relatively small set of liquid/liquid data, while the correlation of Sakiadis and Johnson was fitted to a much larger set of liquid/liquid, solid/liquid, and gas/liquid data. Because it covers a much larger range of properties and represents a later effort by the same research group, the correlation of Sakiadis and Johnson is preferred. However, it should be remembered that all the data were taken on laboratory-scale columns. Also, the ability of the correlation of Sakiadis and Johnson to describe flows of gas through liquid is doubtful, since the flows described used much larger bubbles and much smaller columns than will be found in heat exchange systems. The correlations of Minard and Johnson and of Sakiadis and Johnson are presented in section 9.

Letan (1976) recommends the semiempirical correlation of Richardson and Zaki (1954) to describe the hindered upward velocity of a swarm of drops. Letan defines flooding as the point at which, for a given flow rate of the dispersed phase, the velocity of the continuous phase cannot be increased without gross entrainment of the dispersed phase and at which the holdup (volume fraction of dispersed phase) is at a maximum. An algebraic method is presented for determining the flow rates and holdup at flooding and at the chosen operating condition.

A better picture of the relationship between the flow rates and holdup is achieved by combining the rise velocity equations of Richardson and Zaki (1954) with the graphical presentation of Mertes and Rhodes (1955).

A schematic operating diagram for a system with 4-mm-diameter drops is presented in Fig. 3.1. The drop number is the ratio of the superficial velocity of the dispersed phase to the terminal velocity of a single drop. The liquid number is the ratio of the superficial velocity of the continuous phase to the terminal velocity. It is clear that for a fixed continuous-phase superficial velocity there is a maximum dispersed phase flow rate. The locus of all these points is the flooding curve. Also, for any given ratio R of dispersed to continuous phase flow rates that is set by the system's energy balance, there exists a single flooding point where the line of constant R crosses the flooding curve. Finally, for a given operating point on the chart, it is possible to predict the effect of changing either or both of the flow rates.

Like the correlations of Minard and Johnson and of Sakiadis and Johnson, use of the correlation of Richardson and Zaki by either the method of Letan or of Mertes and Rhodes is validated only in liquid/liquid systems in smaller-diameter columns.

In designing a spray column, the superficial velocities of the two phases are unknown, but the ratio of the volumetric flow rates (and therefore the ratio of the

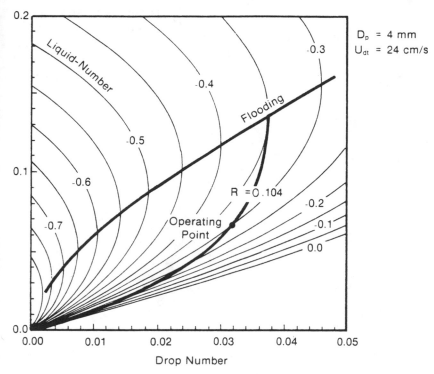

Figure 3.1 Schematic operating diagram for counter-current flow illustrating relationships among dispersed phase flow rate, continuous phase flow rate, and holdup.

superficial velocities) is set by the energy balance on the heat exchanger. Knowing the ratio of the superficial velocities and the physical properties of the two phases, we can solve for the superficial velocity of the continuous phase at flooding as a function of drop size using any of the previously mentioned methods. These superficial velocities represent the maximum possible throughput. Knowing the superficial velocity and volumetric flow rate of the continuous phase, we divide to obtain the minimum permissible cross-sectional area of the heat exchanger.

The maximum allowable superficial velocities in the liquid/liquid section as determined by the three correlations are plotted as a function of drop size in Fig. 3.2, while the corresponding minimum cross-sectional areas for a 50-MW$_t$ heat exchanger are plotted in Fig. 3.3. Drop size in the liquid/liquid section is determined by the dispersed-phase distributor design and may be set by the designer. A review of methods for drop size prediction is given by Horvath (1978).

It is clear from the preceding figures that the superficial velocities at flooding predicted by Richardson and Zaki are up to 50% higher than those predicted by Sakiadis and Johnson, and that the predictions of Minard and Johnson are essentially independent of drop size. However, little weight should be given to the

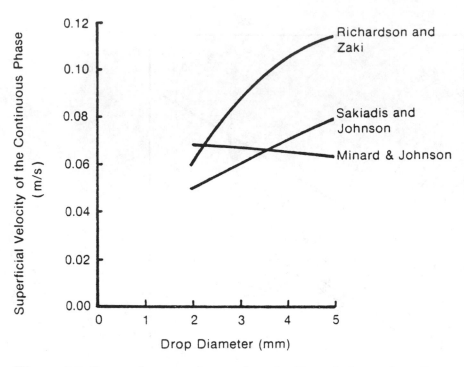

Figure 3.2 Spray column continuous-phase flooding velocity vs. drop diameter.

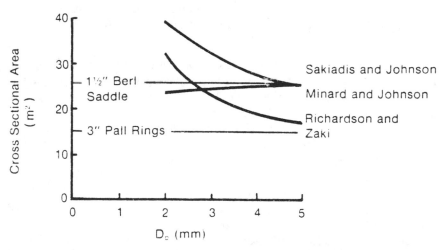

Figure 3.3 Minimum cross-sectional area vs. drop diameter.

prediction of Minard and Johnson because their work is superseded by that of Sakiadis and Johnson, and because the dependence on drop size is at odds with theory and the observations of all other researchers. It is obvious that considerable uncertainty exists in the prediction of the flooding velocity. This is not critical if the preheater and boiler are contained in the same vessel, since the cross-sectional area required by the vapor flow is larger than that required by the liquid. However, the uncertainty will be important if separate preheaters and boilers must be used.

Each of the authors recommends a safety factor for design. Letan recommends the operational holdup be set at 90% of the flooding holdup. Since the slope of holdup versus superficial velocity at flooding is very steep, this amounts to an increase in cross-sectional area of only 5%. Sakiadis and Johnson recommend increasing the cross-sectional area by 25%, and Minard and Johnson recommend a 50% safety margin.

3.2 Backmixing and Packings

It is preferable to operate the heat exchanger as a counter-current device in order to obtain the maximum possible driving force for heat exchange. Backmixing also tends to flatten the continuous-phase temperature profile of Fig. 2.4 and make the system behave as a mixed tank instead of a plug flow device. Backmixing tends to keep the continuous phase at the brine outlet temperature. This can drastically reduce the average driving force in the vaporization section, but has little effect on the driving force in the preheat section because of the large difference between the inlet pentane and outlet water temperatures. Backmixing has three basic causes. Each rising drop of pentane carries with it a wake of continuous phase. Because there is little convective interchange of heat or mass between the bulk phase and the wake, cold brine is moved from the bottom of the column to the top of the preheat section, decreasing the continuous-phase temperature and driving force higher up in the column. Secondly, the dispersed-phase drops tend to channel upward together in the center of the column and avoid the edges. The continuous phase is dragged upward in the center and compensates by flowing downward faster at the sides. This also mixes the continuous phase and lowers the driving force. Finally, the boiling section is violently agitated by the vaporization and very high velocity of the pentane vapor. This turbulence can set up violent eddies and large-scale mixing within both the vaporization and preheat sections.

Conventional tower packings can be used to suppress backmixing. A packing with a large void fraction (open packing) is desirable, so that backmixing will be reduced, but the flow will be impeded as little as possible. An example of an open packing is the three-inch metal Pall ring manufactured by the Norton Co. (Fig. 3.4). A second possibility is a grid packing (Fig. 3.5) a design that is commonly used in cooling towers. Pall ring packings force the drops to rise in tortuous paths, causing them to shed their wakes repeatedly. The packing increases the pressure drop, preventing the dispersed phase from channeling up the center, and the solid walls block the formation of large-scale turbulence. However, by

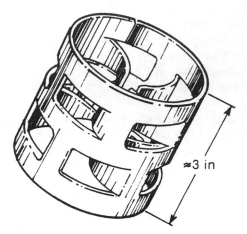

≈3 in

Figure 3.4 Metal pall ring tower packing (Norton Chemical Co.).

impeding the progress of the two phases, the rings make it necessary to increase the size of the column. Grid-type packings offer less resistance to flow and would allow the column to work essentially as a spray column, while preventing the formation of large-scale circulation patterns. By offsetting the grids as the packing grids are stacked on top of each other, the flow can be made to split and recombine repeatedly. The column would behave like many small stirred tanks in series, which is essentially indistinguishable from plug flow.

3.3 Liquid/Liquid Packed Columns

While the correlations for spray columns filled should apply to a large column filled with an open grid packing, a column packing, such as a Pall ring, presents enough resistance to flow that it must be described by different correlations. The

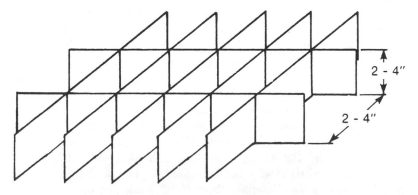

2 - 4"

2 - 4"

Figure 3.5 Grid packing.

larger packings offer less resistance to flow and are less expensive per unit volume and, therefore, are preferred in order to minimize both the size and cost of the tower. Packings should be made of a material, such as ceramic or oxidized metal, that is not preferentially wet by the dispersed phase. As a rule of thumb, the packing pieces should be less than one-eighth the diameter of the column, in order to achieve good distribution and prevent the dispersed phase from tending to migrate to the walls (Treybal, 1973). Since most of the data on liquid/liquid flooding in packed columns were obtained in laboratory-scale columns, most correlations deal with packings smaller than one inch, as opposed to the 3-in. Pall rings which would be most suitable for this application. The only correlation which is valid for both the flow rates and fluid properties in the preheater and for reasonably large packings was derived by Hoffing and Lockhart (1954) (section 9). The largest packing covered by this correlation is a 1-1/2-in. Berl saddle. The cross-sectional area required in the liquid/liquid section for a 1-1/2 in. Berl saddle is shown on Fig. 3.3 and is essentially in the middle of the range for spray towers. If the correlation is applied to 3-in. Pall rings, the predicted cross-sectional area is as low as the lower limiting values predicted for spray towers. It is not probable that the correlation accurately predicts the flooding velocity of the 3-in. Pall ring packing, since the correlation predicts continually diminishing cross-sectional areas as the packing size is increased, instead of reaching a limiting value equal to that of a spray column as would be expected. All that can be conclusively stated is that the cross-sectional area required for a column packed with 3-in. Pall rings would be somewhat less than that for 1-1/2-in. Berl saddles.

3.4 Vapor/Liquid Spray and Packed Columns

When the vapor pressure of the liquid drops reaches the pressure in the vessel, they begin to vaporize. There is a zone where three phases are present: liquid water, liquid pentane, and pentane vapor. This zone is violently agitated and characterized by small drops. There have been no attempts to date to characterize these flows in either packed or spray columns.

In a properly designed column, the entire pentane flow will be vaporized before it reaches the top liquid surface. It is at the point where vaporization is complete that the dispersed-phase volumetric flow rate will be the greatest, because the pentane undergoes an 80-fold expansion on vaporization. If this large flow can move through the column, the lower volumetric flow rate of liquid and vapor pentane will be able to pass through the middle of the column.

The only correlation that has been proposed to predict flooding rates of gas bubbling through a liquid-continuous phase is that of Sakiadis and Johnson (1954). Figure 3.6 shows the calculated cross-sectional areas at flooding as a function of bubble diameter for a 50-MW$_t$ heat exchanger. The area is a strong function of bubble diameter, especially at low diameters. No systematic or quantitative studies of bubble diameter in vaporizing systems have been conducted, although limited visual study of such systems suggests they are in the range of 2 to 5 mm. Bubble diameter is determined by a balance between coalescence and turbulent

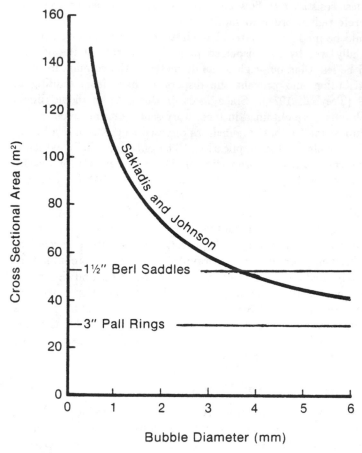

Figure 3.6 Heat-exchange cross-sectional area for packed and spray columns, based on vapor flow.

forces that tend to break up the droplets. Liquid drop size in turbulent systems (Heinz, 1955) and air bubble size in agitated tanks (Calderbank, 1967) have been derived for significantly different systems, and both require knowledge of the mechanical energy input per unit mass, a parameter unknown in the situation being considered. If bubble diameters are of the order of 1 mm, the vessel cross-sectional area required by the vapor section will be much larger than that required by the preheater, while if diameters are larger, the required cross-sectional areas in the two zones will be more similar. It should be remembered that the only gas/liquid data which were used in the fitting of the correlation of Sakiadis and Johnson came from bubbles of 1/4 to 1-1/2 in. in diameter, using a 2-in. diameter column. While this correlation must be used for preliminary evaluation because it is the only one available, its predictions should be used with caution.

Extensive work has been done on gas/liquid flows in packed towers, because packed towers are extensively used in the chemical industry for mass-transfer devices. Sherwood (1938) developed the first correlation of packed-tower flooding velocities using an air/water system. Lobo (1945) modified the method to use experimentally determined packing factors to correlate the flow instead of measured surface-to-volume ratios. Leva (1954) included information on preflooding conditions by including curves of constant pressure drop in the correlation, and by extending the correlation to systems other than air/water. Eckert (1966) refined the method for experimentally determining packing factors for dumped packings.

The plot relating flow ratio and pressure drop to the allowable column diameter is shown in Fig. 3.7. The abscissa is determined from the energy balance and the physical properties of the two phases. When the designer chooses a pressure drop and a column packing material, the gas superficial velocity and column diameter are fixed. In order to minimize both the volume and cost per unit

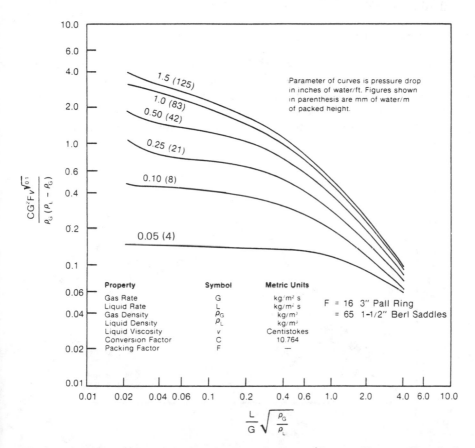

Figure 3.7 Gas-liquid packed tower correlations. (Source: Norton Chemical Process Products Bulletin).

volume of the packing, a large packing is preferable. The calculated column diameters for 3-in. Pall rings and 1-1/2-in. Berl saddles are shown in Fig. 3.6. It is interesting to note that the predicted packed column diameters are less than the spray column diameters. A possible explanation is that at these flow rates in packed columns, the gas is the continuous phase and water is the dispersed phase. The water channels through the vapor, and the pressure drop is greatly reduced. In the model used to describe the spray column, many discrete bubbles of vapor with a very large total surface area are rising through the water. Therefore, there is greater resistance to flow and the required diameter is larger. This also means that at some point in the boiling section of the packed column, the dispersed and continuous phases must reverse. The vapor must be the dispersed phase where boiling is just beginning and the vapor fraction is small, while at the top the vapor may establish itself as the continuous phase. Whether this phase reversal, which is a classic definition of flooding, will lead to unstable operation of the column must be resolved by experiment.

4 HEAT TRANSFER

The considerable literature dealing with heat transfer to single drops, summarized by Sideman (1966), applies mainly to systems where the drops are widely dispersed and do not interact. Since the heat-transfer rate is proportional to the surface area, it is desirable to operate at high holdups where many drops are crowded together. Research on heat transfer in multidrop systems has been

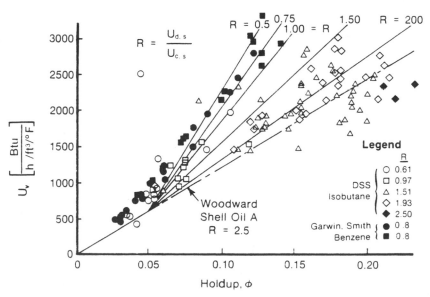

Figure 4.1 Volumetric heat-transfer coefficient vs. holdup with flow ratio as a parameter.

focused on desalination in Israel, with the work of Letan, Kehat, and Sideman, and on geothermal applications in the United States.

4.1 Liquid/Liquid Systems

Data from a large number of spray-tower studies (Suratt, 1977; Garwin and Smith, 1953; and Plass, 1979) is presented graphically and in equation form by Plase and is shown in Fig. 4.1. It is useful to note that the volumetric heat-transfer coefficient increases with holdup for a constant R and decreases with increasing R for a constant holdup. The data can also be fit with the empirical equation:

$$U_v = 45,000 \, (\phi - 0.05)e^{-0.75R}$$

$$+ \, 600 \, \text{Btu/h ft}^{3o}\text{F} \qquad \phi > 0.05$$

$$U_v = 12,000 \, \phi \, \text{Btu/h ft}^{3o}\text{F} \qquad \phi < 0.05$$

where U_v is the volumetric heat-transfer coefficient, ϕ is the dispersed phase holdup, and R is the ratio of the dispersed to continuous phase volumetric flow rates.

The height of the liquid/liquid section may also be described by the method of Letan and Kehat (1968). This model applies to drop Reynolds numbers greater than 30 and laminar flow of the continuous phase in the column. The model involves determination of the temperature distribution throughout the column from energy balances on the fluids in the three zones of the column: the wake-growth zone, the wake-shedding zone, and the mixing zone. In the wake-growth zone, a wake builds up behind the newly formed drop. In this zone, the drop is surrounded by fluid at the bulk continuous phase temperature, and the heat-transfer rate is high. In the wake-shedding zone (the middle region), heat transfer is poor, as the drop is in contact with an attached packet of continuous phase which is essentially in thermal equilibrium with the drop. Heat transfer here is limited by the frequency with which the drops shed their wakes. During the coalescence or transition to boiling zone at the top, the drops again shed their wakes, and the resulting agitation increases the rate of transfer. Nonlinear ordinary differential equations are developed for each zone and are solved simultaneously to yield the temperature distribution, height, and outlet temperature. The equations are described in section 9. The correlation has been successfully compared with experimental data by Letan and Kehat and by Plass. Plass also found good agreement between the correlation of Letan and Kehat and his volumetric heat-transfer coefficient method. Urbanek (1979), however, in experiments on a brine/isobutane preheater/boiler found that the Letan-Kehat model would only fit the observed temperature profile with different constants. Figures 9.3 and 9.4 show that the constants are not yet known with reasonable accuracy. However, as Urbanek merely adjusted the parameters in the model to fit his own data, it is not known whether his parameters actually represent the physical processes that occurred in his experiment.

No data are available for liquid/liquid heat transfer in packed towers, nor can information be reasonably developed from the mass-heat-transfer analogies, as mass transfer generally affects the hydrodynamics of the drops. Mass transfer operations in liquid/liquid packed columns has not been successfully correlated, and commercial equipment is generally designed by scaling up from pilot-plant data. However, it can be said that the heat-transfer performance of packed beds should be superior to that of spray columns because the convoluted path of the drops through the packing ensures frequent wake shedding, and because the packing reduces the unmodeled reductions in performance due to backmixing. Therefore, a conservative method of estimating heat transfer in liquid/liquid packed columns is to estimate the holdup by the method of Pratt (Treybal, 1973) which is shown in section 9, and then to estimate the heat-transfer coefficient from the graph by Plass.

4.2 Boiling Systems

Little is known about direct-contact vaporization. Some experimental observations are available, but there is no theory or method of correlation to describe the results in the region of interest. Sideman (1964) analyzed the vaporization of single drops of pentane rising through water. In his study, the drops grew into single large bubbles. However, in heat exchangers where many drops are vaporizing simultaneously, the resulting turbulence breaks the large bubbles into many small bubbles. In a later study, Sideman and Gat (1966) measured volumetric heat-transfer coefficients and column heights required to vaporize pentane in a laboratory-scale spray column. The data are presented in Fig. 4.2.

Measured volumetric heat-transfer coefficients are in the range 8,000-20,000 $kJ/m^3 \ h^{\circ}C$ (5,000-12,500 $Btu/ft^3 \ h^{\circ}F$). The coefficients decrease rapidly with increasing driving force and are also a function of the ratio of the mass flow rates. As the ratio of the dispersed to continuous phase flow rate is increased, the heat-exchange coefficients rise steeply, pass through a maximum, and then slowly decrease.

A considerable amount of testing of direct-contact preheater/boilers has been carried out for the DOE geothermal program at the University of Utah and at the East Mesa, Utah, test site (Suratt and Hart, 1977; Sims, 1976; Blair, 1976; Deeds, 1976). Data were successfully correlated by the equation

$$St = \frac{UA}{(\dot{m}C_p)_{d,p,1}} = 2.0 \ Ja \ Pr \left(\frac{\dot{m}_c}{\dot{m}_d}\right)^{0.15}$$

where

$$Ja = h_{fg}/C_{p,v}(T_{c,ave} - T_{sat}),$$
$$Pr = (C_p \mu/k)_{d,v},$$
$$U = \text{heat-transfer coefficients,}$$
$$A = \text{heat-transfer area,}$$

$\dot{m}$ = flow rate, and

C_p = heat capacity.

A graph of the correlation is presented in Fig. 4.3. The correlation is useful mainly for evaluating the relative sizes of heat exchangers working with different fluids, and for suggesting the effect of different conditions on the heat-transfer coefficient. Unfortunately, because no methods are available to predict the surface area available in the heat exchanger, the equation cannot be used to predict the heat-transfer coefficient independently. A second limitation on the correlation is that it is derived from data where the driving force for heat transfer is an order of magnitude greater than in solar pond systems. This is of concern because in direct-contact boiling the heat-transfer coefficient is influenced by the magnitude of the driving force. Finally, a typical solar pond heat exchanger would have a JaPr product of approximately 30, which is well to the right of the bulk of the data on the plot.

No heat transfer data are available at the flow ratio and continuous phase superficial velocities which are typical of the pond heat exchanger. However, 20,000 kJ/m^{3}h°C (12,500 Btu/ft^{3}h°F) will be used as the estimate of the heat transfer coefficient in designing boilers (Jacobs and Boehm, 1980). It is encouraging to note that in no case did Sideman require a depth greater than 0.2 m (6 in.) to completely vaporize the pentane. Therefore, it is probable that the heat-exchanger height will be determined primarily by the preheat section.

5 HEAT-EXCHANGE SYSTEM DESIGN

Because of the wide variation in the predictions of column cross-sectional area, three different heat-exchanger configurations will be discussed. The preferred system appears to be the packed column, as this yields both a small cross-sectional area and reduced backmixing. A second choice would be a spray column with grid packing and large vapor bubbles. The least desirable design would be a spray column with small vapor bubbles and a cross-sectional area required for vapor flow that was much larger than that required in the liquid/liquid section.

5.1 Packed Towers

The most desirable combination is the column packed with Pall rings. The required vapor-section, cross-sectional area is 30 m^2, if a pressure drop of 1-in. water/ft of column packing is assumed. This corresponds to a superficial velocity which is 80% of that at flooding. The calculated cross-sectional area required by the liquid is in the range of 15-26 m^2. Therefore, if the columns are straight sided and sized to meet the vapor flow rate, the column will be big enough to handle the liquid phase throughput. A single column with a 30-m^2 cross-sectional area would have a diameter of 6.2 m (21 ft). This would be too large to ship by truck and would have to be site fabricated (an expensive process). Therefore, it is preferable to build three smaller columns with a 3.6-m (11.7-ft) diameter which will be run in parallel to handle the heat-exchange load.

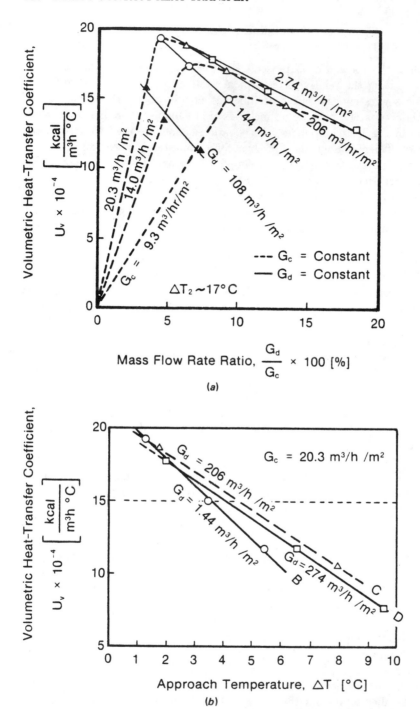

Figure 4.2 Volumetric heat transfer coefficient plotted against mass flow ratio at constant driving force and against driving force.
(a) coefficient vs. mass flow ratio at constant driving force
(b) coefficient vs. driving force

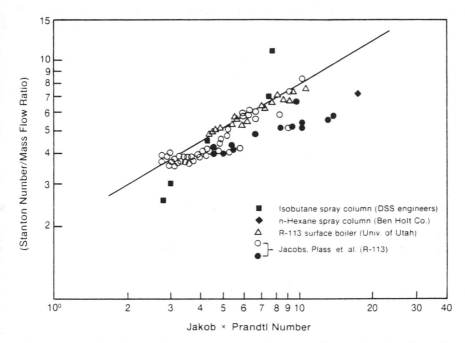

Figure 4.3 Nondimensional vaporization heat transfer vs. Jakob x Prandtl numbers.

To determine the liquid/liquid preheater height, the Pratt's method is used to estimate the holdup in the packing at the design conditions ($\phi = 0.07$), and the volumetric heat-transfer coefficient is conservatively estimated from the graph of Plass ($U_v = 28,000$ W/m^{30}C [1500 Btu/ft^{3}h$^{\circ}$F]). The preheater duty is 11 MW$_t$, and the log mean driving force for counter-current flow is 15.5°C; therefore, the required volume is 25.3 m^3, and the required height is 0.84 m (2.8 ft). As there are large uncertainties in the calculations, a 50% safety factor gives us a 1.2-m (4-ft) tall preheat section.

The boiler section is sized by using a volumetric heat-transfer coefficient of 20,000 W/m^{30}C (12,500 Btu/ft^{3}h$^{\circ}$F) and an average driving force of 5°C. The volume required is 30 m^3 and the height is 1.3 m (4.26 ft). Again, using a 50% safety factor we get an actual boiler height of 2 m (6.4 ft). We then have a heat-exchange section comprised of three units, each with a diameter of 3.6 m (12 ft), active height of 3.2 m (10.5 ft), and a total height of 5.6 m (18.5 ft). One of the modules is illustrated in Fig. 2.3.

It is important to note some of the important assumptions inherent in this design. Large uncertainties exist in the calculation of the cross-sectional area of the liquid/liquid section, but as the sizing is set by the vapor flow, this is not a major drawback. Likewise, the calculations of the heat-transfer coefficients in the preheater are quite uncertain, but as the assumptions are conservative, this is

Table 5-1 Equipment Cost Estimates for 50-MW$_t$ Heat Exchangers (mid-1980 dollars

	Preheater		Boiler		
	Vessel	Packing and Internals	Vessel	Packing and Internals	Total
Packed column 3-in. Pall rings	50,300	69,000			119,30●
Spray column 5-mm bubbles combined preheater/boiler	99,100	116,700			215,80●
5-mm bubbles separate preheater/boiler	28,900	15,700	84,500	52,400	181,50●
1.5-mm bubbles combined preheater/boiler	183,300	206,600			389,90●
1.5-mm bubbles separate preheater/boiler	28,900	15,700	169,000	33,900	247,50●
Shell-and-tube HEX					2,000,00●

again not critical. The vapor section is sized using the correlation of Eckert for vapor liquid columns and is probably reasonably accurate. The uncertainty in the heat transfer calculations in the vaporization section is important because no data are available in the operating region under consideration. Data are not available at the superficial velocities under consideration, and no data are available for heat transfer in packed towers. While the cost of such packed-tower direct-contact heat exchangers is low enough compared to shell-and-tube exchangers that almost any necessary degree of over-design can be afforded (Table 5.1), it is necessary to understand the operation to design an exchanger which will yield the expected performance and ensure that the exchanger will perform stably under all operating conditions.

The most important uncertainties are not those inherent in the calculations but those in the processes that were not considered. For example, some degree of backmixing will exist in the continuous phase, even with Pall ring packing. This will decrease the available driving force and increase the required volume. The major reason that the column diameter at the top is smaller in the packed tower than in the spray tower is that the vapor phase is continuous in the top of the packed tower. However, at the beginning of the boiler section the liquid is the continuous phase. The point where the phases reverse is a flooding point, characterized by large pressure drops. It is not known whether the transition will cause large-scale flow instabilities in the column, such as slugging, or whether it will have no effect at all on the overall column performance. In portions of the boiling section where the vapor is the continuous phase, it is uncertain how the liquid pentane drops will rise through the column. It is possible that the local downward velocity of the water will be greater than the upward velocity of the drops. This could cause accumulation of a layer of liquid pentane at the bottom of the boiling section, which could in turn lead to liquid/liquid flooding or formation of large pockets of pentane vapor. Alternatively, the life span of a liquid drop in the boiling section may be so short that these concerns are unwarranted. In any case, research on the mechanism of fluid flow in the boiling section is needed before such a column could be designed with confidence.

5.2 Spray Towers

The system configuration in a spray-tower system will be strongly dependent on the equilibrium bubble size in the boiling section. Because there is no method available for predicting bubble size, we will create two designs, one for a system with 5-mm bubbles and one for a system with 1.5-mm bubbles.

For a spray column with a grid packing open enough not to affect the flooding calculations, the cross-sectional area of the vaporization section is 45 m^3 at flooding for 5-mm bubbles. Using a 25% safety factor, the total cross-sectional area is 56 m^2. If we assume a drop size of 3.5 mm in the liquid/liquid section, the preheater area at flooding is in the range of 20-30 m^2. Therefore, the area is set by the vapor throughput.

From a knowledge of the boiler duty, heat transfer coefficient, and driving force, we calculate the boiling depth as 0.67 m. With a 50% safety factor, the boiler height is 1 m (3.3 ft). Using the correlation of Richardson and Zaki to estimate the holdup in the preheater, we find it to be only 2.2%. This is because we have spread the small liquid dispersed phase flow over a large area and have reached the point where the drops rise without interference.

This holdup corresponds to a heat transfer coefficient of only 4,200 W/m^3^{o}C (265 Btu/ft^{3}h^{o}F). The volume of the preheater is now 143 m^3, yielding a height of 2.6 m. Using a 50% safety factor, we have an actual preheater height of 3.9 m. The overall active height of the heat exchanger is now 4.9 m (16 ft). In order to shop-fabricate these units, we have 5 units with a 3.8-m diameter (12.5 ft) and a total height of 7.3 m (24 ft). It is useful to note that the increase in the total area required for the vapor flow did not increase boiler volume, but increased the volume of the preheat section dramatically. The major uncertainties in this design are in the degree of backmixing, the calculation of the vaporization heat transfer coefficient, the flooding rate at 5-mm bubble diameter, and of course, what the actual bubble diameter will be. No data are available on vaporization heat transfer coefficients at these operating conditions. The predictions of the gas/liquid flooding velocity in large diameter columns by the method of Sakiadis and Johnson are suspect, even if the average bubble size is known.

The low holdup in the preheater lowered the volumetric heat transfer coefficient and led to a relatively large and expensive unit. The size and cost may be reduced by building a system with separate preheaters. Using the same methods described earlier, we find that the preheating load can be carried out in two parallel vessels, each with an active height of 3.3 m (11 ft), a total height of 5.3 m (17 ft), and a diameter of 4.1 m (13.5 ft). By regulating the pressure in the preheater, the pentane is prevented from flashing. The penetane then leaves the preheater and is injected into the brine just withdrawn from the pond in several relatively shallow tanks, where it is vaporized. The boiler section consists of five vessels in parallel, each with an active height of 1.8 m (6 ft), a total height of 4.3 m (14 ft), and a diameter of 3.6 m (12 ft). While this system has a somewhat lower capital cost (Table 5.1), it will be more complex due to the piping connecting the boiler and preheater. In this case it is necessary to be able to accurately predict the heat transfer coefficients and holdups in the liquid/liquid section of the exchanger as well as in the boiler. Because the preheater and boiler are separate, the preheater is not necessarily oversized as it is in the combined system.

The situation with a 1.5-mm diameter bubble is an exaggeration of that at 5 mm. The cross-sectional area required by the vapor phase at flooding is now 100 m^2, compared with 20-30 m^2 for the preheater. If the preheater and boiler are combined in a single vessel, the result is 10 vessels, each with an active height of 4.6 m (15 ft), a total height of 7 m (23 ft), and a diameter of 3.6 m (12 ft). If separate preheaters and boilers are used, we again have two preheaters with a total height of 5.3 m and a diameter of 4.1 m. The boiling is carried out in 10 vessels in parallel, each with an active height of 1 m (3.3 ft), a total height of 4.3 m (14 ft), and a diameter of 3.6 m (12 ft).

5.3 Cost Estimate

Table 5.1 presents the direct (fabricated) cost estimates for the five heat exchanger designs described in Secs. 5.1 and 5.2, and for a conventional shell-and-tube heat exchanger designed for the same duty. The packed-column exchanger is least expensive because the phase reversal allows the vapor to flow through a smaller cross section, and because the smaller cross section leads to a denser packing of drops and a higher volumetric heat-transfer coefficient in the preheat section. Because of the greater cross-sectional area required for vapor flow when the bubbles are small, the cost of the spray-column systems is larger than that of the packed tower. The difference may be lessened if separate preheaters and boilers are used, eliminating the effect of the enlarged vapor section cross-sectional area on the preheater.

The columns were costed by the methodology described by Pikulik and Diaz (1977). The vessels were assumed to be plain carbon steel with 7/16-in. walls. (It is possible that these vessels could be built from fiber-reinforced plastic at a large reduction in cost.) The conventional heat exchanger was sized using an overall heat transfer coefficient of 450 W/m^2°C (80 Btu/ft^{2}h°F) for the preheater and 790 W/m^2°C (140 Btu/ft^{2}h°F) in the boiler (Bell, 1973). The heat exchangers were calculated to cost $215/m^2o(20/ft^2)$, a representative figure in 1980 dollars suggested by Stearns-Roger Services.

6 RESEARCH REQUIREMENTS

Several issues need to be addressed before direct-contact preheater/boilers can be designed with confidence. Because the packed tower appears most promising, research should be done on flow patterns in the vaporization section. It should be verified that the gas phase does indeed become the continuous phase in the top of the packing. The flow behavior at the point of phase inversion should be studied to be sure that the local flooding will not disrupt the operation of the rest of the column. Also it should be determined whether the correlation of Eckert adequately describes the liquid/vapor flow. It should be determined whether the vaporizing pentane has a tendency to accumulate in large pockets and slug through the column, as well as whether liquid pentane tends to accumulate in pockets. Next, it should be determined whether salt water brines have a tendency to create a stable foam at the liquid/vapor surface, as such a foam would eventually be carried over into the turbine.

In addition to the fluid flow questions, work is required on vaporization heat transfer coefficients. The only heat transfer data available describe spray-tower operation. However, the heat transfer coefficient is a function of flow ratio and driving force, and no data are available at the conditions under which the column operates. Also, no theory yet exists which gives an explanation for the type of variations observed. Such an understanding would allow more confident scaleup of bench- and pilot-scale experience to full-scale units. Also, experiments need to be conducted to determine what effect the presence of packing has on the spray-column heat transfer results.

The most important questions in spray-column systems are: what is the equilibrium size of the bubbles formed in the vaporization section, what is the maximum allowable vapor superficial velocity, and does the correlation of Sakiadis and Johnson adequately describe flooding in the system? These questions are critical because the size, configuration, and cost of the heat exchanger are primarily determined by the allowable vapor superficial velocity.

7 CONCLUSIONS

The use of a direct-contact preheater/boiler instead of a shell-and-tube unit can significantly reduce the heat-exchange costs of a solar pond power plant. The size and cost of direct-contact heat exchangers is related to both the heat transfer and hydrodynamics of the system. The height of an exchanger is determined by the driving force available for heat exchange, the interfacial area, and the enthalpy change of the dispersed phase. The cross-sectional area is determined by the volumetric flow rates of the two phases. Methods exist for predicting the height and cross-sectional areas of direct-contact heat exchangers, but large uncertainties exist. The liquid/liquid preheater section of the exchanger is reasonably well understood, but it is the boiling section which usually determines the diameter of a combined preheater/boiler column. Flow characteristics of packed columns with vaporization are not understood. No heat-transfer data are available on the vaporization of drops of a dispersed organic phase at the driving forces and flow rates which will be encountered in this system. The prediction of flooding velocities by the method of Sakiadis and Johnson requires a knowledge of the bubble diameter, a parameter that cannot be predicted. In addition, validity of the correlation for gas/liquid flows in large diameter columns is suspect. Research is required on the flow patterns in packed columns, flooding characteristics and bubble size in spray columns, and heat transfer coefficients in the boiling sections of both packed and spray columns. Packed- and spray-column heat exchangers were designed and costed which covered the full range of possible designs. In all cases, the exchangers would offer significant cost savings over shell-and-tube heat exchangers.

8 REFERENCES FOR SECTIONS 1 THROUGH 7

Bell, K. J. 1973. "Thermal Design of Heat Exchangers, Condensers, and Reboilers." *Chemical Engineers Handbook*, Section 10, 5th Edition. Edited by R. H. Perry and C. H. Chilton. New York: McGraw Hill.

Blair, C. K.; Boehm, R. F.; Jacobs, H. R. 1976. *Heat Transfer Characteristics of a Three-Phase Volume Boiling Direct-Contact Heat Exchanger*. DOE Report IDO/1523-1. University of Utah.

Calderbank. 1967. *Mixing*. New York: Academic Press, Vol. 2.

Deeds, R. S.; Jacobs, H. R.; Boehm, R. F. 1976. *Heat Transfer Characteristics of a Surface Type Direct Contact Boiler*. DOE Report DIO/1523-2. University of Utah.

Eckert, J. S.; Foote, E. H.; Walter, L. F. 1966 (Jan.). "What Affects Packing Performance." *Chemical Engineering Progress*. Vol. 62 (No. 1): p. 59.

Garwin, L.; Smith, B. D. 1953. "Liquid-Liquid Spray-Tower Operations in Heat Transfer." *Chemical Engineering Progress*. Vol. 49 (No. 11): pp. 591-601.

Heinz, J. O. 1955 (Sept.). "Fundamentals of the Hydrodynamic Mechanism of Splitting in Dispersion Process." *American Institute of Chemical Engineers Journal*. Vol. 2 (No. 3): pp. 289-295.

Hoffing, E. H.; Lockhart, F. J. 1954. "A Correlation of Flooding Velocities in Packed Columns." *Chemical Engineering Progress*. Vol. 47: p. 423.

Horvath, M.; Steiner, S.; Harland, S. 1978 (Feb.). *Canadian Journal of Chemical Engineering*. Vol. 56: pp. 9-19.

Jacobs, H. R.; Plass, S. B.; Hansen, A. C.; Gregory, R. 1977. "Operational Limitations of Direct-Contact Boilers for Geothermal Applications." ASME Paper No. 77-HT-5. 1976 ASME/AIChE National Heat Transfer Conference.

Jacobs, H. R.; Boehm, R. F. 1980 (Dec.). "Direct-Contact Binary Cycles." *Sourcebook on the Production of Electricity from Geothermal Energy*. Edited by J. Kestin. U.S. Department of Energy; pp. 413-470. Available from U.S. Government Printing Office, Washington, DC 20402.

Jayadev, T. S.; Edesess, M. 1980 (Apr.). *Solar Ponds*. SERI/TR-731-587. Golden, CO: Solar Energy Research Institute. Available from: NTIS, Springfield, VA 22161.

Letan, R. 1976. "Design of a Particulate Direct-Contact Heat Exchanger: Uniform Countercurrent Flow." ASME Paper 76-HT-27. ASME/AIChE Heat Transfer Conference.

Letan, R.; Kehat, E. 1968. "The Mechanism of Heat Transfer in a Spray Column Heat Exchanger." *American Institute of Chemical Engineers Journal*. Vol. 14 (No. 3): pp. 398-405.

Leva, M. 1954. *Chemical Engineering Progress Symposium Series*. Vol. 50 (No. 10): p. 57.

Lobo, W. E.; Friend, L.; Hashmall, F.; Zenz, F. A. 1945. "Limiting Capacity of Dumped Tower Packings." *Transactions of the AIChE*. Vol. 41: p. 693.

Mertes, T. S.; Rhodes, H. B. 1955 (Sept.). "Liquid-Particle Behavior: Part I." *Chemical Engineering Progress.* Vol. 51 (No. 9): pp. 429-432.

Minard, G. W.; Johnson, A. I. 1952 (Feb.). "Limiting Flow and Holdup in a Spray Extraction Column." *Chemical Engineering Progress.* Vol. 48 (No. 2): pp. 62-74.

Norton Chemical Process Products. "Design Information for Packed Towers." Norton Bulletin DC-11. Akron, OH 44309.

Pikulik, A.; Diaz, H. E. 1977 (Oct. 10). "Cost Estimating for Major Process Equipment." *Chemical Engineering.* pp. 106-122.

Plass, S. G.; Jacobs, H. R.; Boehm, R. F. 1979. "Operational Characteristics. of a Spray Column Direct-Contact Preheater." *American Institute of Chemical Engineers, Symposium Series–Heat Transfer.* San Diego. Vol. 75 (No. 189): pp. 227-234.

Richardson, J. F.; Zaki, W. N. 1954. "Sedimentation and Fluidization, Part I." *Transactions of the Institute of Chemical Engineers.* Vol. 32: pp. 35-53.

Sakiadis, B. C.; Johnson, A. I. 1954 (June). "Generalized Correlation of Flooding Rates." *Industrial and Engineering Chemistry.* Vol. 46 (No. 6): pp. 1229-1238.

Sherwood, T. K.; Shipley, G. H.; Holloway, F. A. L. 1938. *Ind. Eng. Chem.* Vol. 30 (No. 7): p. 765.

Sideman, S.; Taitel, Y. 1964. "Direct-Contact Heat Transfer with Change of Phase." *International Journal of Heat and Mass Transfer.* Vol. 7: pp. 1273-1289. London: Pergamon Press.

Sideman, S. 1966. "Direct-Contact Heat Transfer in Immiscible Liquids." *Advances in Chemical Engineering.* Vol. 6: p. 207. New York: Academic Press.

Sideman, S.; Gat, Y. 1966. "Direct-Contact Heat Transfer with Change of Phase: Spray-Column Studies of a Three-Phase Heat Exchanger." *American Institute of Chemical Engineers Journal.* Vol. 12 (No. 3): pp. 296-303.

Sims, A. F. 1976. "Geothermal Direct-Contact Heat Exchange." Final Report, ERDA Contract No. E(04-3) 1116. Pasadena, CA: The Ben Holt Co.

Suratt, W. B.; Hart, G. K. 1977. "Study and Testing of Direct-Contact Heat Exchangers for Geothermal Brines." DOE Report ORO-4893-1. Ft. Launderdale, FL: DES Engineers, Inc.

Treybal, R. E. 1973. "Liquid-Liquid Systems." *Chemical Engineers Handbook,,* Section 21, 5th Edition. Edited by R. H. Perry and C. H. Chilton. New York: McGraw Hill.

Urbanek, M. W. 1979. "Development of Direct Contact Heat Exchangers for Geothermal Brines—Final Report, Oct. 4, 1977–June 30, 1978." U.S. Government Report No. LBL-8558; available from National Technical Information Service.

Wright, J. D. 1981 (June). *An Organic Rankine Cycle Coupled to a Solar Pond by Direct-Contact Heat Exchange—Selection of a Working Fluid.* SERI/TR-631-1122. Golden, CO: Solar Energy Research Institute. Available from NTIS, Springfield, VA   22161.

9 FLOODING AND HOLDUP CORRELATIONS

Minard and Johnson (1952) determined the form of the i r correlation by performing a force balance on the two phases and determined the constants and exponents from experimental data. The data were taken in a column 32 in. tall with a diameter of 8-5/8 in. The continuous and dispersed phases were both liquids with density differences between 0.1 and 0.6 g/cm^3. Continuous phase viscosity varied from 0.9 to 36 cp, and drop diameters ranged from 0.55 to 1.27 cm. The form of the correlation is

$$U_c^{1/2} = \left[-1.8\ D_p^{0.056}\ \eta_c^{-0.075}\ \left(\frac{\rho_d'}{\rho_c'} \right)^{0.5} \right] U_d^{1/2}$$

$$+ \left[47\ \eta_c^{-0.07}\ \Delta_\rho'^{0.14}\ \rho_c'^{0.50} \right]$$

where

U = superficial velocity (ft^3/ft^2h),

D_p = drop diameter (in.),

η_c = continuous-phase velocity (cp),

c = continuous phase, and

d = dispersed phase,

Sakiadis and Johnson (1954) developed the form of their correlation by writing a force balance on the two phases. The constants and exponents were derived by fitting the equations to a large amount of liquid/liquid, solid/liquid, and gas/liquid flow data, gathered from experiments carried by a variety of investigators. The ability of the correlation to accurately predict the behavior of gas/liquid systems is questionable, as the gas bubble diameter was on the same order of magnitude as the column diameter. The final form of the correlation is

$$1 + 1.8\left[\left(\frac{\rho_d}{\rho_c}\right)^{1/4}\left(\frac{U_d}{U_c}\right)^{1/2}\right] = 0.565\, D_p^{1/4}\left[\frac{U_c^2\rho_c}{g_c\Delta\rho}\,\mu_c^{1/4}\right]^{-1/4}$$

where

D_p = particle diameter (in.),

g_c = acceleration due to gravity (4.17E8 ft/h^2),

U = superficial velocity (ft^3/ft^2h),

ρ = density (lb/ft^3),

μ = viscosity (cp),

c = continuous phase, and

d = dispersed phase.

Letan (1974) surveyed the methods of predicting settling velocities in systems where holdup was important and determined that the method of Richardson and Zaki (1954) correlated a wide range of data found in the literature. In 1976 Letan described a design methodology for spray columns that used the correlation of Richardson and Zaki.

Richardson and Zaki correlated the hindered settling or rise velocity V_s of a particle with its terminal velocity V_T at zero holdup ($\phi = 0$) and with the unhindered Reynolds number Re$_o$.

$$V_s = V_T(1 - \phi)^m$$

where

$m = 3.65$ \qquad for Re$_o < 0.2$

$$m = 4.35 \, \text{Re}_o^{-0.03} - 1 \quad \text{for } 0.2 < \text{Re}_o < 1$$

$$m = 4.45 \, \text{Re}_o^{-0.1} - 1 \quad \text{for } 1 < \text{Re}_o < 500$$

$$m = 1.39 \qquad\qquad \text{for } 500 < \text{Re}_o .$$

Defining the flooding velocity as

$$\frac{\partial \phi}{\partial V_c}\bigg|_{V_d} = \infty ,$$

where V_c and V_d are the continuous and dispersed phase superficial velocities, the following quadratic equation relates holdup at flooding to the flow ratio and drop properties:

$$(m + 1)(1 - R)\phi_f^2 + (m + 2) \, R\phi_f - R = 0 ,$$

where $R = V_d/V_c$, and ϕ_f is the holdup at the flooding conditions. It is then suggested that the column be operated so that the holdup is at least 10% below flooding:

$$\phi = 0.9\phi_f .$$

The superficial velocity of the continuous phase is then

$$V_c = \frac{V_T \phi (1 - \phi)^{m+1}}{R(1 - \phi) + \phi} .$$

Mertes and Rhodes (1955) developed theoretical upper and lower bounds on the hindered settling (slip) velocity V_s of a swarm of particles

low estimate $\qquad \dfrac{V_s}{V_T} = 1 - 1.209 \, \phi^{2/3}$

high estimate $\qquad \dfrac{V_s}{V_T} = \dfrac{1 - \phi}{(1 - \phi) + 1.209\phi^{2/3}} .$

where V_T is the terminal velocity of a single particle, and ϕ is the holdup.

An equation was also derived that related holdup, particle velocity number (V_s/V_T), continuous-phase throughout number $(L = U_c/V_T)$, and dispersed-phase throughput number $(D = V_D/V_T)$:

$$D = \frac{V_D}{V_T} = \phi \left[\frac{V_s}{V_T} + \frac{1}{1 - \phi} \frac{U_c}{V_T} \right] .$$

By substituting in the high or low particle number predictions derived by Mertes and Rhodes on the semiempirical correlations of Richardson and Zaki, the behavior of the system is described.

A convenient method of presenting the results is to plot holdup versus dispersed-phase throughput number, with the liquid-phase throughout number as a parameter (Fig. 9.1). Allowing the dispersed-phase throughput number to take on negative values, the plot shows the cocurrent downflow, cocurrent upflow

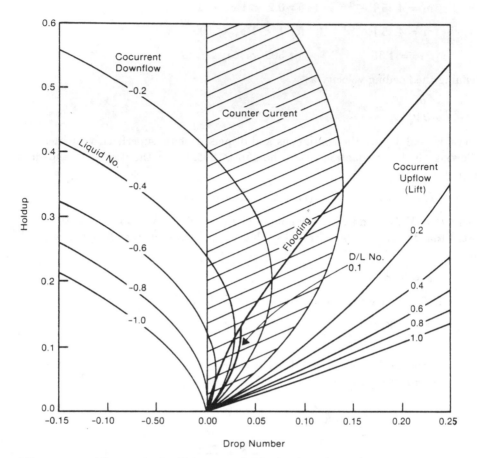

Figure 9.1 The method of Mertes and Rhodes (1955).

(pneumatic lift), and counter-current flow. The flooding line, which connects the points where

$$\frac{\partial \phi}{\partial D}\bigg|_L = \infty$$

divides the counter-current region into two parts—the n-phase (dense packing) and p-phase (dispersed packing). In the dense packing zone, the holdup decreases with increased dispersed-phase throughput, while in the dispersed zone, holdup increases with increased dispersed-phase throughput. Although operation in the dense region would be desirable, since high holdups result in increased interfacial area and heat transfer, such operation is possible only in columns where the dispersed-phase flow rate can be controlled by the coalescence rate at the outlet. This is not possible in vaporizing systems. Additionally, densely packed systems often exhibit phase reversal where the dispersed phase coalesces while rising through the column. The graphical method may be used to size columns. The ratio U_D/U_c is set by the energy balance—a line of constant U_D/U_c can be constructed on the graph. The intersection with the flooding line is the flooding point for the system. The margin of safety can then be specified, i.e., operate at 90% of flooding throughput. It is useful to note that because of the steepness of the line of constant L/D near the flooding point, small changes in throughput result in large reductions in holdup, and therefore in the volumetric heat transfer coefficient. Also, for any given drop size and ratio of flow rates, the holdup can be calculated. Therefore, this plot can be used to evaluate the effect of variations in operating conditions.

Hoffing and Lockhart (1954) studied packed-tower flooding in liquid/liquid systems using a 6-in.-diameter column. Flooding velocities were correlated with the physical properties of the two phases and with an experimentally determined area/void fraction parameter $(a/F^{1.2})$ similar to the packing factor defined by Eckert (1966). The correlation is presented in figure form (Fig. 9.2). Given the flow ratio $R = U_D/U_c$, the abscissa is read to give $f(R)$ and the following equation solved to yield the continuous phase superficial velocity:

$$U_c = \left[\frac{f(R)\,\Delta\rho^{0.5}}{3.33E - 5\,\rho_d^{0.22}\,\rho_c^{0.1}\,\mu_d^{0.08}\,\mu_c^{0.1}\left(\dfrac{\sigma}{\sigma_{w-a}}\right)^{0.5}\left(\dfrac{a}{F^{1.2}}\right)^{0.67}} \right]^{1.25} R^{0.25}$$

where

μ = viscosity in centipoise,

ρ = density in g/cm^3,

σ = actual interfacial tension, and

σ_{w-a} = surface tension of water in air.

The correlation was also tested against data from earlier studies. The largest packing (smallest $a/F^{1.2}$) described by the correlation is the 1-1/2-in. Berl saddle. This packing is much less open than a modern high-capacity packing such as the Pall ring.

Letan and Kehat (1968) developed a model of heat transfer in a spray tower. As drops enter the column, wakes begin to grow behind the drops. Heat transfer in this region is rapid as the wakes approach the temperature of the drop. After a short distance, the drops rise steadily through the column, periodically shedding their wakes. As the wakes are essentially in thermal equilibrium with the drop, and the majority of the heat transfer is from drop to wake, heat transfer in this region, which occupies most of the column, is relatively slow and limited by the rate at which wakes are shed. When the drops reach the top, they completely shed their wakes and coalesce. Differential equations can be written to describe the heat transfer rate in each zone. The equations can then be solved simultaneously to give the temperature difference between the two phases at the top of the

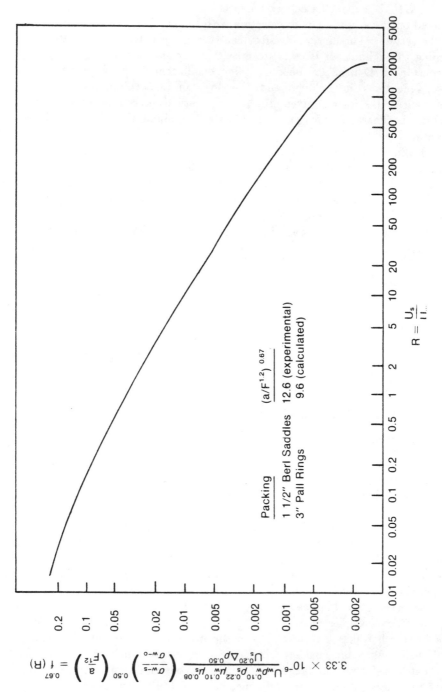

Figure 9.2 Liquid/liquid flooding correlation of Hoffing and Lockhart.

column.

$$\frac{T_{d,out} - T_{c,in}}{T_{d,in} - T_{c,out}} = \left[(1 + S) \exp(\alpha_1 L) - S \exp(\alpha_2 L) \right] \exp\left(-\frac{m}{r} \right)$$

where

$$S = \frac{\alpha_1 + \left[\frac{m}{R} \right] - \left[\frac{r}{p} \right] \left[\frac{m}{M} \right] \left[\exp\left[\frac{M}{r} \right] - 1 \right]}{\alpha_2 - \alpha_1},$$

$$\alpha_{1,2} = \frac{-m}{2} \left[\left[\frac{1}{M} + \frac{1}{r} - \frac{1}{p} \right] \pm \sqrt{\left[\frac{1}{M} + \frac{1}{r} + \frac{1}{p} \right]^2 - \frac{4}{Mr}} \right],$$

$$\rho = 1 + \frac{MR}{r},$$

G = volumetric flow rate,

L = column height (cm),

m = volume of wake elements shed per volume of drop and unit length

of column (cm^{-1}),

M = ratio of wake-to-drop volume,

$r = (\rho C_p)_d / (\rho C_p)_c$,

R = ratio of volumetric flow rates G_d / G_c, and

T = temperature (°C).

The variation of M with holdup is shown in Fig. 9.3. The number of elements of wake shed per drop volume and unit length of column (m) are shown as a function of holdup in Fig. 9.4. The equations may also be solved to give the column height, but only in the special case of $Rr = 1$.

Pratt et al. (Treybal, 1973) studied holdup in liquid/liquid packed towers. For commercial packings larger than 1/2 in., and at low values of the dispersed-phase superficial velocity (U_D), the holdup (ϕ) varies linearly with U_D up to $\phi = 0.1$. With a further rise in U_D, the holdup increases sharply. At a higher value of U_D there is an upper transition point, drops of the dispersed phase begin to coalesce, and U_D can be increased without an increase in holdup. At a still higher superficial velocity, the system floods.

Drops of the dispersed phase reach a constant characteristic diameter as they travel up through the packing. For any fluid system, there is a critical packing size above which the equilibrium drop size will be minimum and independent of packing size. The critical packing size d_p (usually greater than 1/2 in.) is given by

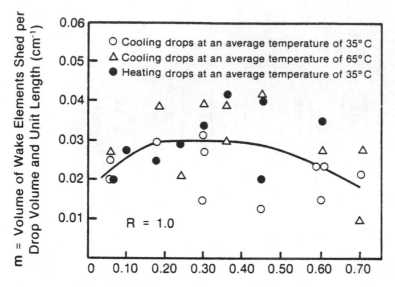

Figure 9.3. Holdup at bottom of wake shedding zone, as a function of wake elements shed.

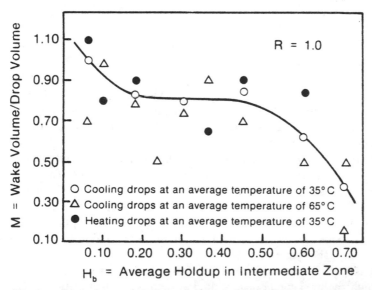

Figure 9.4. Average holdup in the intermediate zone, as a function of Wake and Drop Volume.

$$d_{pack,c} = 2.42 \left[\frac{\sigma g_c}{\Delta \rho g} \right]^{0.5},$$

where

$d = $ ft,

$\sigma = $ interfacial tension in lb_f/ft,

$g_c = $ gravitational constant $(4.18 \times 10^8 \text{ lb}_m \text{ ft}/lb_f \text{ h}^2)$,

$\Delta \rho = $ difference in density (lb_m/ft^3), and

$g = $ acceleration of gravity $(4.18 \times 10^8 \text{ ft}/h^2)$.

For packings larger than $d_{pack,c}$, the equilibrium drop diameter is

$$d_p = 0.92 \left[\frac{\sigma g_c}{\Delta \rho g} \right]^{0.5} \left[\frac{V_K \epsilon \phi}{V_D} \right],$$

where ϵ is the packing void fraction and V_K is a characteristic drop velocity (ft/h) which is obtained from Fig. 9.5. In Fig. 9.5, V_T is the drop terminal velocity (ft/h) and T is the tower diameter in feet. Because the drop diameter enters into the correlation through the terminal velocity in the correlation, several iterations are required to pick a drop diameter and characteristic velocity.

The following equation may then be utilized to determine the holdup:

$$\frac{U_D}{\phi} + \frac{U_c}{1 - \phi} = \epsilon V_K (1 - \phi) .$$

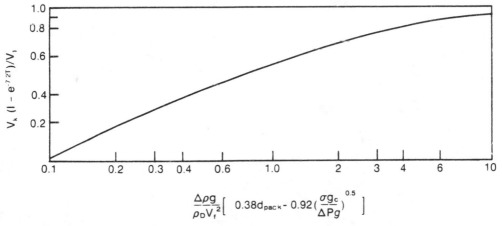

Figure 9.5 Characteristic drop velocity for packed towers (Pratt, 1955).

The interfacial area (a) is then determined by

$$a = \frac{6\epsilon\phi}{d_p} \, .$$

REFERENCES FOR SECTION 9

Eckert, J. S.; Foote, E. H.; Walter, L. F. 1966 (Jan.). "What Affects Packing Performance." *Chemical Engineering Progress.* Vol. 62 (No. 1): p. 59.

Hoffing, E. H.; Lockhart, F. J. 1954 (Feb.). "A Correlation of Flooding Velocities in Packed Columns." *Chemical Engineering Progress.* Vol. 47 (No. 2): pp. 94-103.

Letan, R.; Kehat, E. 1968 (May). "The Mechanism of Heat Transfer in a Spray-Column heat exchanger." *AIChE J.* Vol. 14 (No. 3): pp. 398-405.

Letan, R. 1974. "On Vertical Dispersed Two-Phase Flow." *Chemical Engineering Science,* Vol. 29: pp. 621-624.

Letan, R. 1976. "Design of a Particulate Direct-Contact Heat Exchanger: Uniform, Counter-Current Flow." ASME Paper 76-HT-27. 1976 ASME/AIChE Heat Transfer Conference.

Mertes, T. S.; Rhodes, H. B. 1955 (Sept.). "Liquid-Particle Behavior: Part 1." *Chemical Engineering Progress.* Vol. 51 (No. 9): pp. 429-432.

Mertes, T. S.; Rhodes, H. B. 1955 (Sept.). "Liquid-Particle Behavior: Part 2." *Chemical Engineering Progress.* Vol. 51 (No. 11): pp. 517-522.

Minard, G. W.; Johnson, A. I. 1952 (Feb.). "Limiting Flow and Holdup in a Spray Extraction Column." *Chemical Engineering Progress.* Vol. 48 (No. 2): pp. 62-74.

Richardson, J. F.; Zaki, W. N. 1954. "Sedimentation and Fluidization: Part 1." *Transactions of the Institute of Chemical Engineers.* Vol. 32: pp. 35-53.

Sakiadis, B. C.; Johnson, A. I. 1954 (June). "Generalized Correlation of Flooding Rates." *Industrial and Engineering Chemistry.* Vol. 46 (No. 6): pp. 1229-1238.

Treybal, R. E. 1973. "Liquid/Liquid Systems." *Chemical Engineers Handbook.* Section 21, 5th Edition. Edited by R. H. Perry and C. H. Chilton. New York: McGraw-Hill.

DESIGN OF A DIRECT CONTACT LIQUID-LIQUID HEAT EXCHANGER

R. Letan

1 SPECIFICATIONS FOR EXAMPLE

The theoretical background for the following example is presented in detail in Chapter 6 "Liquid-Liquid Processes". The example deals with sea-water which has to be heated by kerosene in a direct contact heat exchanger. The operation is to be carried out in a countercurrent manner to achieve close approach of temperatures at both inlet-outlet ends. The kerosene is to be dispersed as droplets into the sea-water. For the operation and design of the heat exchanger the following parameters have to be referred to:

Operating Parameters: flow rates of both liquids, inlet and outlet temperatures of both liquids, holdup of the dispersed liquid, diameter of droplets.
Geometric Parameters: diameter of the column proper, length of the column proper, the number and size of nozzles in the disperser.

The example herein considered relates to sizing. Therefore, the operating conditions must be either specified or prescribed.

Operating Conditions:

Sea-Water: specified flow rate $Q_c = 1.66 \times 10^{-3}$ m³/s
specified inlet temperature $T_{ci} = 20°$ C
prescribed outlet temperature $T_{co} = 70°$ C

Kerosene   specified inlet temperature $T_{di} = 75°$ C
disperse packing of
droplets of diameter $d = 3.5 \times 10^{-3}$ m

Physical Properties of Liquids:

Sea-Water: density $\rho_c = 1000$ kg/m³
specific heat $c_{pc} = 4200$ J/kg·° C
viscosity $\mu_c = 5 \times 10^{-4}$ kg/m·s

Kerosene: density $\rho_d = 800$ kg/m³
specific heat $c_{pd} = 2100$ J/kg·° C

2 OPERATING PARAMETERS

2.1 Flow Rate Ratio

At steady state, with physical properties constant

$$(Q \cdot \rho \cdot c_p)_d \cdot (T_{di} - T_{do}) = (Q \cdot \rho \cdot c_p)_c \cdot (T_{co} - T_{ci})$$

The flow rate ratio is defined as:

$$R = \frac{Q_d}{Q_c} = \frac{V_d}{V_d}$$

The ratio of heat capacities is:

$$r = \frac{(\rho \cdot c_p)_d}{(\rho \cdot c_p)_c}$$

Therefore

$$R = \frac{1}{r} \frac{(T_{co} - T_{ci})}{(T_{di} - T_{do})}$$

For the same approach of temperatures at both ends of the column

$$T_{di} - T_{co} = T_{do} - T_{ci}$$

and therefore,

$$T_{di} - T_{do} = T_{co} - T_{ci}$$

That leads to

$$Rr = 1$$

and in the presently specified case:

$$r = 0.4$$

$$R = 1/r = 2.5$$

$$\underline{R = 2.5}$$

2.2 Flow Characteristics of a Single Droplet

The size of the dispersed droplets is an independent variable which may be optim-ized. In the case presented herein the size is specified as,

$$\underline{d = 3.5 \times 10^{-3}\ m}$$

The terminal velocity of the droplet is obtained from the drag coefficient - Rey-nolds number relation

$$C_D \cdot Re_o^2 = \frac{4}{3} \cdot d^3 \cdot \frac{\rho_c \cdot (\rho_c - \rho_d) \cdot g}{\mu_c^2}$$

The drag coefficient correlation for rigid spheres applies reasonably well also to spherical droplets. For larger $(d > 4 \times 10^{-3}\ m)$ and distorted droplets more appropriate correlations have to be applied.

Thus, using the standard $C_D(Re_o)$ curve of a rigid sphere provides for the present case:

$$Re_o = 1000 \quad \text{and } C_D = 0.44$$

Then the terminal velocity is obtained:

$$U_T = \frac{Re_o}{d} \frac{\mu_c}{\rho_c}$$

for $d = 3.5 \times 10^{-3}$ m, $\rho_c = 1 \times 10^3$ kg/m^3, $\mu_c = 5 \times 10^{-4}$ kg/m·s

$$\underline{U_T = 0.143\ m/s}$$

2.3 Flow Characteristics of the System

Flow characteristics of a particulate system are defined by the relationship between slip velocity and holdup. The relationship is unique for a system of specified properties. Semi-empirical expressions like the one by Richardson and Zaki ("Liquid-Liquid Processes", Eq. (4)) will provide a satisfactory approximation:

$$U_S = \frac{V_c}{1-\phi} = U_T (1 - \phi)^{(m-1)}$$

where for $Re_o > 500$, m = 2.4.

Therefore in our case the slip velocity - holdup relationship is

$$U_S = 0.143 \, (1 - \phi)^{1.4}$$

and is also related to the superficial velocities in the countercurrent flow as follows:

$$U_T \, (1 - \phi)^{m-1} = \frac{V_d}{\phi} + \frac{V_c}{(1 - \phi)}$$

However, $V_d = R \cdot V_c$ and therefore:

$$0.143 \cdot (1 - \phi)^{1.4} = V_c \left[\frac{2.5}{\phi} + \frac{1}{1-\phi} \right]$$

2.4 Flooding and Operational Holdup

Flooding represents the state of disruption of the stable flow conditions. The holdup at flooding in a disperse packing of droplets corresponds to the maximum operable holdup and the combination of flow rates. On the flooding curve:

$$\left. \frac{\partial V_c}{\partial \phi} \right|_{V_d} = 0$$

Differentiation of the superficial velocity-holdup equation yields a relationship between ϕ, and V_d. Substituting RV_c for V_d and combining again with the holdup - superficial velocity relationship results in:

$$\phi_f \simeq 0.35$$

for holdup at flooding. The operational holdup has to be at least 10% lower:

$$\underline{\phi \leq 0.315}$$

2.5 Superficial Velocities

Superficial velocities of both liquids are calculated at the operational holdup. Thus:

$$V_c = U_T \, \frac{\phi \cdot (1-\phi)^{(m+1)}}{R(1-\phi) + \phi}$$

For $U_T = 0.143$ m/s, $\phi = 0.315$, $R = 2.5$ and $m = 1.4$:

$$V_c = 8.95 \times 10^{-3} \text{ m/s}$$

and

$$V_d = R \cdot V_c = 22.4 \times 10^{-3} \text{ m/s}$$

2.6 Temperatures

The inlet temperatures are specified

$$T_{ci} = 20\,^\circ C \quad \text{and} \quad T_{di} = 75\,^\circ C$$

The outlet temperature of the sea water is prescribed, and the outlet temperature of the kerosene is calculated at $R = 2.5$:

$$T_{co} = 70\,^\circ C \quad \text{and} \quad T_{do} = 25\,^\circ C$$

3 GEOMETRIC PARAMETERS

3.1 Diameter of the Column

The diameter of the column proper is obtained by using the flow rate and superficial velocity of the dispersed or continuous liquid:

$$D = 2 \left(\frac{Q_c}{\pi V_c} \right)^{\frac{1}{2}}$$

For $Q_c = 1.66 \times 10^{-3}$ m^3/s, $V_c \simeq 9 \times 10^{-3}$ m/s and

$$D = 0.485\ m$$

$$\underline{D \simeq 0.5\ m}$$

3.2 Length of the Column

To calculate the length of the column proper for the specific case of $Rr = 1$, the appropriate equations ("Liquid-Liquid Processes", Eq. (34) - (39)) take the following form:

$$L = \left[\left[\left(\frac{T_{do} - T_{co}}{T_{di} - T_{co}} \right) \cdot \exp{(MR)} - 1 \right] \frac{r}{m \cdot S} - \frac{1}{\alpha_1} \left(\frac{1}{S} + 1 \right) \right]$$

where S and α_1 are expressed by Eqs. (37) and (36) respectively in Chapter 6, "Liquid-Liquid Processes".
At $\phi = 0.315$, $M = 0.8$, and $m = 3.0$ m^{-1}.
For these conditions, $S = -1.11$, $\alpha_1 = -8.75$ m^{-1}. Substituting for $R = 2.5$, $r = 0.4$ and the respective temperatures provides the column proper length:

$$\underline{L = 8.2\ m}$$

Disperser

Nozzles of diameter

$$d_N = 1.5 \times 10^{-3} \; m$$

are required for the specified size of droplets $d = 3.5 \times 10^{-3}$ m. For uniformly sized droplets the kerosene velocity through the nozzles is:

$$V_N = 0.5 \; m/s$$

Therefore, the number of nozzles in the plate must be:

$$N = Q_c / \left[\frac{\pi}{4} \cdot d_N^2 \cdot V_N \right]$$

For $Q_c = 1.66 \times 10^{-3}$ m^3/s, $d_N = 1.5 \times 10^{-3}$ m and $V_N = 0.5$ m/s we obtain the number of nozzles as:

$$N = 1900$$

The surface of the nozzles is then

$$S_N = \frac{\pi}{4} \cdot d_N^2 \cdot N \simeq 3.4 \times 10^{-3} \; m^2$$

If the disperser is located in an enlarged conical bottom, it may be designed of the same diameter as the column proper, D. Thus,

$$D_N = 0.5 \; m$$

Then the relative drilled surface is:

$$S_N/S_D = 1.74\%$$

4 VOLUMETRIC HEAT TRANSFER COEFFICIENT

The performance of a direct contact heat exchanger is usually assessed by means of a volumetric heat transfer coefficient defined as:

$$U_V = \frac{(Q \cdot \rho \cdot c \cdot \Delta T)_c}{\left[\frac{\pi}{4} \cdot D^2 \cdot L \right] \cdot \Delta T_m}$$

where ΔT_m is the logarithmic mean temperature difference. For the present design:

$$U_V = 4.45 \times 10^4 \; W/m^3 \; K$$

5 SUMMARY

Operating Conditions:

Flowrates: sea-water $Q_c = 1.66 \times 10^{-3}$ m^3/s
kerosene $Q_d = 4.15 \times 10^{-3}$ m^3/s
flow rate ratio $R = 2.5$

Temperatures: sea-water, inlet $T_{ci} = 20\,^\circ$C, outlet $T_{co} = 70\,^\circ$C
kerosene, inlet $T_{di} = 75\,^\circ$C, outlet $T_{do} = 25\,^\circ$C

Design:

Diameter of column proper	$D = 0.5$ m
Length of column proper	$L = 8.2$ m
Disperser: nozzle diameter	$d_N = 1.5 \times 10^{-3}$ m
number of nozzles	$N = 1900$

NOMENCLATURE

C_D drag coefficient
c_p specific heat capacity
D diameter of column proper
d diameter of droplet
d_N diameter of nozzle
g gravitational acceleration
L length of column proper
M ratio of wake to droplet volumes
m wake elements shed per unit length of column and per unit volume of droplet
N number of nozzles
Q volumetric flow rate
R flow rate ratio, Q_d/Q_c
r heat capacity ratio $(\rho \cdot c_p)_d/(\rho \cdot c_p)_c$
S surface
Re_o Reynolds number of a single droplet
T temperature
U_S slip velocity
U_T terminal velocity
U_V volumetric heat transfer coefficient
V superficial velocity
μ dynamic viscosity
ρ density
ϕ holdup

SUBSCRIPTS

c	continuous
d	dispersed
f	flooding
i	inlet
o	outlet

THERMAL AND HYDRAULIC DESIGN OF DIRECT-CONTACT SPRAY COLUMNS FOR USE IN EXTRACTING HEAT FROM GEOTHERMAL BRINES

Harold R. Jacobs

ABSTRACT

This Appendix outlines the current methods being used in the thermal and hydraulic design of spray column type, direct contact heat exchangers. It provides appropriate referenced equations for both preliminary design and detailed performance analysis. The design methods are primarily empirical and are applicable for use in the design of such units for geothermal application and for application with solar ponds. Methods of design, for both preheater and boiler sections of the primary heat exchangers, for direct contact binary power plants are included.

Based on a report submitted to the U.S. Department of Energy, Contract No. DE-AS07-76ID 01523 with the Department of Mechanical and Industrial Engineering, The University of Utah, Salt Lake City, Utah, April, 1985, Revised June, 1985.

1 INTRODUCTION

The spray column has been widely studied in the chemical industry for many years due to its inherent simplicity as a counter-current device for heat or mass transfer. Developments were enhanced in the 1960's due to increased interest in desalination systems (Saline Water Conversion Engineering Data Book, 1971). More recently, in the 1970's, Jacobs and Boehm (1980) suggested their use for extracting heat from moderate temperature geothermal brines. They and a number of other investigators have carried out a wide range of studies under U.S. Department of Energy funding for nearly 10 years. This work culminated in the construction of the 500 kW$_e$ Geothermal Direct Contact Binary Cycle Power Plant at East Mesa, California, by Barber-Nichols Engineering under U.S.D.O.E. funding, (Olander, et al., 1983). The 500 kW$_e$ direct contactor was designed by the present author as a combined working fluid preheater-boiler. The working fluid was isobutane with the continuous fluid being the immiscible geothermal brine.

Based upon the relative success of this technique, a number of other applications have been spawned. Most closely related is the use of a modified spray type direct contactor for the extraction of heat from a salt-stratified solar pond. The low temperature design conditions for a solar pond dictate the use of pentane as the working fluid if it is desired to utilize the working fluid vapor to generate electricity.

Although both geothermal and solar pond applications have the same ultimate purpose, to generate electricity from a moderate to low temperature source and to obtain the energy exchange at small approach temperature differences, many source-related characteristics cause significant differences in their design. For the geothermal application, it has generally been conceded that the most economical design is to utilize as much heat as possible from each unit mass of geothermal brine. This leads to near equal mass flow rates of the working fluid and brine. For the case of solar ponds, with much lower peak temperatures and concerns about returning too cold a fluid back to the pond, the mass flow rate of brine far exceeds that of the working fluid.

For a combined boiler-preheater, it is clear that for high-pressure, high-temperature vapor generation (such as for geothermal applications) the heat duty of the preheater can greatly exceed that of the boiler. For solar pond applications, where the vapor is generated at temperatures as low as 67 ° C, the boiler duty can be two to three times that of the preheater. Thus, design philosophy can be considerably different. Nevertheless, in this Appendix, an attempt is made to provide information for general purpose design of spray column type direct contactors. As nearly as possible, the information provided herein is current and provides the best available information.

2 SPRAY COLUMN DESCRIPTION

A spray column is one of the oldest known devices for contacting two immiscible fluids in order to transport either heat or some chemical substance from one fluid to the other. It is basically an empty vertical column with injection devices for each fluid and outlets for each fluid. In most common applications, each fluid is in a liquid phase; however, for use with binary power cycles, a single column can include a liquid-liquid preheating zone and a boiling or evaporation zone.

When only liquid-liquid heat exchange is desired, the column must have a disengagement zone at both the top and bottom of the column (see Fig. 5.1). A properly designed column needs only to control the flow rates of the two liquid streams to insure a pseudo-steady operation.

A column in which both preheating and boiling takes place requires that there be a disengagement section at the bottom of the column, a level control device to insure that the column is not completely flooded with liquids, and a vapor reservoir at the top of the column for the generated vapors. The vapor reservoir must be sufficiently large to insure that a liquid phase does not exit as a mist with the vapor mixture. Thus, mist eliminators may also be required (see Fig. 5.2).

3 DESIGN OF DISPERSED PHASE FLUID INJECTORS

In order to design a direct contact heat exchanger of the spray column type, it is necessary to design a distributor which can produce regular uniform-sized drops of one of the two fluids. Normally this is the lighter fluid. Thus, for geothermal or solar applications, this would be the hydrocarbon working fluid. This is achieved by designing a distributor which uses a perforated plate of a material not wetted by the fluid to be dispersed to form the drops. Typically, punched holes which leave a slight nozzle at the surface exposed to the continuous phase are used. For geothermal applications, a mild steel plate pickled in sulfuric acid provides clean jets and well-formed drops.

Actual design of the holes is not critical as long as the flow rate through them is equal to the jetting velocity, but less than the critical jetting velocity. Exceeding or equaling the jetting velocity is important due to the fact that lower velocities can lead to situations where not all of the nozzles are flowing and due to the fact that drops formed when V_J is exceeded are very regular in size. Regular size drops are important in order to predict column performance. Steiner and Hartland (1983) recommend that the jetting velocity be calculated from

$$V_j = \frac{2\sigma}{\rho_d \, d_n} \left(1.07 - 0.75 \, \frac{\Delta\rho d_n^2 g}{4\sigma} \right) \tag{1}$$

where σ is the interfacial tension[*] and dn is the nozzle diameter.

[*]Interfacial tension is not surface tension. It can be predicted by Antonoff's rule which states that for two saturated liquid layers in equilibrium, the interfacial tension is equal to the difference between the two individual surface tensions of the two mutually saturated phases under a common

It is also necessary not to exceed the critical jetting velocity. Above this velocity the length of jet decreases dramatically followed quickly by automization of the dispersed phase. This requires a large pressure drop across the nozzle and is generally undesirable for spray columns. The critical velocity and corresponding critical drop diameter can be calculated according to Skelland and Johnson (1974) as follows

$$d_{jc} = \frac{d_n}{0.485K^2 + 1} \quad \text{for } K < 0.785 \tag{2}$$

or

$$d_{jc} = \frac{d_n}{1.51K + 0.12} \quad \text{for } K \geq 0.785 \tag{3}$$

where

$$K = \frac{d_n}{\sqrt{\sigma/\Delta\rho g}} \tag{4}$$

$$V_{jc} = 2.69 \frac{d_{jc}^2}{d_n^2} \frac{\sigma}{d_{jc}\left(0.514\rho_d + 0.472\rho_c\right)} \tag{5}$$

At this critical velocity, Treybal (1963) recommends the following equations for the critical drop size

$$d_{DC} = \frac{2.07 d_n}{0.485 E_o^+ 1} \quad \text{for } E_o < 0.615 \tag{6}$$

and

$$d_{DC} = \frac{2.07 d_n}{1.51 E_o^{1/2} + 0.12} \quad \text{for } E_o \geq 0.615 \tag{7}$$

where E_o is the Eötvös number, defined as

$$E_o = \frac{\Delta\rho g d_n^2}{\sigma} \tag{8}$$

For conditions between the jetting velocity and the critical jetting velocity, the following correlation is recommended (Steiner and Hartland, 1983) to determine the drop size.

$$d_P = d_{jc} \left(2.06 \frac{V_{jc}}{V_n} - 1.47 \ln \frac{V_{jc}}{V_n}\right) \tag{9}$$

vapor or gas, $\sigma = \sigma_{ds} - \sigma_{cs}$. If the above values are not known saturated-phase surface tensions can be used. Accuracy within 15% is claimed for organic-water and organic-organic systems for the latter estimate. Saline Water Conversion Engineering Data Book, 2nd Edition (1971) gives values for organic fluids and water and brine under air. These properties should be used.

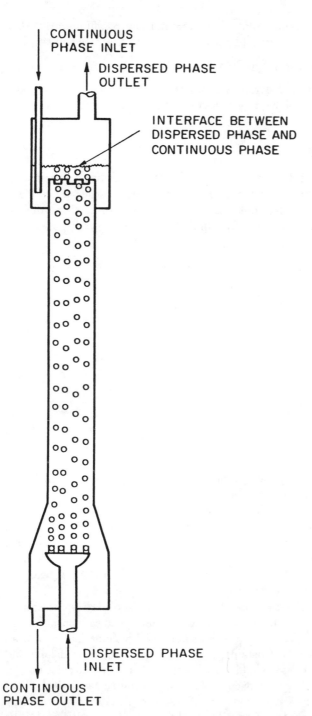

CONTINUOUS
PHASE INLET

DISPERSED PHASE
OUTLET

INTERFACE BETWEEN
DISPERSED PHASE AND
CONTINUOUS PHASE

DISPERSED PHASE
INLET

CONTINUOUS
PHASE OUTLET

Figure 5.1 Direct contact spray tower for liquid-liquid heat exchange.

Steiner and Hartland (1983) recommend maintaining a minimum Weber number (defined with nozzle velocity and the density of the dispersed phase, i.e., $W_e = \dfrac{V_n^2 d_n \rho_d}{\sigma}$) greater than two to prevent seeping along the surface and secure drop formation on all openings. Experience in the laboratory indicates that nozzle or perforation spacing should not be closer than 1.5 d_D to insure that jet or drop coalescence does not occur.

4 BEHAVIOR OF DROPS

Drops formed from jets or nozzles may behave differently according to their density, interfacial tension, volume, and whether heat or mass transfer takes place between them and the surrounding continuous phase. For a drop rising due to gravity in an immiscible liquid, there are five dimensionless groups that govern the motion of the drop:

Reynolds number $\qquad\qquad Re = \dfrac{\rho_c d_D V_D}{\mu_c}$ $\qquad\qquad$ (10)

Eötvös number $\qquad\qquad E_o = \dfrac{\Delta \rho g d_D^2}{\sigma}$ $\qquad\qquad$ (11)

M - group $\qquad\qquad M = \dfrac{g \mu_c^4 \Delta \rho}{\rho_c^2 \sigma^3}$ $\qquad\qquad$ (12)

Viscosity ratio $\qquad\qquad K_1 = \mu_d / \mu_c$ $\qquad\qquad$ (13)

and Density ratio $\qquad\qquad \gamma = \rho_d / \rho_c$ $\qquad\qquad$ (14)

For any particular liquid-liquid combination M, K, γ are constant in an isothermal system. Thus, Grace (1983) correlated drop behavior by plotting Re versus E_o for constant values of M for a large number of liquid pairs. K_1 and γ play a small role in the results. Figure 5.3 categorizes drops into three regimes: the Spherical regime, the Ellipsoidal regime and the Spherical Cap regime. An approximate curve is shown which separates the former two regimes. (Experiments conducted at the University of Utah at high values of Re for E_o of near one indicate that the spherical regime exists longer than shown.)

The spherical regime contains that region where drops are spherical, or nearly so. For spherical drops, little or no internal circulation takes place.

Somewhat larger drops obtain, on a mean time basis, a shape like that of an oblate ellipsoid of revolution. The instantaneous shape may depart radically and undergo wobbling which, of course, would cause significant internal circulation.

When $E_o \geq 40$ for all $M \leq 10^2$ droplets have a leading surface which looks spherical, but the rear may be flat or concave. These drops may move randomly and their behavior is hard to correlate. Thus, they should be avoided in the design of a spray column.

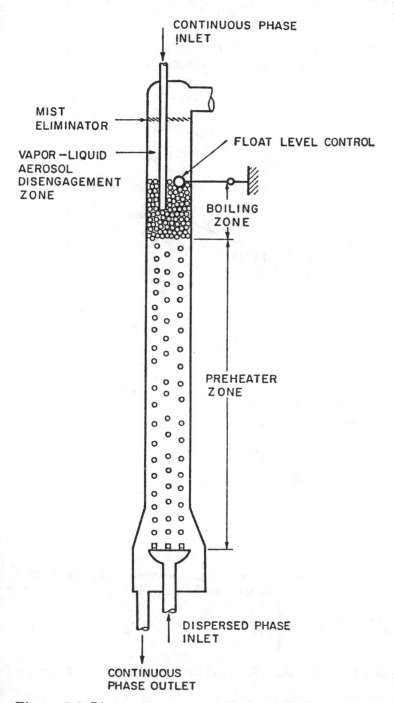

CONTINUOUS PHASE INLET

MIST ELIMINATOR

VAPOR—LIQUID AEROSOL DISENGAGEMENT ZONE

FLOAT LEVEL CONTROL

BOILING ZONE

PREHEATER ZONE

DISPERSED PHASE INLET

CONTINUOUS PHASE OUTLET

Figure 5.2 Direct contact spray tower for preheating and boiling dispersed phase.

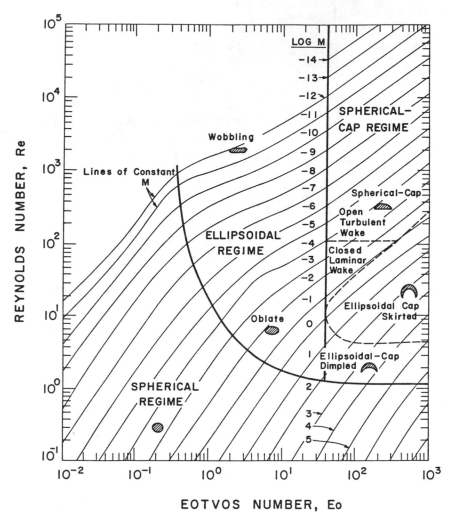

Figure 5.3 Drop characterization map.

For spherical drops, the terminal velocity in a quiescent fluid can be calculated by a simple balance of the gravity force by the drag yielding

$$f \ V_T^2 = \frac{4}{3} \ d_D \left| \left(\frac{\rho_d - \rho_c}{\rho_c} \right) \right| g \tag{15}$$

If it is assumed that the drops behave like rigid smooth spheres, then f varies as follows

$$Re_c < 0.1 \qquad\qquad\qquad f = 24/Re_c$$

$$2 < Re_c < 500 \qquad\qquad f = 18.5/Re_c \qquad (16)$$

$$500 < Re_c < 2 \times 10^5 \qquad Rf = 0.44$$

$$2 \times 10^5 < Re_c \qquad\qquad Rf = 0.2$$

More recently Rivkind and Ryskin (1976) have proposed for the drag coefficient:

$$f = \frac{1}{K+1} \left[K\left(\frac{24}{Re_c} + \frac{4}{Re_c^{2/3}} \right) + \frac{14.9}{Re_c^{0.78}} \right] \qquad (17)$$

which accounts for the relative motion of the interface due to the differences in fluid viscosities. It should be noted, however, that the presence of contaminants such as found in geothermal brines or salt pond brines tend to make the interface more immobile. However, data is missing on the influence of surfactants and impurities.

For ellipsoidal drops Grace (1985) recommends that the terminal drop velocity be calculated from

$$V_T = \frac{\mu_c}{\rho_c d_D} M^{-0.149} (J - 0.857) \qquad (18)$$

where

$$J = 0.94 H^{0.757} \text{ for } 2 < H < 59.3 \qquad (19)$$

$$J = 3.42 H^{0.441} \text{ for } H > 59.3 \qquad (20)$$

with

$$H = \frac{4}{3} \, Eo \, M^{-0.149} \left[\frac{\mu_c}{0.0009} \right]^{-0.14} \qquad (21)$$

In the above equations μ_c is in kg/msec.

5 VELOCITY OF DROPS IN SWARMS

Drops in a spray column, depending upon the holdup, may move in dense swarms. As the drops get closer together they interact changing not only their own velocities but also that of the continuous phase. Steiner and Hartland (1983) recommend

$$\frac{G_c}{A} = \frac{\left[\dfrac{k \Delta \rho d_D g}{\rho_c} \right]^{1/2} (1 - \epsilon)^{1+n/2}}{\left(1 + \epsilon^{1/2} \right)^{n/2} \left[1 + \dfrac{R(1-\epsilon)}{\epsilon} \right]} \qquad (22)$$

to predict the superficial velocity of the continuous phase. This compares with

$$\frac{G_c}{A} = \frac{V_T \, \epsilon \, (1 - \epsilon)^{m+1}}{(1 - \epsilon)R + \epsilon} \tag{23}$$

which has been used by Letan, et al. (1968) and by Jacobs and Boehm (1980). In Equation 22, recommended values for k and n are 2.725 and 1.834 respectively. The value of m in Equation 23 is a function of the drop Reynolds number based on terminal velocities, as follows:

$$
\begin{aligned}
Re_c &< 0.2 & m &= 3.65 \\
0.2 &< Re_c < 1.0 & m &= 4.35 \, Re_c^{-0.03} - 1 \\
1 &< Re_c < 500 & m &= 44.5 \, Re_c^{-0.1} - 1 \\
Re_c &\geq 500 & m &= 1.39
\end{aligned} \tag{24}
$$

In the above equations, ϵ is the holdup.

In low holdup situations, the relative velocity between the drops and the continuous phase is equal to the terminal rise velocity of a single drop.

$$V_r = V_{D_T} = \frac{G_c}{A(1 - \epsilon)} + \frac{G_d}{A\epsilon} \tag{25}$$

However, at higher holdup the close proximity of adjacent drops influences the terminal velocity of a typical drop within the swarm, thus, $V_r \neq V_{D_T}$. Kumar (1980) recommends for this situation that

$$V_r = \left[\frac{2.725 \, \Delta\rho d_D g}{\rho_c} \left(\frac{1 - \epsilon}{1 + \epsilon^{1/2}} \right)^{1.834} \right]^{1/2} \tag{26}$$

It was this expression together with Equation 25 from which Equation 22 was developed. Comparison of Equation 26 with experimental data from other investigators indicate mixed results especially at low holdup. Nevertheless, it is recommended by Steiner and Hartland (1983). Deviation from $V_r = V_T$, however, is small unless dense packing is achieved. Thus, Equation 23 is preferred unless high holdups are encountered.

6 HOLDUP OF THE DISPERSED PHASE

The holdup is the fraction of the total volume occupied by the dispersed phase. In an isothermal spray column, it is constant along the column length. However, when heat transfer occurs, the holdup can vary along the column length. Whereas for many applications the variation may be small, in a geothermal application using isobutane as the working fluid, changes in holdup must be taken into account in the preheater much less the boiler (if a boiler-preheater combination is used). This is due to the fact that density changes of 25% can readily occur in the preheater.

If the continuous phase flow rate is held constant and the dispersed phase flow rate is gradually increased, a condition known as flooding can eventually occur. (Similarly, flooding will eventually occur if the dispersed phase flow rate is held constant and the continuous phase flow is increased.) At this point, it is impossible to continue passing more of the dispersed fluid through the column and a fraction would then be washed out of the column with the continuous phase or the drops might all coalesce, changing either the drop size or which fluid is dispersed. As this cannot be readily predicted, flooding is to be avoided. We, thus, must be able to predict the flooding point.

There is not much data on flooding in the literature (Steiner and Hartland, 1983). Based on the relationship of Richardson and Zaki (1954), Letan (1976) recommends that

$$(m + 1)(1 - R)\epsilon_f^2 + (m + 2)R\epsilon_f - R = 0 \tag{27}$$

be used to calculate the holdup at flooding where m is given in Equation 24 and $R = G_d/G_c$, the ratio of volumetric flow rates. For stable flow operation

$$\epsilon < 0.9 \, \epsilon_f \tag{28}$$

is advised.

It should be noted, that it is possible to operate at a holdup greater than the flooding point. This type of operation occurs in what is called the dense packing regime. Operating below the so-called flooding point is called the dispersed packing regime. Theoretically, it is possible to operate a spray column in each regime for a given pair of flow rates. Although considerable work has been directed toward dense packing, in practice, it is difficult to achieve. It does have advantages over the common operation with loose packing. The interfacial area can be three to five times higher and back mixing is considerably reduced. The dense dispersion can be controlled so that its interface is 30–50 mm above the dispersed phase distributor. However, if the jets from the distributor plate enter into the dense packing, it can lead to coalescence of the drops. Once coalescence starts, it continues through the column height and operation cannot be maintained. This coalescence is the main danger in operating in the dense packing regime.

Not much is known about either the hydrodynamics or the heat transfer in the dense packing regime. Thus, although it is possible to operate in this manner, prudent designers have stuck to disperse packing operating. Further, research to counteract coalescence by the adding of surfactants to the drops could result in practical dense phase operation of a spray column. The influence of the surfactants on heat and mass transfer would also have to be studied.

7 AXIAL MIXING

Spray columns are designed with the intent to facilitate countercurrent contact between two immiscible fluids. The degree to which this is achieved depends upon the design of the injectors for the two fluids. The design for the dispersed phase is reasonably straightforward and is essentially a perforated plate covering a manifold. The limiting factors on nozzle jet velocity have already been discussed. The injection of the continuous phase is more difficult and flaws in its design are probably the leading cause of axial circulation in a liquid-liquid spray column. However, it should be noted that little research on internal back mixing or axial mixing has been done.

In the analyses of spray columns, models used are normally one-dimensional and transient or steady state. Back mixing can be introduced by introducing single dispersion coefficients, E_c, and E_d which may be correlated by comparing column operating results against different operating parameters. Principally, it is assumed that an additional flux exists in the opposite direction to the main flow of each of the phases. The magnitude of the back flow is assumed to be proportional to the negative gradient of the parameter in question (temperature, concentration, velocity, etc.). For example, for mass transfer, the conservation equations take the form (Steiner and Hartland, 1983)

$$\frac{\partial x}{\partial t} = V_c \frac{\partial x}{\partial z} + E_c \frac{\partial^2 x}{\partial z^2} - \frac{k_c a}{1 - \epsilon} \left(x - x^* \right) \tag{29}$$

for the continuous phase, and

$$\frac{\partial y}{\partial t} = -V_d \frac{\partial y}{\partial z} + E_d \frac{\partial^2 y}{\partial z^2} + \frac{k_c a}{\epsilon} \left(x - x^* \right) \tag{30}$$

for the dispersed phase. The values for E_c and E_d will be dependent on column geometry, injector design, etc., as well as the fluids being used. These points are discussed by Steiner and Hartland (1983). They carried out experiments in a spray column without any dispersed phase present. They noted strong circulation in the continuous phase only. Other back mixing can be caused by the fact that the drops carry wakes that are periodically shed as they rise. As pointed out by Letan and Kehat (1968), in a sufficiently tall column, this effect can even out along the length of the column. In their heat transfer work this resulted in long columns operating with a nearly constant value of the volumetric heat transfer along the column length for columns with length to diameter ratios, L/D_c, greater than eight.

In small diameter columns, bulk circulation can occur due to the fact that drops concentrate in the middle of the column. In larger diameter columns, this can lead to more difficulties as it has been proposed that the axial dispersion is proportional to the diameter raised to a power $(E \sim D^n)$. However, it is believed that this is again caused by the difficulty in designing an appropriate continuous phase inlet. Such was the problem, it is believed, with the 500 kW$_e$ spray column direct contactor at East Mesa. However, changes in other operating conditions were also made at the same time as the injector was changed. Thus, no one knows for sure whether the injector modification alone led to improved operation. However, testing with continuous phase nozzle design prototypes at the University of Utah in its six inch diameter unit, seemed to indicate significant reduction in back mixing. Thus, it is believed that appropriate design can significantly reduce back mixing.

No general rule is available to design the continuous phase injector. The following rules of thumb are proposed:

1) The continuous phase injector should not release a strong jet of fluid. A local strong jet produces recirculation regions about it. If the fluid is injected downward, strong axial recirculation cells may develop. Further, a strong jet can lead to local flooding and potential breakup of the drops.

2) Lateral release of the continuous phase is desirable; however, again, strong flows can lead to drop breakup. While this may be tolerated in a boiler-preheater combination where the nozzle is in the boiler region it cannot be tolerated in the preheater where countercurrent flow is strongly desired.

3) Multiple-spaced inlets can insure low speeds and small velocities, thereby, minimizing any jet lengths and thus recirculation zones. No other guides are available, although examples of inlets are provided in the literature.

8 HEAT TRANSFER CORRELATIONS

8.1 Liquid-Liquid Heat Transfer

Direct contact heat transfer between the drops of the dispersed phase and the continuous phase is complex. It depends not only on the thermal properties of each phase, but also on the dynamics of the drops themselves. As was noted earlier, the drops can be spherical, ellipsoidal or cap-shaped depending upon their Reynolds number, their Eötvös number and their M-group. To date, most experiments for spray columns were carried out with drops whose diameters were less than 7.5 mm with most heat transfer studies being carried out for drops with 2.0 mm $\leq d_D \leq$ 4.0 mm. In this range for use with geothermal brines or water as the continuous phase, the drops are nominally spherical with the possibility of being in the fluctuating ellipsoidal regime. For hexane, pentane or isobutane, experience at the University of Utah indicates that the drops are spherical. In the presence of impurities, such as occur in geothermal brines, there is further justification to presume this behavior. This is due to the fact that impurities tend to immobilize the interface. Thus, it is believed that little or no circulation will take place in drops of light hydrocarbons in water, or brine for diameters less than 4 mm. Further, in swarms, the terminal velocities of drops are decreased which makes it less probable that they are in the fluctuating ellipsoidal regime.

When the drops have low thermal conductivity, as is the case with hydrocarbons, it is likely that the governing resistance to heat transfer is internal to the drops. However, until the recent work of Jacobs and Golafshani (1985, 1985), it was generally assumed that external resistances were governing. Letan (1968, 1976) and coworkers postulated models based purely on the hydrodynamics of the drops. Noting that at moderate Re_c, $Re_c > 500$, the drops periodically shed their wakes, Letan and Kehat (1968) postulated that the reduced heat transfer as compared to single drops was related to the shedding frequency and wake size. They argued that for long columns most of the heat transfer occurred in the wake shedding regime. Allowing for some empirical constants in the model, they were able to fit their own and some other data. Further, for long columns they could justify the use of a constant volumetric heat transfer coefficient. As their experiments were conducted at small temperature differences between the incoming fluids, they did not worry about temperature dependent fluid properties.

Following the lead of Letan and earlier investigators, Plass, Jacobs and Boehm (1979) ran a series of experiments to determine a volumetric heat transfer coefficient, U_v. They correlated both their own data and that of other investigators for organic fluids dispersed in water or geothermal brine. They claim an accuracy of $\pm 20\%$ for the following correlation

$$U_v = 1.2 \times 10^4 \, \epsilon \, \frac{Btu}{hrft^3 \, °F} \, (for \, \epsilon < 0.05) \tag{31}$$

$$U_v = \left[4.5 \times 10^4 \, (\epsilon - 0.05)e^{-0.57} G_D/G_c + 600 \right] \tag{32}$$

$$\frac{Btu}{hrft^3 \, °F} \, (for \, \epsilon > 0.05)$$

where

$$U_v = \frac{\dot{Q}}{Vol \, LMTD} \tag{33}$$

The above equation was used in estimating the preheater length requirement for the 500 kW$_e$ East Mesa combined boiler/preheater spray column.

For drops originally of 3.0–3.5 mm in diameter, Jacobs and Golafshani (1985) showed that the heat transfer is reasonably well-represented by Equations 31–32 when actual local holdup values are used. However in deriving Equations 31–32, Plass, et al. (1979) used the correlation of Johnson, et al. (1957) for holdup. The use of the latter correlation gave "sometimes agreement" with the data from East Mesa (Letan, 1976). The degree of accuracy in predicting preheater length was approximately $\pm 20\%$ depending upon how the holdup was calculated. The calculated, detailed temperature profiles did not compare as well with the experiments. Jacobs and Golafshani (1985) also investigated a model using the assumption of no drop internal resistance to heat transfer and one where the heat transfer was governed by diffusion within the drop. This latter model showed better agreement, especially when it accounted for drop growth.

For drops less than 4.0 mm in diameter, it is recommended that final preheater spray column sizing be done using the conduction drop growth model of Jacobs and Golafshani (1985) and Jacobs (March 1985). Preliminary sizing can be carried out using Equation 33.

For drops greater than 4 mm in diameter, it is highly probably that the drops will be in the fluctuating ellipsoidal regime. For this regime, both internal and external resistances to heat flow would have to be considered. For single drops, Sideman (1966) gives

$$Nu_d = 50 + 8.5 \times 10^{-3} \, Re_c \, Pr^{0.7} \tag{34}$$

for the external surface coefficient for oscillating drops for $150 < Re_c < 700$. A maximum deviation of 12% was reported. No reliable expression is available for $Re > 700$.

The internal resistance to heat transfer can be calculated by Equation 28 of Golsfshani and Jacobs (1985). It gives

$$Nu_d = 0.00375 \; \frac{Re_d Pr_d}{1 + \mu_d/\mu_c} \tag{35}$$

This equation is reported to be accurate to ±20%.

It is not clear from the studies conducted to date whether increases in the surface heat transfer due to internal circulation will offset reduced surface area by going to larger drops. Before settling on a drop size for a given applications, however, such a study is warranted.

8.2 Direct Contact Boiling Heat Transfer

The sizing of the boiler for the 500 kW$_e$ unit at East Mesa was also done using an estimated value for the volumetric heat transfer coefficient. Based on all available data for light hydrocarbons and freons boiling in water, it was observed that

$$U_v = 48,000\epsilon \; \frac{Btu}{hrft^3 \; {}^\circ F} \tag{36}$$

where ϵ is estimated at the value just below the boiler in the preheater section. Although this yielded a reasonable estimate and appears to agree well with the 500 kW$_e$ facilities operation, it has no basis in the physics of the boiling phenomenon. However, no correlation yet proposed does. Further, Walter (1981), in a recent Ph.D. dissertation at the University of Tennessee concludes, "There appears to be no way to calculate a heat transfer coefficient". For the lack of anything better, Equation 36 is recommended.

9 DESIGN APPROACH FOR GEOTHERMAL APPLICATION

Based on laboratory and field experience with the 500 kW$_e$ unit at East Mesa, it is clear that we can safely design a DCHX for approach or pinch temperatures of 2.5°C (4.5°F). The pinch temperature is a necessary parameter in carrying out the thermodynamic analyses to select the optimum direct contact binary cycle for a given geothermal resource. It will set the flow rates of the two fluids, brine and working fluid once the working fluid is selected. The complete power system can be chosen utilizing the computer program DIRGEO described by Jacobs and Boehm (1980) and Riemer, et al. (1976).

On the basis of the system thermodynamic analyses, the mass flow rates of brine and working fluid would be available for a given geothermal resource. The direct contactor pressure, working fluid boiling point, and the inlet and outlet temperatures for both the brine and working fluid would be established. One could now, utilizing this information, proceed to design the direct contactor.

The first thing that must be done is to decide on the drop size for the dispersed phase. Typically the nozzle diameter, d_n, will be from one half to two thirds as large as d_D. As the equation for the drop size, Equation 9, depends upon the critical jetting velocity as well as the critical jet diameter, which in turn depend on d_n, we must select d_n first. For light hydrocarbons in brine values of d_n less than 1.58 mm (1/16 inch) should result in drops from 3.0 to 3.2 mm ($\sim$1/8 inch) in diameter. Such drops should remain nearly spherical with little internal circulation. A larger nozzle diameter will result in fluctuating drops whose behavior at high holdup could lead to an unstable column. This, of course, would need to be examined for the actual fluids selected in light of Fig. 5.3.

After selecting d_n, the jetting velocity, V_j, can be calculated from Equation 1. This is the minimum velocity for the injection nozzles. Next the critical jetting velocity and jet diameter are determined from Equations 2–5. The critical jetting velocity, V_{jc}, cannot be exceeded. At the critical jetting velocity, the drops formed will be given by either Equation 6 or 7 depending upon the Eo number. One can operate the injector at any velocity between the jetting velocity and the critical jetting velocity at long as the Weber number is greater than two. This is necessary to prevent seeping.

Knowing the desired drop diameter, Equation 9 can be used to calculate the nozzle velocity, V_n. As long as the criteria mentioned above is satisfied, we will have selected the nozzle size. We can now proceed to determine the required number of perforations, or nozzles, in the dispersed phase distributor.

As the total mass flow rate of the disperse phase is known, as well as its temperature and pressure, we can calculate the volumetric flow rate, $G_d = \dfrac{\dot{m}_d}{\rho_d}$. The number of nozzles required is

$$n = \frac{G_d}{\dfrac{\pi}{4} d_n^2 V_n} \qquad (37)$$

The number of nozzles, of course, must be a whole number. We round off the calculation to the nearest one, making sure we do not cause a problem by exceeding the value of V_{jc}. Using the whole number, we calculate a new value of V_n from Equation 37. We then recalculate d_D from Equation 9. This is then the drop diameter.

Depending upon whether or not the drops are spherical or wobbling ellipsoidal, which can be checked roughly from Fig. 5.3, we are ready to calculate the drop terminal velocity using either Equation 15 or 18, as is appropriate.

Next the flooding holdup, ϵ_f, should be calculated from Equation 27. The design value of holdup for the column should be selected as 0.9 ϵ_f or less as pointed out in Equation 28. Correcting for the amount of mass of geothermal brine that is vaporized in the boiler with the working fluid, the mass entering the preheater is determined prior to the actual calculation of ϵ_f. Knowing the exit conditions of the brine, the volumetric flow rate G_c is determined. We now know both G_d and G_c. Their ratio $G_d/G_c = R$. Using Equation 24 and calculating Re_c for the drop we obtain m. ϵ_f is the small positive root of Equation 27.

Having established the holdup for the bottom of the preheater, we can now calculate the superficial velocity of the continuous phase at the bottom of the column above the dispersed phase injector. Either Equation 22 or 23 can be used. Equation 23 is sufficient unless the holdup is very high, i.e., > 0.35. The superficial velocity is defined as the mean velocity of the continuous phase across the entire column. Thus,

$$D_C = \left[\frac{4}{\pi} \frac{G_c}{(G_c/A)} \right]^{1/2} \tag{38}$$

gives the needed diameter of the column just above the dispersed phase injector.

If the column was isothermal, selecting the diameter just calculated would insure its holdup being constant along its entire length. With heat transfer, the holdup may vary. This is due to the fact that the density of the brine and selected working fluid can vary considerably with temperature if the column pressure is sufficiently high. If the working fluid density varies, the drop diameters will vary and thus their terminal velocity, etc. No where along the column length should the holdup exceed $0.9\epsilon_f$. Thus, we should next calculate the conditions at the top of the preheater in a manner similar to what was done at the bottom. The column diameter should be whichever is larger.

Having determined the column diameter, we should next check the overall size of the dispersed phase distribution plate. The nozzle holes should not be placed on a center to center distance less than 1.5 d_D. With this type of layout, the overall distributor plate injector should not be larger in diameter than the column. This will insure that the drops can rise vertically.

The columns shown in Figs. 5.1 and 5.2 both have a conical section in which the dispersed phase injector is located. Jacobs and Boehm (1980) and Treybal (1963) indicate that the cone half angle should be about $15°$. The distributor plate should be so located that the annular ring of open space around it should equal the cross-sectional area of the column. This will insure that there is no undercarry of the dispersed phase and that there is good separation of the phases.

We are now ready to proceed with the calculation of the length of the preheater part of the column. If the thermodynamic properties of the working fluid liquid and brine do not vary significantly with temperature, it is possible to calculate the length of the preheater from

$$L_{P.H.} = \frac{\dot{m}_d (h_{bp} - h_{in})_d}{U_v (\frac{\pi}{4} D_C^2) LMTD} \tag{39}$$

where h_{bp} is the enthalpy of the working fluid liquid at the boiling point, h_{in} is its enthalpy at the inlet, U_v is the volumetric heat transfer coefficient and LMTD is the log mean temperature difference across the preheater assuming counterflow. Unfortunately for fluids like isobutane, the specific heat varies considerably with temperature as was noted in the design of the 500 kW$_e$ unit at East Mesa. Thus, it is necessary to evaluate the heat transfer in a number of steps along the preheater length. This was done by hand using Equations 31–33 for the 500 kW$_e$ unit (Olander, et al., 1983). For design studies it is recommended that the steady state computer program described in Jacobs (March, 1985) be used to determine the preheater length. The computer program is easily modified to include a variety of methods for estimating the heat transfer rate. For spherical drops, it is recommended that the variable diameter drop conduction model described by Jacobs and Golafshani (1985) be used to determine the preheater length. For fluctuating ellipsoidal drops, Equations 34 and 35 can be used.

The length of the boiler section can be calculated using the value of U_v given in Equation 36. Due to the nature of the equation as discussed, it is not warranted to divide the boiler into segments. Thus the length of the boiler should be calculated as

$$L_b = \frac{\dot{m}(h_{exit} - h_{b.p.})_d}{\frac{\pi}{4} D_C^2 U_v \, LMTD}$$

where LMTD is based on counterflow temperatures across the boiler.

Based on results of the 500 kW$_e$ design, the length of the spray column should start at the top of the conical section housing the dispersed phase distributor.

It is recommended that the continuous phase injector or injectors be located in the middle of the boiler section to insure maximum heat transfer rates.

A disengagement section at the top of the column needs to be included as shown in Fig. 5.2. This should be sufficiently large to insure no liquid carryover with the exiting vapor.

A sensitivity analysis should be made to determine possible input variations and the time constant for the column such as was done by Golafshani and Jacobs (1985) for the 500 kW$_e$ unit. The controls for the column should be designed based on this information and the resulting information indicating possible preheater and boiler length excursions.

Following the above procedures it is possible to design a highly reliable spray column for geothermal power applications. The techniques posed appear to be conservative based on experience with the 500 kW$_e$ East Mesa unit. Further refinements will require additional laboratory studies as mentioned in the discussions in this manual.

10 APPLICATIONS OF THE METHOD TO ISOBUTANE–WATER SYSTEMS

Isobutane has been shown (Jacobs and Boehm, 1980) to be an excellent choice as a working fluid for geothermal brines when the brine temperature is above 300 °F (~149 °C). This section presents some calculations using the methods discussed in this Appendix to show the influence of various parameters on spray column design.

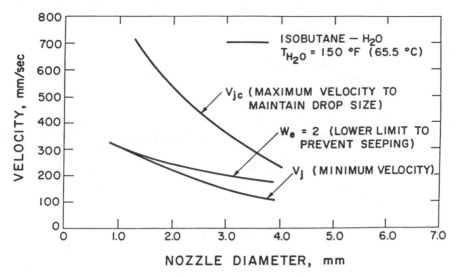

Figure 5.4 Typical limiting velocities for changes in nozzle diameter.

10.1 Establishing Drop Size

In designing a spray column, it is necessary to choose the drop size in order to establish the column hydrodynamics. Figure 5.4 was developed for the isobutane-H_2) system utilizing Equations 1–5. For a small nozzle diameter, it is clear that a wide range of nozzle velocities is possible between the working limits of V_j and V_{jc}. However, as the nozzle diameters increase, the range of permissible velocities decrease. Thus, it would appear safer to design using nozzles of < 3.0 mm in diameter if one wished to provide for significant variations in velocity. However, since the area of the nozzles is proportional to their diameter squared, near equivalent volume flow changes may occur for much smaller changes in velocity for the larger nozzles.

If one considers the actual drop sizes as a function of the nozzle velocities, it is clear from Fig. 5.5 that nearly the same variation in drop diameters is possible over all the nozzle sizes shown. However, the approximate nozzle diameter limit for non-circulating drops is less than 2.5 mm (7/64 inch) and at velocities near the jetting velocity. As it is most easy to hydrodynamically design spray columns for situations where the drops behave as rigid spheres, typical nozzle diameters of 1.5–2.0 mm are normally chosen. Plugging or partial plugging of some holes can lead to velocities exceeding the critical jetting velocity. Reductions in working fluid flow rate can cause drop diameters to exceed the limit for rigid sphere behavior. Thus, fluctuations in spray column operation can occur even without recirculation. Nonetheless, the advantages of spray column direct contact heat exchange make them attractive to pursue. It should also be noted that the limits for rigid drop behavior and critical jetting velocity are experimentally established and are, in general, conservative.

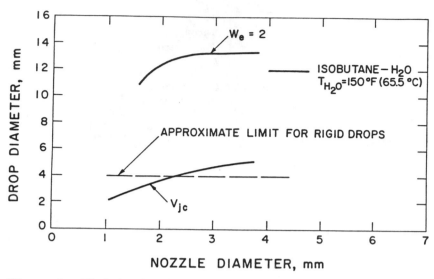

Figure 5.5 Variation in drop sizes produced by nozzles within operating velocity limits.

Figure 5.6 shows the terminal velocities calculated using Equation 15 for rigid drops and Equation 18 for fluctuating drops. For all drops formed between the limits on Weber number and V_{jc} for the nozzle dimensions considered, isobutane drops in H_2O fall within either the spherical or fluctuating ellipsoidal regime shown in Fig. 5.3 as log M is approximately -12.16 and $.7 \leq E_o \leq 2.6$. It should further be noted that the terminal velocity is higher for drops with internal circulation by up to 50%. Thus, it should be possible to increase the throughput by increasing drop size. However, as the path of larger drops is not vertical, coalescence is more likely to occur. Therefore, at this time, it does not appear prudent to significantly increase drop size without further experimental data on drop hydrodynamics, especially for drop swarms.

10.2 Establishing Flooding Limits

Data on flooding in spray columns has primarily been determined in small diameter columns where change of phase of the volatile fluid does not occur. This has been noted in the preceding sections, and is the case for Equation 27. This correlation was developed for the spherical drop regime. Equation 23 presents a correlation for the continuous phase superficial velocity also for the rigid sphere regime. Utilizing these two correlations, the nondimensional, continuous phase, superficial velocity as a function of holdup, ϵ, for a range of the volumetric flow ratio, R, is shown in Fig. 5.7. Note the flooding limit, ϵ_f, and $0.9\epsilon_f$ and $0.8\epsilon_f$ are also shown. The range of R is consistent with the use of DCHX systems with moderate temperature geothermal brines and solar ponds. For this range, the flooding condition

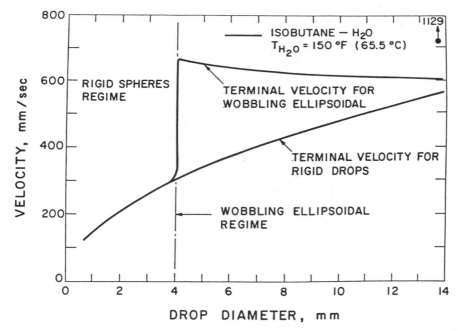

Figure 5.6 Terminal velocities for rigid spheres and fluctuating drops according to Figure 5.3.

corresponds to a nearly maximum value of continuous phase superficial velocity. The peak value of ϵ_f varies from nearly 0.34 at an R of 2.25 to 0.25 for $R = 0.5$. In each case, the superficial velocity only undergoes a small change in order to reduce to $0.8\epsilon_f$ or $0.9\epsilon_f$.

The advantage for heat transfer of operating close to the flooding point is shown in Fig. 5.8. For the case of small values of R, as is the case for solar ponds, the correlation of Plass, et al. (1979), indicates volumetric heat transfer coefficients of nearly 8,000 Btu/hrft^3°F (148.8 kW/m^{3}K) when operated near flooding. Of course, it should be noted that in the liquid preheating regime, such a low value of R leads to low utilization of the heating capacity of the continuous phase brine. In typical geothermal applications, it is desired to utilize a considerable portion of the thermal capacity of the hot liquid brine. With an organic working fluid such as isobutane, a typical value for R would be around two. The correlation of Reference 15 indicates a decrease in U_v with increasing R, but an increase with ϵ. Since the rate of increase of U_v with increasing ϵ is less for higher values of R, operating further away from the flooding point does not, as significantly, affect U_v as it would at lower R, more typical of solar pond applications.

Figure 5.8 can be used for preliminary design applications provided we keep in mind the data from which it was determined. The limitations are: (a) the dispersed phase is an organic fluid; (b) the continuous phase is brine or water; (c)

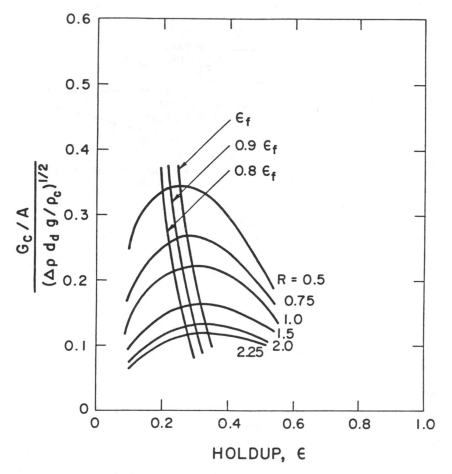

Figure 5.7 Superficial continuous phase velocity as a function of holdup for spherical drops.

the drops are less than 5.0 mm in diameter; and (d) the correlation is good to only within ±20%.

11 EXAMPLE OF DCHX DESIGN

Consider the design of a DCHX preheater-boiler for brine entering at a rate of 85,840 lb_m/hr. Isobutane at a flow rate of 91,434 lb_m/hr enters the DCHX at 86°F. These mass flow rates correspond to 327.9 gpm of IC_4 and 192 gpm of brine. The boiling point of the IC_4 is 244.5°F (corresponding to a vapor pressure of 400 psi), and the temperature of the brine entering the preheater is 251°F. The brine exits the column at 129.5°F.

The change in enthalpy of the isobutane across the preheater is 10.332 × 10^6 Btu/hr. In order to size the column, we must first choose the size drops we

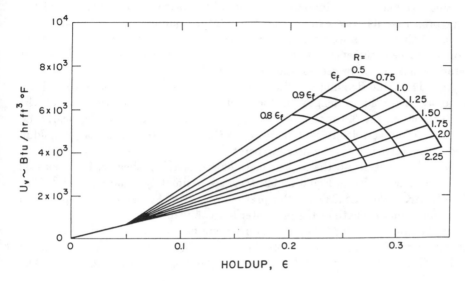

Figure 5.8 Volumetric heat transfer coefficient as a function of holdup from Plass et al., 1979.

desire. Let us choose 3.5 mm diameter at entrance. This size can be achieved for nozzle diameters less than 1.95 mm (see Fig. 5.5). The maximum velocity, V_{jc}, will be 540 mm/sec (1.772 ft/sec) for this size nozzle. Let us choose a smaller nozzle diameter. One-sixteenth inch diameter holes can be drilled (1.587 mm) to form nozzles. For this size nozzle, the critical jetting velocity is 630 mm/sec and the critical drop diameter $d_{DC} = 2.5$ mm. Using Equations 2 and 9, we can solve for d_{jc} and V_n, the nozzle velocity. The equations yield $d_{jc} = 1.46$ mm and $V_n = 430$ mm/sec (1.47 ft/sec). This velocity is in the midpoint range between the V_{jc} and $W_e = 2.0$ limits. The number of holes required for the distributor plate can be calculated from Equation 37. The nearest whole number yields 23,328 nozzles. The area of the nozzles is 0.497 ft^2 or 0.0462 m^2.

Next we must calculate the terminal velocity of the drops. As they are in the spherical regime (see Fig. 5.6), the terminal velocity can be calculated from Equation 15. The terminal velocity at the bottom of the column is 280 mm/sec or 0.919 ft/sec. Before determining the superficial velocity of either phase, we must next calculate the holdup at flooding.

Using the terminal velocity, the Reynolds number of a drop is calculated. This is necessary to establish the Re_c regime. As the Re_c is greater than 500, $Re_c = 2212$, we obtain $m = 1.39$. From Equation 27, we next find the value of the holdup at flooding, ϵ_f. Given the ratio of flow rates $R = 1.708$ and $m = 1.39$, ϵ_f is 0.326. If we choose to design for $0.9\epsilon_f$, then $\epsilon = 0.2934$. From Equation 23, we can calculate the superficial velocity at the bottom of the column. The chosen conditions yield $G_c/A = 0.0784$ ft/sec or 23.88 mm/sec.

Since we know the volumetric flow rate of the brine G_c, the diameter of the column proper at its base can be calculated from Equation 38, to be 2.80 ft or 0.855 m. Although this would be an appropriate diameter for the DCHX column for conditions near the entry of the working fluid, large changes in the density of isobutane can occur at pressures near 400 psi as it is heated to its saturation temperature. Thus, the flooding conditions will have to be checked at several locations along the column length in the preheater section. At the top of the preheater section, the 3.5 mm drops will have grown to a diameter of nearly 4.0 mm. The terminal velocity will increase and generally the deviation in holdup will be to one further from the flooding point.

In order to establish the approximate length of the preheater, Equations 32 and 39 can be used. From Equation 32, U_v is approximately 4,650 Btu/hrft^3°F or 86.5 kW/m^{3}K. The LMTD for the preheater is 21.8°F. Thus, from Equation 39, the approximate length of the preheater is 16.55 ft or 5.04 m.

In the boiler section, 5.85×10^6 Btu/hr will be transferred to the isobutane. Equation 36 yields a $U_v = 14.083$ Btu/hrft^3°F. The LMTD across the boiler is 31.5°F. Thus, the boiler volume is 13.19 ft^3. The boiler length is 2.14 ft or 0.65 m.

We now have a preliminary sizing of the column proper. The diameter is 2.8 ft (0.855 m) and the combined length of the preheater and boiler is 18.7 ft (5.70 m). Using these preliminary dimensions, the steady state computer program described by Jacobs (March, 1985) can be used to refine the size for the preheater. Such a calculation yields an overall length of 19 ft (5.8 m).

For a vapor-mist disengagement section, a height of two diameters is suggested (5.6 ft or 1.71 m). Two rows of chevron mist eliminators located mid-way in the disengagement section would eliminate any droplet carry-over. The chevrons should be two inches high and inclined at 60° from the vertical.

The brine injector would be located near the midpoint of the boiler section. An injector might be designed to distribute the brine horizontally from a single tube. It should yield a horizontal velocity no greater than G_c/A at the wall of the column. This would require the injector distribution over a height of 7.4 inches or 18.9 cm. A design such as used for the 500 kW$_e$ East Mesa DCHX unit described by Olander, et al. (1983), would be satisfactory.

The isobutane distribution plate, as noted, would require 23,328 1/16 inch diameter holes spaced over 6.16 ft^2. This is equivalent to one hole every 0.038 square inches, or 0.22 inches or 5.59 mm between centers. The distributor should be located in the conical frustrum section below the column proper where the diameter is 3.96 ft. Ideally, the frustrum 1/2 angle would be 15°. This will allow for the brine to pass by the IC$_4$ distributor with no increase in the continuous phase superficial velocity. The brine exit would be located on the pressure head below the column in a manner similar to that shown on the schematic of Fig. 5.1.

This design example used flow conditions identical to those for the East Mesa DCHX observed on November 12, 1980 (Olander, et al., 1983). The East Mesa DCHX had a diameter of 3.67 ft. Under these flow conditions the holdup, ϵ, for the East Mesa DCHX was only 0.227, which was only 70% of the flooding value. The resulting U_v calculated from Equation 32 would have been 2,718 Btu/hrft^3 °F. The experiment indicated a U_v of 2,390 Btu/hrft^3 °F. The comparison is within the ±20% claimed by Plass, et al. (1979).

12 NOMENCLATURE

A	Internal cross-sectional area of the column
d_D	Drop diameter
d_{DC}	Critical drop size according to Treybal
d_{jc}	Drop diameter at critical jetting velocity
d_n	Nozzle diameter
$\Delta\rho$	$\lvert(\rho_c - \rho_d)\rvert$
D_C	Column diameter
E_c, E_d	Dispersion coefficients, empirically determined constants used in one dimensional conservation equations to account for backmixing
E_o	Eötvös number, defined by Equation 8 and 11
ϵ	Void fraction, local global fraction of volume occupied by the dispersed phase
ϵ_f	Void fraction at the flooding point
f	Drag coefficient defined by Equation 16 for rigid spheres and Equation 17 for surface mobile spheres
g	Gravitational constant
G_c	Volumetric flow rate of continuous phase
G_c/A	Superficial velocity of the continuous phase
G_D	Volumetric flow rate of dispersed phase
h_{bp}	Enthalpy of liquid at the boiling point
h_{exit}	Enthalpy at exit from the column
h_{in}	Inlet value of enthalpy
H	Defined by Equation 21
J	Defined by Equation 19 and 20
k, n	Constants in Equation 22
K	Defined by Equation 4
K_1	Defined by Equation 13
L	Length of column
LMTD	Log mean temperature difference for countercurrent flow
$L_{P.H.}$	Preheater length
m	Constant in Equation 23 defined by Equation 24
μ_c	Dynamic viscosity of continuous phase
$M-group$	Defined by Equation 12
n	The number of nozzles required in the dispersion plate in Equation 37
Nu_d	Nusselt Number defined on the basis of drop diameter
Pr	Prandtl number
Q/Vol	Total local heat transfer between phases per unit volume

ρ_c	Continuous phase density
ρ_d	Dispersed phase density
R	G_D/G_C
Re	Reynolds number, defined by Equation 10
σ	Interfacial tension
U_v	Volumetric heat transfer coefficient defined by Equation 33
V_{cj}	Maximum velocity in nozzles to insure near uniform drop formation
V_j	Jetting velocity, minimum velocity in nozzle to insure all nozzles are flowing
V_n	Actual nozzle velocity corresponding to d_n, actual nozzle diameter and flow rate of dispersed phase
V_T	Terminal drop velocity for surface mobile spheres calculated from Equation 18
We	Weber number, $\dfrac{V_n^2 d_n^2 \rho_d}{\sigma}$

REFERENCES

Golafshani, M., H. R. Jacobs. *Stability of a Direct Contact Spray Column Heat Exchanger,* ASME/AIChE National Heat Transfer Conference, Denver, CO (1985).

Grace, J. R. "Hydrodynamics of Liquid Drops in Immiscible Liquids," Chapter 38, *Handbook of Fluids in Motion,* N. P. Cheremisinoff and R. Gupta, Editors, Ann Arbor Science, The Butterfield Group, Ann Arbor, MI, pp. 1003-1025 (1983).

Jacobs, H. R. *Stability Analysis of Direct Contact Heat Exchangers Subject to System Perturbations,"* Final Report-Task 2, U.S. Dept. of Energy Contract DE-AS07-76IDO 1523, Modification A014 (March 1985).

Jacobs, H. R., R. F. Boehm. "Direct Contact Binary Cycles," Section 4.2.6 *Sourcebook on the Production of Electricity from Geothermal Brines,* J. Kestin, Editor, Published by U.S. Dept. of Energy, Washington, D.C., DOE/RA/4051-1, pp. 413-471 (March 1980).

Jacobs, H. R., M. Golafshani. *"A Heuristic Evaluation of the Governing Mode of Heat Transfer in a Liquid-Liquid Spray Column,"* ASME/AIChE National Heat Transfer Conference, Denver, CO (1985).

Johnson, A. I., G. W. Mirand, C. J. Huang, J. H. Hansuld, V. M. McNamara. "Spray Extraction Tower Studies," *AIChE Journal* 3:101-110 (1957).

Kumar, A. D., K. Vohra, S. Hartland. *Canadian J. Chemical Engineering,* 53:158 (1980).

Letan, R. *Design of a Particle Direct Contact Heat Exchanger: Uniform Countercurrent Flow,* ASME Paper 76-HT-27, ASME/AIChE Heat Transfer Conference (1976).

Letan, R., E. Kehat. "The Mechanism of Heat Transfer in a Spray Column Heat Exchanger," *AIChE Journal,* 14(3):398-405 (1968).

Olander, R., S. Oshmyanshu, K. Nichols, D. Werner. *Final Phase Testing and Evaluation of the 500 kW_e Direct Contact Pilot Plant at East Mesa,* U.S. Dept. of Energy Report DOE/SF/11700-TI, Arvada, CO (December 1983).

Plass, S. B., H. R. Jacobs, R. F. Boehm. "Operational Characteristics of a Spray Column Type Direct Contact Preheater," *AIChE Symposium Series,* 189:227-234 (1979).

Richardson, J. F., W. N. Zaki. "Sedimentation and Fluidization, Part I," *Transactions of the Inst. of Chemical Engineers,* 32:35-53 (1954).

Riemer, D. H., H. R. Jacobs, R. F. Boehm. *Analysis of Direct Contact Binary Cycles for Geothermal Power Generation (Program DIRGEO),* University of Utah, U.S. Dept. of Energy Report IDO/1549-5 (1976).

Riemer, D. H., H. R. Jacobs, R. F. Boehm, D. S. Cook. *A Computer Program for Determining the Thermodynamic Properties of Light Hydrocarbons,* University of Utah, U.S. Dept. of Energy

Report IDO/1549-3 (1976).

Riemer, D. H., H. R. Jacobs, R. F. Boehm. *A Computer Program for Determining the Thermodynamic Properties of Freon Refrigerants*, University of Utah, U.S. Dept. of Energy Report IDO/1549-4 (1976).

Riemer, D. H., H. R. Jacobs, R. F. Boehm. *A Computer Program for Determining the Thermodynamic Properties of Water*, University of Utah, U.S. Dept of Energy Report IDO/1549-2 (1976).

Rivkind, V. Y., G. M. Ryskin. "Flow Structure in Motion of a Spherical Drop in a Fluid Medium at Intermediate Reynolds Number," *Fluid Dynamics*, 1:5-12 (1976).

Saline Water Conversion Engineering Data Book, 2nd Edition, published by the M. W. Kellogg Company, Piscataway, NJ, for the U.S. Office of Saline Water, Contract No. 14-30-2639 (November 1971).

Sideman, S. "Direct Contact Heat Transfer Between Immiscible Liquids," *Advances in Chemical Engineering*, Vol. 6, Academic Press, New York, NY (1966).

Skelland, A. H. P., K. R. Johnson. *Canadian J. Chemical Engineering*, 52:732 (1974).

Steiner, L., S. Hartland. "Hydrodynamics of Liquid-Liquid Spray Columns," Chapter 40 *Handbook of Fluids in Motion*, N. P. Cheremisinoff and R. Gupta, Editors, Ann Arbor Science, The Butterfield Group, Ann Arbor, MI, pp. 1049-1092 (1983).

Treybal, R. E. *Liquid Extraction*, 2nd Edition, McGraw-Hill, New York, NY (1963).

Walter, D. B. *An Experimental Investigation of Direct Contact Three-Phase Boiling Heat Transfer*, Ph.D. Dissertation, The University of Tennessee, Knoxville, TN (1981).

THERMAL DESIGN OF
WATER-COOLING TOWERS

John C. Campbell

1 INTRODUCTION

Water-cooling towers are used for the removal of a major portion of the heat generated in industrial plants. The basic principle of operation is the direct contact of the heat-laden water with atmospheric air, resulting in cooling the water by evaporating a portion of it. Since the latent heat of vaporization of water is in the order of 4,000 times the specific heat of the cooling air, the sensible heat transferred by the latter is generally insignificant, and the process may be considered to be the removal of heat by means of mass transfer.

The basic heat transfer equations are fairly simple, since the only primary fluids involved are air and water. Since the transfer is primarily by evaporation of water, the temperature difference between the two streams cannot be used to determine the driving force; instead *enthalpy difference* is generally used. The problem of accurate evaluation of the magnitude of the effective surface through which the heat and mass transfer occurs adds a complication not usually present in indirect contact heat transfer. This is due to the fact that in many types of cooling tower packing the fill faces provides only part of the transfer surface, the rest being the surface of the water droplets. The number, size, and shape of these droplets vary widely with such variables as fill arrangement, type of fill, operating conditions, and air velocity. The difficulty of accurate evaluation of total transfer surface is circumvented by using the transfer coefficient and corresponding area as a single term. The familiar heat transfer equation

$$Q = (u)(A)(\Delta T) \tag{1}$$

is thus modified to

$$Q = (K_g)(a)(v)(\Delta h) \tag{2}$$

Substitution $(L)(C_p)(T_1 - T_2)$ for Q, and simplifying by assuming that $C_p = 1.0$, the integral form of Equation (2) is the famous Merkel [1] equation:

$$\frac{KaV}{L} = \int_{T_2}^{T_1} \frac{dT}{h_w - h_a} \tag{3}$$

The term KaV/L has been given several names; popular ones are "tower characteristic" and "number of diffusion units." Calculated using Equation (3), it is representative of the *required* heat and mass transfer surface for given thermal conditions.

A companion equation is convenient to express the *capability* of a specific fill design and arrangement:

$$\frac{KaV}{L} = C + M \left[\frac{L}{G}\right]^{-n} \tag{4}$$

The constants C, M, and n are normally developed from experimental data.

Equations (3) and (4) are the basic tools generally used for cooling tower thermal design. The former, in single integral form, is applicable to counter-current flow, and can be evaluated numerically by the Tchebycheff method [2], or by other similar approximation methods. For crossflow, double integration is required, and numerical evaluation is best achieved by aid of a computer. For convenience, elaborate sets of curves have been prepared for both counterflow and crossflow, relating required characteristic to L/G ratio for a wide range of thermal conditions [3,4]. These curves, often termed *demand* curves, are particularly useful to those who have only occasional need for design calculations and do not have computer assistance.

For given thermal requirements, a solution is readily achieved by the following procedure:

(a)   Select a suitable type and arrangement of fill, and establish the allowable L/G ratio. Direct solution can be made by determining the intersection of the applicable *demand* curve and the fill *characteristic* curve.

(b)   From heat balance and psychrometric data establish properties and volumetric flow rates of inlet and exit air streams.

(c)   Choose a cooling tower size and shape appropriate for the fill arrangement and air and water flow rates.

(d)   Compute air velocities and pressure losses through the cooling tower.

(e)   Design equipment for providing the necessary flow rates and distributions of the air and water streams.

Many combinations of

> type of cooling tower
> fill type
> active volume
> packed height
> air flow rate

will yield workable cooling towers for a given heat removal rate. The skilled designer will pare this multitudinous number down to the one selection of greatest value to the user, considering first cost, operating power, ground area, flexibility, maintenance costs, and other pertinent factors.

The following examples* are presented to clarify the foregoing brief description.

2   MECHANICAL DRAFT COOLING TOWER

Problem: Design a cooling tower suitable for the following conditions:

Water circulation rate, gpm	40,000
Inlet water temperature, °F	104
Outlet water temperature, °F	86
Wet-bulb temperature, °F	77
Dry-bulb temperature, °F	95
Elevation	sea level
Cost of power for fans, $/HP	2,000
Cost of power for water pumps, $/foot of pumping head	22,000
Maximum ground area, L × W, feet	200 × 60

(a)   The design conditions indicate that a counterflow cooling tower requiring moderately low fan and pump power may be optimum; therefore this illustration of one of many possible selections will follow these guidelines. A film type fill requiring relatively low packed height and low resistance to air flow will be used; and is identified as No. J-5. The characteristic curve for this fill is shown in Figure 1. This curve may be superimposed on the counterflow demand curves applicable to the design wet-bulb temperature of

---

*The units used in these examples conform to industrial practice.

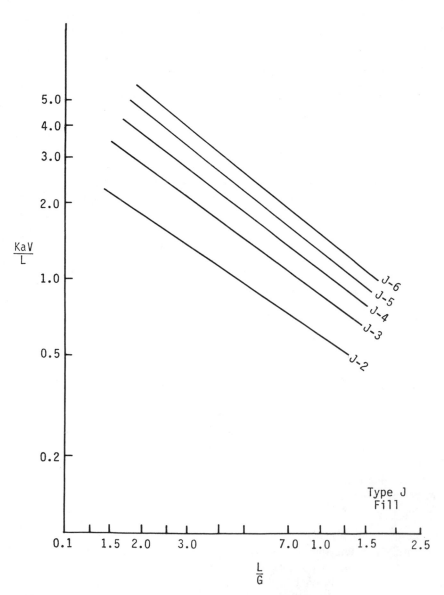

Figure 6.1 Characteristic curves for type J film fill.

77°F and cooling range of 18°F, as shown in Figure 2, establishing the allowable L/G ratio of 1.41.

(b)   The properties and volumetric flow rates of the inlet and exit air streams are determined from equations and psychrometric data as follows:

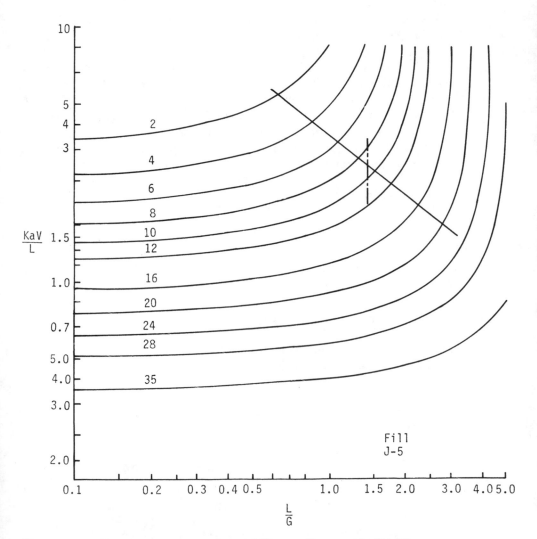

Figure 6.2 Characteristic curve J-5 and "demand" curve 104/86/77.

$L = \text{(gpm) (lb/gal)} = (40,000)\,(8.280) = 331,200 \text{ 16/min}$
$G = (L)/(L/G) = 331,200/1.41 = 234,900 \text{ lb/min dry air}$
$Q = (L)\,(T_1-T_2)\,(C_p) = (331,200)\,(104\text{-}86)\,(1.00) = 5,961,600 \text{ Btu/min}$
$\Delta h = Q/G = 5,961,600/234,900 = 25.38 \text{ Btu/lb dry air}$
h_1 @ inlet conditions = $\underline{40.57}$ Btu/lb dry air
h_2 @ exit conditions = 65.95 Btu/lb dry air
Volumetric flow rate, $ACFM = (G)\,(V)$

Inlet air: $ACFM_1 = (234,900) (14.563) = 3,420,850$
Exit air: $ACFM_2 = (234,900) (14.891) = 3,498,000$

(c)   The economic advantage of construction standardization normally dictates the *bay* sizes to be used by the thermal designer; for wood framed counterflow cooling towers, column spacings of 6′ or 8′ are popular. For this example, assume the designer will use 8′ × 8′ spacing. For large heat removal rates, fairly large cell sizes will usually be cost effective. Based on these guidelines, and on the stated ground area restriction, try a cell width of 6 bays, or 48′. Since the value of fan power is fairly high, assume moderately low air velocity to ensure low resistance to air flow, say in the order of 300 to 400 feet per minute through the fill inlet plane. A rough estimate of a trial tower length can now be made:

$$\text{estimated length, ft} = \frac{ACFM_1}{(\text{est. } u_1)(W, \text{ ft})}$$

$$= \frac{3,420,850}{(\sim 350)(48)} = 203.6 \text{ ft}$$

This slightly exceeds the 200′ plot length limitation, and leads to the selection of four 48′ long cells. Thus this selection will be four in-line cells, each $48'W \times 48'L$, with a total active plan area of 9,216 square feet.

(d)   The design air velocities through the fill, based on nominal plan area, will be:

Fill inlet: $u_1 = 3,420,850/9,216 = 371.19 \text{ ft/min}$
Fill exit: $u_2 = 3,498,000/9,216 = 379.56 \text{ ft/min}$

For convenience, pertinent air properties and flow rates are tabulated below:

	INLET	EXIT	AVERAGE
WBT, °F	77.0	96.6	86.8
DBT, °F	95.0	96.6	95.8
V, ft^3/lb dry air	14.563	14.891	14.727
ρ, lb mix./ft^3 mix.	0.06976	0.06974	0.06975
θ, ρ/ρ_s	0.9314	0.9311	0.93125
$ACFM$, total	3,420,850	3,498,000	3,459,400
$ACFM$ per cell	855,200	874,500	864,850
u through fill, ft/min	371.19	379.56	375.37

The major pressure losses to cooling tower air flow are:

(1)   tower inlet
(2)   fill
(3)   distribution system
(4)   mist eliminators
(5)   plenum and structure

(6)   water spray
(7)   contractions, expansions, and changes in direction
(8)   net velocity pressure at exit
(9)   fan entrance

The summation of these is the total pressure differential at design conditions.

(1) tower inlet

Assume this example tower is equipped with air inlet louvers of a design characterized by the manufacturer's empirical equation

$$\Delta P = \frac{(5.0)(u)^{1.82}(\theta)}{10^7} \tag{5}$$

Also assume a design louver face velocity of 800 ft/min.

$$\Delta P = \frac{(5.0)(800)^{1.82}(0.9314)}{10^7} = \underline{0.0895}'' \ H_2O$$

(2) fill

The pressure drop through the fill is a function of:

air velocity: $u_a = 375.37$ ft/min.
water loading: gpm/ft^2 = 40,000/9,216 = 4.34
relative air density: $\theta_a = 0.093125$

Using the manufacturer's data, Fig 3, and correcting for deviation from standard air density,

$$\Delta P = (0.1490)(0.93125) = \underline{0.1388}'' \ H_2O$$

(3) distribution system

This air pressure loss is primarily due to the energy of the water spraying countercurrently to the air stream. The water distribution piping presents some additional restriction. Manufacturer's data, Fig 4, is applicable to this example. At u_2 = 379.56 ft/min and a water loading of 4.34 gpm/ft^2, the corrected pressure drop is

$$\Delta P = (0.0210)(0.9311) = \underline{0.0196}'' \ H_2O$$

(4) mist eliminators

The pressure drop across the type of mist eliminators used in the example cooling tower is given by the manufacturer's equation

$$\Delta P = \frac{(0.39)(u)^{1.82}\theta}{10^6} \tag{6}$$

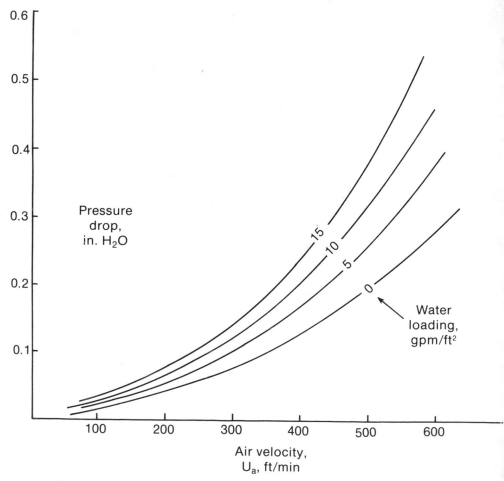

Figure 6.3 Pressure drop data for film fill No J-5.

where u is the *net* air velocity. For this example the mist eliminator frames and supports block 7.0 percent of the nominal face area.

$$\textit{net face area} = (0.93)\,(9{,}216) = 8{,}570 \;\; \textit{ft}^2$$

$$\textit{u net} = 3{,}498{,}000/8{,}570 = 408.13 \;\; \text{ft/min.}$$

$$\Delta P = \frac{(0.39)(408.13)^{1.82}(0.93125)}{10^6} = \underline{0.0205} \;\; '' \; H_2O$$

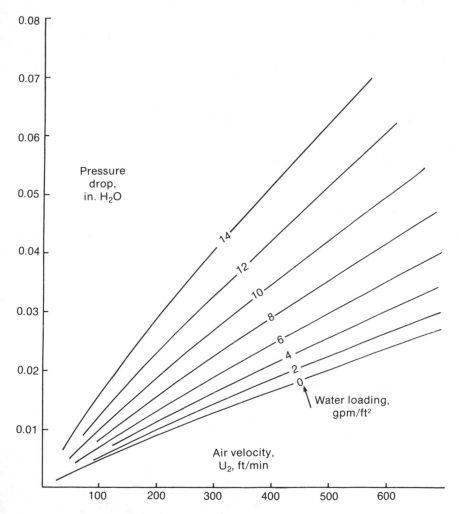

Figure 6.4 Distribution system pressure drop.

(5) plenum and structure

The applicable equation for the example tower is

$$\Delta P = (1.70)\left(\frac{u_2}{4005}\right)^2 (\theta_2) \tag{7}$$

$$= (1.70)\left(\frac{379.56}{4005}\right)^2 (0.9311) = \underline{0.0142} \text{ '' } H_2O$$

(6) water spray

The air pressure loss due to the water spray, exclusive of the distribution system spray accounted for in item (3), is a function of the average air travel below the fill, the water loading, and the average air velocity through this "rain" area:

$$\Delta P = (0.0031) \left(\frac{u_{louv.}}{2 \times 10^3} \right)^{1.85} (\frac{W}{4})(\frac{gpm}{ft^2})^{0.40}(\theta_1) \tag{8}$$

$$= (0.0031) \left(\frac{800}{2 \times 10^3} \right)^{1.85} \left[\frac{48}{4} \right] \left(4.34 \right)^{0.40} (0.9314)$$

$$= \underline{0.0114 '' \ H_2O}$$

(7) contractions, expansions, and changes in direction

For the example tower, these flow abnormalities result in a total pressure drop equal to 1.50 velocity heads, referred to the average face velocity:

$$\Delta P = (1.50) \left[\frac{u_a}{4005} \right]^2 (\theta_a) \tag{9}$$

$$= (1.50) \left[\frac{375.37}{4005} \right]^2 (0.93125) = \underline{0.0123 '' \ H_2O}$$

(8) net velocity pressure at exit

Assume that 30 ' diameter fans will be used for air movement, and that the velocity recovery stacks selected will recover, or "regain", 75% of the velocity pressure difference between the fan and the stack exit plane.

$$net \ A_f = (\frac{\pi}{4})[(D_f)^2 - (D_h)^2]$$

$$= (\frac{\pi}{4})[(30.0)^2 - (6.5)^2] = 673.7 \ ft^2$$

$$u_f = ACFM_2/A_f = 874,500/673.7 = 1298.1 \ ft/min$$

$$VP_f = \left[\frac{u_f}{4005} \right]^2 (\theta_2) = \left[\frac{1298.1}{4005} \right]^2 (0.9311) = 0.0978'' \ H_2O$$

$$A_s = (\frac{\pi}{4})(D_s)^2 = (\frac{\pi}{4})(33.0)^2 = 855.3 \ ft^2$$

$$u_s = ACFM_2/A_s = 874,500/855.3 = 1022.4 \ ft/min$$

$$VP_s = \left[\frac{u_s}{4005} \right]^2 (\theta_2) = \left[\frac{1022.4}{4005} \right]^2 (0.9311) = 0.0607'' \ H_2O$$

$$VP \ recovered = (0.75)(VP_f - VP_e)$$
$$= (0.75)(0.0978 - 0.0607) = 0.0278'' \ H_2O$$

$$net \ VP = VP_f - VP_{rec.}$$
$$= 0.0978 - 0.0278 = \underline{0.0700} \ '' \ H_2O$$

(9) fan entrance

An ideally designed inlet bell will virtually eliminate fan entrance loss, while a very poor design will result in a loss approaching the velocity pressure at the fan. The fan inlet bells on the example cooling tower are designed for a loss equal to 14 percent of VP_f:

$$\Delta P = (0.14)(0.0978) = \underline{0.0137} \ '' \ H_2O$$

Summary of pressure losses

(1)	tower inlet	$0.0895''$ H_2O
(2)	fill	$0.1388''$ H_2O
(3)	distribution system	$0.0196''$ H_2O
(4)	mist eliminators	$0.0205''$ H_2O
(5)	plenum and structure	$0.0142''$ H_2O
(6)	water spray	$0.0114''$ H_2O
(7)	contractions, expansions, etc.	$0.0123''$ H_2O
(8)	net velocity pressure at exit	$0.0700''$ H_2O
(9)	fan entrance	$\underline{0.0137} \ ''$ H_2O
	Total	$0.3900''$ H_2O

(e)   The height of the example cooling tower is 30 feet, measured from basin curb to fan deck. This dimension is determined by:

(1)   sufficient space for air inlet louvers, fill, fill support structure, spray chamber, mist eliminators, and plenum chamber, and
(2)   acceptably uniform distribution of air and water streams, and
(3)   manufacturer's standards.

The velocity recovery fan stack extends 20 feet above the fan deck, for an overall height of 50 feet.

The fans and speed reduction gears chosen for the tower have overall efficiencies of 78% and 96%, respectively. The required driver-output power is calculated from the equation

$$HP/cell = \frac{(ACFM_2)(TP)}{(6356)(\eta_f)(\eta_g)}$$

$$= \frac{(874,500)(0.3900)}{(6356)(0.78)(0.96)} = \underline{71.7}$$

Normally the cooling tower would be equipped with either 75 or 100 horsepower motors for driving the fans, the larger size enabling operation of about 12 percent above the specified heat removal capability.

3 NATURAL DRAFT COOLING TOWER

Problem: Design a natural draft cooling tower for the following conditions:

Water circulation rate, gpm	400,000
Inlet water temperature, °F	110
Outlet water temperature, °F	80
Wet-bulb temperature, °F	60
Dry-bulb temperature, °F	72
Elevation	sea level
Cost of power for water pumps, $/foot of pumping head	230,000
Maximum ground area, $L \times W$, feet	700 × 700

(a) This example will describe the thermal design of a counterflow natural draft tower. For convenience, the fill selected will be the same as used in the Example 1 mechanical draft type. The applicable characteristic and "demand" curves, shown in Fig 5, intersect at $L/G = 1.40$. In this type of tower, the relatively large volume of the rain area below the fill contributes significantly to the heat and mass transfer surface. Previous experience with similar designs indicates that, at the specified thermal conditions, this added surface will increase the allowable L/G by 9.0 percent. Therefore a net effective ratio of $1.40 \times 1.09 = 1.526$ will be used.

(b) The properties and volumetric flow rates of the inlet and exit air streams are determined as in Example 1:

$L = (400,000)(8.269) = 3,307,600$ lb/min
$G = 3,307,600/1.526 = 2,167,500$ lb/min
$Q = (3,307,600)(100 - 80)(1.00) = 99,228,000$ Btu/min
$\Delta h = 99,228,000/2,167,500 = 45.78$ Btu/lb dry air
h_1 @ inlet conditions = 26.46 Btu/lb dry air
h_2 @ exit conditions = 72.24 Btu/lb dry air

$ACFM_1 = (2,167,500)(13.578) = 29,430,300$
$ACFM_2 = (2,167,500)(15.098) = 32,724,900$

(c) Air flows through the natural draft cooling tower due to the existence of a pressure differential ΔP_s between inlet and outlet. ΔP_s is the product of *air density difference* $\Delta\rho$ and *effective stack height* H_e:

$$\Delta P_s = (0.1924)(\Delta\rho)(H_e)(g/g_c) \tag{11}$$

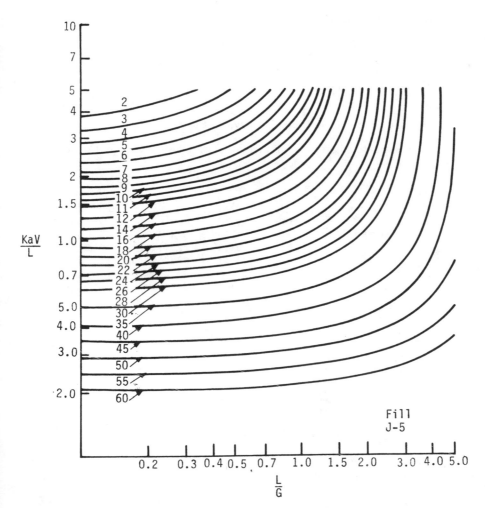

Figure 6.5 Characteristic curve for fill J-5 and 110/80/60 demand curve.

At equilibrium, ΔP_s will equal the total air flow resistance of the system $\sum \Delta P_R$.

The air density difference is a relatively small number, so the stack will be correspondingly high; natural draft towers are much taller than mechanical draft types of similar capacity.

A convenient thermal design procedure consists of assuming a cooling tower size and shape, computing $\sum \Delta P_R$, and then checking the required net effective stack height using Equation (11). If the ratio of height to diameter H/D is not suitable, different selections are calculated to determine the one with suitable shape and greatest value to the owner.

In this example, assume acceptable H/D ratios are between 1.20 and 1.50. For a first try, assume that the air velocity at the fill inlet plane will be in the order of 360 feet per minute.

$$\text{est. } AF_1 = ACFM_1 \, / \sim 360$$

$$= 29{,}430{,}300 \, / \sim 360 = \sim 81{,}800 \text{ ft}^2$$

$$\text{est. } DF_1 = \left[\left(AF_1 \right) \left(\frac{4}{\pi} \right) \right]^{\frac{1}{2}} = \sim 322 \text{ ft}$$

The assumed shape of the trial design is shown in Fig. 6.6.

(d)   Nominal diameters, areas, and face velocities are shown in the following tabulation:

	Dia., ft	Area, ft^2	ACFM	u, ft/min
Base	340	90,792	---	---
Fill inlet	322	81,433	29,430,300	361.40
Fill exit	319	79,923	32,724,900	409.46
Eliminator exit	317	78,924	32,724,900	414.64
Throat	223	39,057	32,724,900	837.87
Tower exit	260	53,093	32,724,900	616.37

Pertinent air properties and flow rates at tower inlet and exit are:

	INLET	OUTLET	AVERAGE
WBT, °F	60	100.28	80.14
DBT, °F	72	100.28	86.14
w	0.00837	0.043588	0.025979
V	13.578	15.098	14.338
ρ	0.074265	0.069124	0.071695
θ	0.991289	0.922667	0.956978
ACFM	29,430,300	32,724,900	31,077,600

The major resistances to air flow are similar in scope to those tabulated for the Example 1 cooling tower.

(1) tower inlet

Assume an inlet height of 34 feet, with air blockage due to support structure of 10%.

$$\text{Gross inlet area} = (34)(\pi)[(340 + 322)/2] = 35{,}355 \text{ ft}^2$$

$$\text{Net inlet area} = (0.90)(35{,}355) = 31{,}820 \text{ ft}^2$$

$$\text{Inlet air velocity} = 29{,}430{,}300/31{,}820 = 924.90 \text{ ft/min}$$

Using a pressure drop equal to 1.08 velocity heads:

$$\Delta P = (1.08) \left(\frac{924.90}{4005} \right)^{2} (0.991289) = \underline{0.05710 \, '' \, H_2O}$$

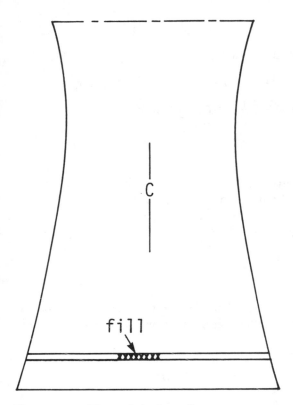

Figure 6.6 Natural draft cooling tower.

(2) fill

The average plan area of the fill is $(81{,}433 + 79{,}923)/2 = 80{,}678$ ft^2. The corresponding water loading is $400{,}000/80{,}678 = 4.958$ gpm/ft^2. The average air velocity is 385.4 ft/min. From Fig 3,

$$\Delta P = (0.1602)(0.956978) = \underline{0.15331}\text{ '' } H_2O$$

(3) distribution system

At the level of the distribution system the inside diameter of the veil is 318 feet.

$$A = (\frac{\pi}{4})(318)^2 = 79{,}423 \text{ ft}^2$$

$$u = 32{,}724{,}900/79{,}423 = 412.04 \text{ ft/min}$$

$$water\ loading = 400{,}000/79{,}423 = 5.036 \text{ gpm/ft}^2$$

From Fig 4, $\Delta P = (0.0235)(0.922667) = \underline{0.02168}\text{ '' } H_2O$

(4) mist eliminators

Assume the mist eliminator design is the same as in Example 1.

$$\text{net face area} = (0.93)(78,924) = 73,399 \text{ ft}^2$$

$$u_{net} = 32,724,900/73,399 = 445.85 \text{ ft/min}$$

$$\Delta P = \frac{(0.39)(445.85)^{1.82}(0.922667)}{10^6} = \underline{0.02386 \text{ '' } H_2O}$$

(5) plenum and structure

The resistance of the plenum and structure, including the friction of the inner wall of the veil, is equal to 0.25 velocity head, based on the *throat* velocity:

$$\Delta P = (0.25) \left[\frac{837.87}{4005} \right]^2 (0.92267) = \underline{0.01010 \text{ '' } H_2O}$$

(6) water spray

The air pressure loss due to the resistance of the falling water droplets below the fill is given by Equation 8, except that the average air travel is a function of DF, rather than W:

$$\Delta P = (0.0031) \left[\frac{u_{inlet}}{2 \times 10^3} \right]^{1.85} (\frac{DF_1}{4})(\frac{gpm}{ft^2})^{0.40}(\theta_1)$$

$$= (0.0031) \left[\frac{924.90}{2 \times 10^3} \right]^{1.85} \left[\frac{322}{4} \right] \left[\frac{400,000}{81,433} \right]^{0.40} (0.991289)$$

$$= \underline{0.11226 \text{ '' } H_2O}$$

(7) contractions, expansions, and changes in direction

For the example tower, these losses total 1.30 velocity heads, based on the average velocity through the fill:

$$\Delta P = (1.30)(\frac{385.4}{4005})^2(0.956978) = \underline{0.01152 \text{ '' } H_2O}$$

(8) net velocity pressure at exit

Velocity pressure recovery for the particular hyperbolic shape used in this example is 73 percent of the difference between throat and exit:

$$throat \ VP = (\frac{837.87}{4005})^2(0.922667) = 0.04038'' H_2O$$

$$exit\ VP = (\frac{616.37}{4005})^2(0.922667) = 0.02185'' \text{ H}_2\text{O}$$

$$recovered\ VP = (0.73)(0.04038 - 0.02185) = 0.01353'' \text{ H}_2\text{O}$$

$$net\ VP\ @\ exit = 0.04038 - 0.01353 = \underline{0.02685}\ '' \text{ H}_2\text{O}$$

Summary of pressure losses

(1)	tower inlet	0.05710" H$_2$O
(2)	fill	0.15331" H$_2$O
(3)	distribution system	0.02168" H$_2$O
(4)	mist eliminators	0.02386" H$_2$O
(5)	plenum and structure	0.01010" H$_2$O
(6)	water spray	0.11226" H$_2$O
(7)	contractions, expansions, etc.	0.01152" H$_2$O
(8)	net velocity pressure at exit	0.02685 " H$_2$O
	TOTAL = sumΔP_R	0.41668 " H$_2$O

(e)   The required effective stack height may now be computed, using Equation 11.

$$\rho_1 = 0.074265 \text{ lb/ft}^3$$

$$\rho_2 = 0.069124 \text{ lb/ft}^3$$

$$\Delta\rho = 0.005141 \text{ lb/ft}^3$$

$$H_e = \frac{\sum \Delta P_R}{(0.1924)(\Delta\rho)(g/g_c)}$$

$$= \frac{0.41668}{(0.1924)(0.005141)(1.00)} = 421.3 \ ft$$

The required *overall* height is the sum of the effective height and the height of the air inlet is approximately:

$$H = 421.3 + 34.0 = 455.3 \ ft$$

The assumed size and shape depicted in Fig 6 will yield cooling capacity slightly in excess of that specified, consistent with good design practice. However an additional trial or two would readily determine a size and shape to satisfy the equality $\Delta P_s = \sum \Delta P_R$. The H/D ratio of the example tower is within the allowable limits previously stated.

4 SUMMARY

The thermal calculations presented in Examples 1 and 2 cover, in each case, only one of many possible selections. Present day practice enables the design engineer, with computer assistance, to investigate all practicable possibilities and choose the particular cooling tower that will be the best buy for the potential owner.

5 NOMENCLATURE

SYMBOL	DESCRIPTION	UNITS
a	area of transfer surface per unit of tower volume	sq ft/cu ft
A	area	sq ft
A_f	area of stack at plane of fan	sq ft
A_h	area of fan hub seal disc	sq ft
A_s	area of fan stack at exit	sq ft
$ACFM$	actual cubic feet per minute	$ACFM$
AF	flow area at fill inlet	sq ft
C	a constant	dimensionless
C_p	specific heat at constant pressure	Btu/lb per °F
D	diameter	ft
D_f	diameter of stack at plane of fan	ft
D_h	diameter of fan hub seal disc	ft
D_s	diameter of fan stack at exit	ft
DBT	dry-bulb temperature	°F
DF	diameter of fill inlet plane	ft
g	acceleration due to gravity	ft/sec^2
g_c	conversion factor	32.2 $lb_m \cdot ft$ ($lb_f \cdot sec^2$)
gpm	gallons per minute	gpm
h	enthalpy	Btu/lb dry air
h_a	enthalpy of air - water vapor mixture at wet-bulb temperature	Btu/lb dry air
h_w	enthalpy of air - water vapor mixture at bulk water temperature	Btu/lb dry air
H	height	ft
H_e	net effective stack height	ft
H/D	height/diameter ratio	dimensionless
HP	horsepower	HP
K	overall enthalpy transfer coefficient	lb per hr per sq ft per lb water per lb dry air
K_g	enthalpy transfer coefficient of gas film	lb per hr per sq ft per lb water per lb dry air
KaV/L	tower characteristic	dimensionless
L	length	ft
L	water flow rate	lb/hr
L/G	water/air mass flow ratio	lb water/lb dry air
M	a constant	dimensionless
n	a constant	dimensionless
Q	heat load	Btu/hr or Btu/min
T	water temperature	°F
TP	total pressure referred to atmospheric	inches of water

u	velocity	ft/min
u_f	velocity at plane of fan	ft/min
u_{louv}	velocity through louvers	ft/min
U	overall heat transfer rate	Btu per lb per sq ft per °F
V	effective cooling tower volume	cu ft
v	specific volume of air	cu ft mixture per lb dry air
VP	velocity pressure	inches of water
VP_f	velocity pressure at fan	inches of water
VP_e	velocity pressure at stack exit	inches of water
w	humidity	lb water vapor per lb dry air
W	width	ft
WBT	wet-bulb temperature	°F
Δ	difference	Δ
Δh	enthalpy difference	Btu per lb dry air
$\Delta\rho$	density difference	lb mixture per cu ft mixture
ΔP	pressure differential	inches of water
ΔP_R	air flow resistance	inches of water
ΔP_S	pressure differential between inlet and outlet	inches of water
ΔT	temperature difference	°F
η	efficiency	dimensionless
η_f	fan efficiency	dimensionless
η_g	gear efficiency	dimensionless
θ	density ratio ρ/ρ_e	dimensionless
ρ	air density	lb mixture per cu ft mixture
ρ_e	standard air density	lb per cu ft
Σ	summation	

SUBSCRIPTS

1   inlet
2   exit
a average

REFERENCES

(1)   *Verdunstungskühlung* by F. Merkel, V.D.I. Forschungsarbsiten, No. 275, Berlin, 1925.

(2)   *Acceptance Test Code for Water-Cooling Towers,* CTI Code ATC-105, Houston, Texas, June 1982 issue, Appendix III-D, p. 18.

(3)   *Cooling Tower Institute Performance Curves,* Cooling Tower Institute, Houston, Texas, 1967.

(4)   *Kelly's Handbook of Crossflow Cooling Tower Performance,* Neil W. Kelly & Associates, Kansas City, MO, 1976.

AUTHOR INDEX